# Undergraduate Texts in Mathematics

# Undergraduate Texts in Mathematics

**Undergraduate Texts in Mathematics** are generally aimed at third- and fourth-year undergraduate mathematics students at North American universities. These texts strive to provide students and teachers with new perspectives and novel approaches. The books include motivation that guides the reader to an appreciation of interrelations among different aspects of the subject. They feature examples that illustrate key concepts as well as exercises that strengthen understanding.

More information about this series at http://www.springer.com/series/666

Richard P. Stanley

# Algebraic Combinatorics

## Walks, Trees, Tableaux, and More

Second Edition

 Springer

Richard P. Stanley
Department of Mathematics
Massachusetts Institute of Technology
Cambridge, MA, USA

ISSN 0172-6056　　　　　　　　ISSN 2197-5604　(electronic)
Undergraduate Texts in Mathematics
ISBN 978-3-030-08389-2　　　　ISBN 978-3-319-77173-1　(eBook)
https://doi.org/10.1007/978-3-319-77173-1

Mathematics Subject Classification (2010): Primary 05Exx; Secondary 15-01

Printed on acid-free paper

This Springer imprint is published by the registered company Springer International Publishing AG part of Springer Nature.
The registered company address is: Gewerbestrasse 11, 6330 Cham, Switzerland

*to*
*Kenneth and Sharon*

*Look'd at each other with a wild surmise—*

# Preface to the Second Edition

The primary change from the first edition is the addition of a chapter entitled "A Glimpse of Combinatorial Commutative Algebra" (Chapter 12). Writing this chapter was an interesting challenge. The "standard" applications of commutative algebra to combinatorics require a background in algebraic topology and homological algebra. I wanted to give a substantial application without using these two subjects and without developing a lot of commutative algebra. It was still necessary to present quite a bit of material on simplicial complexes so that the main result (Theorem 12.25) can be adequately appreciated. The result has been that Chapter 12 is the longest chapter in the book. I hope that I have succeeded in giving some of the flavor of the remarkable connections between commutative algebra and the combinatorics of simplicial complexes.

I have also added a section to Chapter 13 (Section 13.8) involving commutative algebra, but at a much simpler level than Chapter 12. A few new exercises have been added throughout the book, and numerous typos and minor inaccuracies have been corrected. There is now too much material for a one-semester course. The instructor therefore has the pleasure of choosing among diverse treasures, while the student has something to look forward to when the course has ended.

Cambridge, MA, USA                                             Richard P. Stanley

# Updated Preface to the First Edition

This book is intended primarily as a one-semester undergraduate text for a course in algebraic combinatorics. The main prerequisites are a basic knowledge of linear algebra (eigenvalues, eigenvectors, etc.) over a field, existence of finite fields, and some rudimentary understanding of group theory and (for Chapter 12 and Section 13.8) ring theory. The one exception is Section 13.6, which involves finite extensions of the rationals including a little Galois theory. Prior knowledge of combinatorics is not essential but will be helpful.

Why write an undergraduate textbook on algebraic combinatorics? One obvious reason is simply to gather some material that I find very interesting and hope that students will agree. A second reason concerns students who have taken an introductory algebra course and want to know what can be done with their new-found knowledge. Undergraduate courses that require a basic knowledge of algebra are typically either advanced algebra courses or abstract courses on subjects like algebraic topology and algebraic geometry. Algebraic combinatorics offers a byway off the traditional algebraic highway, one that is more intuitive and more easily accessible.

Algebraic combinatorics is a huge subject, so some selection process was necessary to obtain the present text. The main results, such as the weak Erdős–Moser theorem and the enumeration of de Bruijn sequences, have the feature that their statement does not involve any algebra. Such results are good advertisements for the unifying power of algebra and for the unity of mathematics as a whole. All but the last two chapters are vaguely connected to walks on graphs and linear transformations related to them. The final chapter is a hodgepodge of some unrelated elegant applications of algebra to combinatorics. The sections of this chapter are independent from each other and the rest of the text. There are also three chapter appendices on purely enumerational aspects of combinatorics related to the chapter material: the RSK algorithm, plane partitions, and the enumeration of labelled trees. Almost all the material covered here can serve as a gateway to much additional algebraic combinatorics. We hope in fact that this book will serve exactly this purpose, that is, to inspire its readers to delve more deeply into the fascinating interplay between algebra and combinatorics.

Many persons have contributed to the writing of this book, but special thanks should go to Christine Bessenrodt and Sergey Fomin for their careful reading of portions of earlier manuscripts.

Cambridge, MA, USA                                                                                                Richard P. Stanley

# Contents

# Basic Notation

| | |
|---|---|
| $\mathbb{P}$ | Positive integers |
| $\mathbb{N}$ | Nonnegative integers |
| $\mathbb{Z}$ | Integers |
| $\mathbb{Q}$ | Rational numbers |
| $\mathbb{R}$ | Real numbers |
| $\mathbb{C}$ | Complex numbers |
| $[n]$ | The set $\{1, 2, \ldots, n\}$ for $n \in \mathbb{N}$ (so $[0] = \emptyset$) |
| $\mathbb{Z}_n$ | The group of integers modulo $n$ |
| $R[x]$ | The ring of polynomials in the variable $x$ with coefficients in the ring $R$ |
| $Y^X$ | For sets $X$ and $Y$, the set of all functions $f : X \to Y$ |
| $:=$ | Equal by definition |
| $\mathbb{F}_q$ | The finite field with $q$ elements |
| $(j)$ | $1 + q + q^2 + \cdots + q^{j-1}$ |
| $(j)!$ | $(1)(2)\cdots(j)$ |
| $\binom{n}{k}$ | $\frac{(n)!}{(k)!(n-k)!}$, for $0 \leq k \leq n$ |
| $\#S$ or $|S|$ | Cardinality (number of elements) of the finite set $S$ |

| | |
|---|---|
| $S \dot\cup T$ | The disjoint union of $S$ and $T$, i.e., $S \cup T$, where $S \cap T = \emptyset$ |
| $2^S$ | The set of all subsets of the set $S$ |
| $\binom{S}{k}$ | The set of $k$-element subsets of $S$ |
| $\left(\!\binom{S}{k}\!\right)$ | The set of $k$-element multisets on $S$ |
| $KS$ | The vector space with basis $S$ over the field $K$ |
| $B_n$ | The poset of all subsets of $[n]$, ordered by inclusion |
| $\rho(x)$ | The rank of the element $x$ in a graded poset |
| $[x^n]F(x)$ | Coefficient of $x^n$ in the polynomial or power series $F(x)$ |
| $x \lessdot y, y \gtrdot x$ | $y$ covers $x$ in a poset $P$ |
| $\delta_{ij}$ | The Kronecker delta, which equals 1 if $i = j$ and 0 otherwise |
| $|L|$ | The sum of the parts (entries) of $L$, if $L$ is any array of nonnegative integers |
| $\ell(\lambda)$ | Length (number of parts) of the partition $\lambda$ |
| $p(n)$ | Number of partitions of the integer $n \geq 0$ |
| $\ker \varphi$ | The kernel of a linear transformation or group homomorphism |
| $\mathfrak{S}_n$ | Symmetric group of all permutations of $1, 2, \ldots, n$ |
| $\iota$ | The identity permutation of a set $X$, i.e., $\iota(x) = x$ for all $x \in X$ |

# Chapter 1
# Walks in Graphs

Given a finite set $S$ and integer $k \geq 0$, let $\binom{S}{k}$ denote the set of $k$-element subsets of $S$. A *multiset* may be regarded, somewhat informally, as a set with repeated elements, such as $\{1, 1, 3, 4, 4, 4, 6, 6\}$. We are only concerned with how many times each element occurs and not on any ordering of the elements. Thus for instance $\{2, 1, 2, 4, 1, 2\}$ and $\{1, 1, 2, 2, 2, 4\}$ are the same multiset: they each contain two 1's, three 2's, and one 4 (and no other elements). We say that a multiset $M$ is *on* a set $S$ if every element of $M$ belongs to $S$. Thus the multiset in the example above is on the set $S = \{1, 3, 4, 6\}$ and also on any set containing $S$. Let $\left(\binom{S}{k}\right)$ denote the set of $k$-element multisets on $S$. For instance, if $S = \{1, 2, 3\}$, then (using abbreviated notation),

$$\binom{S}{2} = \{12, 13, 23\}, \quad \left(\binom{S}{2}\right) = \{11, 22, 33, 12, 13, 23\}.$$

We now define what is meant by a graph. Intuitively, graphs have vertices and edges, where each edge "connects" two vertices (which may be the same). It is possible for two different edges $e$ and $e'$ to connect the same two vertices. We want to be able to distinguish between these two edges, necessitating the following more precise definition. A (finite) *graph $G$* consists of a *vertex set* $V = \{v_1, \ldots, v_p\}$ and *edge set* $E = \{e_1, \ldots, e_q\}$, together with a function $\varphi \colon E \to \left(\binom{V}{2}\right)$. We think that if $\varphi(e) = uv$ (short for $\{u, v\}$), then $e$ connects $u$ and $v$ or equivalently $e$ is *incident* to $u$ and $v$. If there is at least one edge incident to $u$ and $v$, then we say that the vertices $u$ and $v$ are *adjacent*. If $\varphi(e) = vv$, then we call $e$ a *loop* at $v$. If several edges $e_1, \ldots, e_j$ $(j > 1)$ satisfy $\varphi(e_1) = \cdots = \varphi(e_j) = uv$, then we say that there is a *multiple edge* between $u$ and $v$. A graph without loops or multiple edges is called *simple*. In this case we can think of $E$ as just a subset of $\binom{V}{2}$ [why?].

The *adjacency matrix* of the graph $G$ is the $p \times p$ matrix $A = A(G)$, over the field of complex numbers, whose $(i, j)$-entry $a_{ij}$ is equal to the number of edges incident

© Springer International Publishing AG, part of Springer Nature 2018
R. P. Stanley, *Algebraic Combinatorics*, Undergraduate Texts in Mathematics,
https://doi.org/10.1007/978-3-319-77173-1_1

to $v_i$ and $v_j$. Thus $A$ is a real symmetric matrix (and hence has real eigenvalues) whose trace is the number of loops in $G$. For instance, if $G$ is the graph

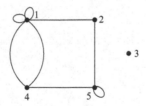

then

$$A(G) = \begin{bmatrix} 2 & 1 & 0 & 2 & 0 \\ 1 & 0 & 0 & 0 & 1 \\ 0 & 0 & 0 & 0 & 0 \\ 2 & 0 & 0 & 0 & 1 \\ 0 & 1 & 0 & 1 & 1 \end{bmatrix}.$$

A *walk* in $G$ of *length* $\ell$ from vertex $u$ to vertex $v$ is a sequence $v_1, e_1, v_2, e_2, \ldots,$ $v_\ell, e_\ell, v_{\ell+1}$ such that:

- Each $v_i$ is a vertex of $G$.
- Each $e_j$ is an edge of $G$.
- The vertices of $e_i$ are $v_i$ and $v_{i+1}$, for $1 \le i \le \ell$.
- $v_1 = u$ and $v_{\ell+1} = v$.

**1.1 Theorem.** *For any integer $\ell \ge 1$, the $(i, j)$-entry of the matrix $A(G)^\ell$ is equal to the number of walks from $v_i$ to $v_j$ in $G$ of length $\ell$.*

*Proof.* This is an immediate consequence of the definition of matrix multiplication. Let $A = (a_{ij})$. The $(i, j)$-entry of $A(G)^\ell$ is given by

$$(A(G)^\ell)_{ij} = \sum a_{ii_1} a_{i_1 i_2} \cdots a_{i_{\ell-1} j},$$

where the sum ranges over all sequences $(i_1, \ldots, i_{\ell-1})$ with $1 \le i_k \le p$. But since $a_{rs}$ is the number of edges between $v_r$ and $v_s$, it follows that the summand $a_{ii_1} a_{i_1 i_2} \cdots a_{i_{\ell-1} j}$ in the above sum is just the number (which may be 0) of walks of length $\ell$ from $v_i$ to $v_j$ of the form

$$v_i, e_1, v_{i_1}, e_2, \ldots, v_{i_{\ell-1}}, e_\ell, v_j$$

(since there are $a_{ii_1}$ choices for $e_1$, $a_{i_1 i_2}$ choices for $e_2$, etc.) Hence summing over all $(i_1, \ldots, i_{\ell-1})$ just gives the total number of walks of length $\ell$ from $v_i$ to $v_j$, as desired.                                                                                    □

We wish to use Theorem 1.1 to obtain an explicit formula for the number $(A(G)^\ell)_{ij}$ of walks of length $\ell$ in $G$ from $v_i$ to $v_j$. The formula we give will depend on the eigenvalues of $A(G)$. The eigenvalues of $A(G)$ are also called simply the *eigenvalues of G*. Recall that a real symmetric $p \times p$ matrix $M$ has $p$ linearly independent real eigenvectors, which can in fact be chosen to be orthonormal (i.e., orthogonal and of unit length). Let $u_1, \ldots, u_p$ be real orthonormal eigenvectors for $M$, with corresponding eigenvalues $\lambda_1, \ldots, \lambda_p$. All vectors $u$ will be regarded as $p \times 1$ *column* vectors, unless specified otherwise. We let $^t$ denote transpose, so $u^t$ is a $1 \times p$ *row* vector. Thus the dot (or scalar or inner) product of the vectors $u$ and $v$ is given by $u^t v$ (ordinary matrix multiplication). In particular, $u_i^t u_j = \delta_{ij}$ (the Kronecker delta). Let $U = (u_{ij})$ be the matrix whose columns are $u_1, \ldots, u_p$, denoted $U = [u_1, \ldots, u_p]$. Thus $U$ is an orthogonal matrix, so

$$U^t = U^{-1} = \begin{bmatrix} u_1^t \\ \vdots \\ u_p^t \end{bmatrix},$$

the matrix whose rows are $u_1^t, \ldots, u_p^t$. Recall from linear algebra that the matrix $U$ *diagonalizes M*, i.e.,

$$U^{-1}MU = \mathrm{diag}(\lambda_1, \ldots, \lambda_p),$$

where $\mathrm{diag}(\lambda_1, \ldots, \lambda_p)$ denotes the diagonal matrix with diagonal entries $\lambda_1, \ldots, \lambda_p$ (in that order).

**1.2 Corollary.** *Given the graph $G$ as above, fix the two vertices $v_i$ and $v_j$. Let $\lambda_1, \ldots, \lambda_p$ be the eigenvalues of the adjacency matrix $A(G)$. Then there exist real numbers $c_1, \ldots, c_p$ such that for all $\ell \geq 1$, we have*

$$(A(G)^\ell)_{ij} = c_1 \lambda_1^\ell + \cdots + c_p \lambda_p^\ell. \tag{1.1}$$

*In fact, if $U = (u_{rs})$ is a real orthogonal matrix such that $U^{-1}AU = \mathrm{diag}(\lambda_1, \ldots, \lambda_p)$, then we have*

$$c_k = u_{ik} u_{jk}.$$

*Proof.* We have [why?]

$$U^{-1}A^\ell U = \mathrm{diag}(\lambda_1^\ell, \ldots, \lambda_p^\ell).$$

Hence

$$A^\ell = U \cdot \text{diag}(\lambda_1^\ell, \ldots, \lambda_p^\ell)U^{-1}.$$

Taking the $(i, j)$-entry of both sides (and using $U^{-1} = U^t$) gives [why?]

$$(A^\ell)_{ij} = \sum_k u_{ik}\lambda_k^\ell u_{jk},$$

as desired.                                                                                □

In order for Corollary 1.2 to be of any use we must be able to compute the eigenvalues $\lambda_1, \ldots, \lambda_p$ as well as the diagonalizing matrix $U$ (or eigenvectors $u_i$). There is one interesting special situation in which it is not necessary to compute $U$. A *closed walk* in $G$ is a walk that ends where it begins. The number of closed walks in $G$ of length $\ell$ starting at $v_i$ is therefore given by $(A(G)^\ell)_{ii}$, so the *total* number $f_G(\ell)$ of closed walks of length $\ell$ is given by

$$f_G(\ell) = \sum_{i=1}^{p}(A(G)^\ell)_{ii}$$

$$= \text{tr}(A(G)^\ell),$$

where tr denotes trace (sum of the main diagonal entries). Now recall that the trace of a square matrix is the sum of its eigenvalues. If the matrix $M$ has eigenvalues $\lambda_1, \ldots, \lambda_p$ then [why?] $M^\ell$ has eigenvalues $\lambda_1^\ell, \ldots, \lambda_p^\ell$. Hence we have proved the following.

**1.3 Corollary.** *Suppose $A(G)$ has eigenvalues $\lambda_1, \ldots, \lambda_p$. Then the number of closed walks in $G$ of length $\ell$ is given by*

$$f_G(\ell) = \lambda_1^\ell + \cdots + \lambda_p^\ell.$$

We now are in a position to use various tricks and techniques from linear algebra to count walks in graphs. Conversely, it is sometimes possible to count the walks by combinatorial reasoning and use the resulting formula to determine the eigenvalues of $G$. As a first simple example, we consider the *complete graph* $K_p$ with vertex set $V = \{v_1, \ldots, v_p\}$ and one edge between any two *distinct* vertices. Thus $K_p$ has $p$ vertices and $\binom{p}{2} = \frac{1}{2}p(p - 1)$ edges.

**1.4 Lemma.** *Let $J$ denote the $p \times p$ matrix of all $1$'s. Then the eigenvalues of $J$ are $p$ (with multiplicity one) and $0$ (with multiplicity $p - 1$).*

*Proof.* Since all rows are equal and nonzero, we have rank$(J) = 1$. Since a $p \times p$ matrix of rank $p - m$ has at least $m$ eigenvalues equal to $0$, we conclude that $J$ has at least $p - 1$ eigenvalues equal to 0. Since $\text{tr}(J) = p$ and the trace is the sum of the eigenvalues, it follows that the remaining eigenvalue of $J$ is equal to $p$.    □

**1.5 Proposition.** *The eigenvalues of the complete graph $K_p$ are as follows: an eigenvalue of $-1$ with multiplicity $p - 1$ and an eigenvalue of $p - 1$ with multiplicity one.*

*Proof.* We have $A(K_p) = J - I$, where $I$ denotes the $p \times p$ identity matrix. If the eigenvalues of a matrix $M$ are $\mu_1, \ldots, \mu_p$, then the eigenvalues of $M + cI$ (where $c$ is a scalar) are $\mu_1 + c, \ldots, \mu_p + c$ [why?]. The proof follows from Lemma 1.4.    □

**1.6 Corollary.** *The number of closed walks of length $\ell$ in $K_p$ from some vertex $v_i$ to itself is given by*

$$(A(K_p)^\ell)_{ii} = \frac{1}{p}((p - 1)^\ell + (p - 1)(-1)^\ell). \tag{1.2}$$

*(Note that this is also the number of sequences $(i_1, \ldots, i_\ell)$ of numbers $1, 2, \ldots, p$ such that $i_1 = i$, no two consecutive terms are equal, and $i_\ell \neq i_1$ [why?].)*

*Proof.* By Corollary 1.3 and Proposition 1.5, the total number of closed walks in $K_p$ of length $\ell$ is equal to $(p - 1)^\ell + (p - 1)(-1)^\ell$. By the symmetry of the graph $K_p$, the number of closed walks of length $\ell$ from $v_i$ to itself does not depend on $i$. (All vertices "look the same.") Hence we can divide the total number of closed walks by $p$ (the number of vertices) to get the desired answer.    □

A combinatorial proof of Corollary 1.6 is quite tricky (Exercise 1.1). Our algebraic proof gives a first hint of the power of algebra to solve enumerative problems.

What about non-closed walks in $K_p$? It's not hard to diagonalize explicitly the matrix $A(K_p)$ (or equivalently, to compute its eigenvectors), but there is an even simpler special argument. We have

$$(J - I)^\ell = \sum_{k=0}^{\ell}(-1)^{\ell-k}\binom{\ell}{k}J^k, \tag{1.3}$$

by the binomial theorem.[1] Now for $k > 0$ we have $J^k = p^{k-1}J$ [why?], while $J^0 = I$. (It is not clear *a priori* what is the "correct" value of $J^0$, but in order for (1.3) to be valid we must take $J^0 = I$.) Hence

---

[1] We can apply the binomial theorem in this situation because $I$ and $J$ *commute*. If $A$ and $B$ are $p \times p$ matrices that don't necessarily commute, then the best we can say is $(A + B)^2 = A^2 + AB + BA + B^2$ and similarly for higher powers.

$$(J - I)^\ell = \sum_{k=1}^{\ell} (-1)^{\ell-k} \binom{\ell}{k} p^{k-1} J + (-1)^\ell I.$$

Again by the binomial theorem we have

$$(J - I)^\ell = \frac{1}{p}((p-1)^\ell - (-1)^\ell)J + (-1)^\ell I. \tag{1.4}$$

Taking the $(i, j)$-entry of each side when $i \neq j$ yields

$$(A(K_p)^\ell)_{ij} = \frac{1}{p}((p-1)^\ell - (-1)^\ell). \tag{1.5}$$

If we take the $(i, i)$-entry of (1.4), then we recover (1.2). Note the curious fact that if $i \neq j$ then

$$(A(K_p)^\ell)_{ii} - (A(K_p)^\ell)_{ij} = (-1)^\ell.$$

We could also have deduced (1.5) from Corollary 1.6 using

$$\sum_{i=1}^{p} \sum_{j=1}^{p} \left( A(K_p)^\ell \right)_{ij} = p(p-1)^\ell,$$

the total number of walks of length $\ell$ in $K_p$. Details are left to the reader.

We now will show how (1.2) itself determines the eigenvalues of $A(K_p)$. Thus if (1.2) is proved without first computing the eigenvalues of $A(K_p)$ (which in fact is what we did two paragraphs ago), then we have another means to compute the eigenvalues. The argument we will give can in principle be applied to any graph $G$, not just $K_p$. We begin with a simple lemma.

**1.7 Lemma.** *Suppose $\alpha_1, \ldots, \alpha_r$ and $\beta_1, \ldots, \beta_s$ are nonzero complex numbers such that for all positive integers $\ell$, we have*

$$\alpha_1^\ell + \cdots + \alpha_r^\ell = \beta_1^\ell + \cdots + \beta_s^\ell. \tag{1.6}$$

*Then $r = s$ and the $\alpha$'s are just a permutation of the $\beta$'s.*

*Proof.* We will use the powerful method of *generating functions*. Let $x$ be a complex number whose absolute value (or modulus) is close to 0. Multiply (1.6) by $x^\ell$ and sum on all $\ell \geq 1$. The geometric series we obtain will converge, and we get

$$\frac{\alpha_1 x}{1 - \alpha_1 x} + \cdots + \frac{\alpha_r x}{1 - \alpha_r x} = \frac{\beta_1 x}{1 - \beta_1 x} + \cdots + \frac{\beta_s x}{1 - \beta_s x}. \tag{1.7}$$

This is an identity valid for sufficiently small (in modulus) complex numbers. By clearing denominators we obtain a polynomial identity. But if two polynomials in $x$ agree for infinitely many values, then they are the same polynomial [why?]. Hence (1.7) is actually valid for *all* complex numbers $x$ (ignoring values of $x$ which give rise to a zero denominator).

Fix a complex number $\gamma \neq 0$. Multiply (1.7) by $1 - \gamma x$ and let $x \to 1/\gamma$. The left-hand side becomes the number of $\alpha_i$'s which are equal to $\gamma$, while the right-hand side becomes the number of $\beta_j$'s which are equal to $\gamma$ [why?]. Hence these numbers agree for all $\gamma$, so the lemma is proved.                                    □

**1.8 Example.** Suppose that $G$ is a graph with 12 vertices and that the number of closed walks of length $\ell$ in $G$ is equal to $3 \cdot 5^\ell + 4^\ell + 2(-2)^\ell + 4$. Then it follows from Corollary 1.3 and Lemma 1.7 [why?] that the eigenvalues of $A(G)$ are given by $5, 5, 5, 4, -2, -2, 1, 1, 1, 1, 0, 0$.

# Notes for Chapter 1

The connection between graph eigenvalues and the enumeration of walks is considered "folklore." The subject of *spectral graph theory*, which is concerned with the spectrum (multiset of eigenvalues) of various matrices associated with graphs, began around 1931 in the area of quantum chemistry. The first mathematical paper was published by L. Collatz and U. Sinogowitz in 1957. A good general reference is the book[2] [28] by Cvetković et al. Two textbooks on this subject are by Cvetković et al. [29] and by Brouwer and Haemers [14].

# Exercises for Chapter 1

NOTE. An exercise marked with (*) is treated in the Hints section beginning on page 245.

1. (tricky) Find a combinatorial proof of Corollary 1.6, i.e., the number of closed walks of length $\ell$ in $K_p$ from some vertex to itself is given by $\frac{1}{p}((p-1)^\ell + (p-1)(-1)^\ell)$.

2. Suppose that the graph $G$ has 15 vertices and that the number of closed walks of length $\ell$ in $G$ is $8^\ell + 2 \cdot 3^\ell + 3 \cdot (-1)^\ell + (-6)^\ell + 5$ for all $\ell \geq 1$. Let $G'$ be the graph obtained from $G$ by adding a loop at each vertex (in addition to whatever loops are already there). How many closed walks of length $\ell$ are there in $G'$? (Use linear algebraic techniques. You can also try to solve the problem purely by combinatorial reasoning.)

---

[2] All citations to the literature refer to the bibliography beginning on page 251.

3. A *bipartite graph* $G$ with *vertex bipartition* $(A, B)$ is a graph whose vertex set is the disjoint union $A \cup B$ of $A$ and $B$, such that every edge of $G$ is incident to one vertex in $A$ and one vertex in $B$. Show by a walk-counting argument that the nonzero eigenvalues of $G$ come in pairs $\pm\lambda$.

   An equivalent formulation can be given in terms of the characteristic polynomial $f(x)$ of the matrix $A(G)$. Recall that the *characteristic polynomial* of a $p \times p$ matrix $A$ is defined to be $\det(A - xI)$. The present exercise is then equivalent to the statement that when $G$ is bipartite, the characteristic polynomial $f(x)$ of $A(G)$ has the form $g(x^2)$ (if $G$ has an even number of vertices) or $xg(x^2)$ (if $G$ has an odd number of vertices) for some polynomial $g(x)$.

   NOTE. Sometimes the characteristic polynomial of a $p \times p$ matrix $A$ is defined to be $\det(xI - A) = (-1)^p \det(A - xI)$. We will use the definition $\det(A - xI)$, so that the value at $x = 0$ is $\det A$.

4. Let $r, s \geq 1$. The *complete bipartite graph* $K_{rs}$ has vertices $u_1, u_2, \ldots, u_r$, $v_1, v_2, \ldots, v_s$, with one edge between each $u_i$ and $v_j$ (so $rs$ edges in all).

   (a) By purely combinatorial reasoning, compute the number of closed walks of length $\ell$ in $K_{rs}$.

   (b) Deduce from (a) the eigenvalues of $K_{rs}$.

5. (*) Let $H_n$ be the complete bipartite graph $K_{nn}$ with $n$ vertex-disjoint edges removed. Thus $H_n$ has $2n$ vertices and $n(n-1)$ edges, each of *degree* (number of incident edges) $n - 1$. Show that the eigenvalues of $H_n$ are $\pm 1$ ($n - 1$ times each) and $\pm(n - 1)$ (once each).

6. Let $n \geq 1$. The *complete p-partite graph* $K(n, p)$ has vertex set $V = V_1 \cup \cdots \cup V_p$ (disjoint union), where each $|V_i| = n$, and an edge from every element of $V_i$ to every element of $V_j$ when $i \neq j$. (If $u, v \in V_i$ then there is no edge $uv$.) Thus $K(1, p)$ is the complete graph $K_p$, and $K(n, 2)$ is the complete bipartite graph $K_{nn}$.

   (a) (*) Use Corollary 1.6 to find the number of closed walks of length $\ell$ in $K(n, p)$.

   (b) Deduce from (a) the eigenvalues of $K(n, p)$.

7. Let $G$ be any finite simple graph, with eigenvalues $\lambda_1, \ldots, \lambda_p$. Let $G(n)$ be the graph obtained from $G$ by replacing each vertex $v$ of $G$ with a set $V_v$ of $n$ vertices, such that if $uv$ is an edge of $G$, then there is an edge from every vertex of $V_u$ to every vertex of $V_v$ (and no other edges). For instance, $K_p(n) = K(n, p)$. Find the eigenvalues of $G(n)$ in terms of $\lambda_1, \ldots, \lambda_p$.

8. Let $G$ be a (finite) graph on $p$ vertices. Let $G'$ be the graph obtained from $G$ by placing a new edge $e_v$ incident to each vertex $v$, with the other vertex of $e_v$ being a new vertex $v'$. Thus $G'$ has $p$ new edges and $p$ new vertices. The new vertices all have degree one. By combinatorial or algebraic reasoning, show that if $G$ has eigenvalues $\lambda_i$ then $G'$ has eigenvalues $(\lambda_i \pm \sqrt{\lambda_i^2 + 4})/2$. (An algebraic proof is much easier than a combinatorial proof.)

9. Let $G$ be a (finite) graph with vertices $v_1, \ldots, v_p$ and eigenvalues $\lambda_1, \ldots, \lambda_p$. We know that for any $i$, $j$ there are real numbers $c_1(i, j), \ldots, c_p(i, j)$ such that for all $\ell \geq 1$,

$$\left(A(G)^\ell\right)_{ij} = \sum_{k=1}^{p} c_k(i, j)\lambda_k^\ell.$$

   (a) Show that $c_k(i, i) \geq 0$.
   (b) Show that if $i \neq j$ then we can have $c_k(i, j) < 0$. (The simplest possible example will work.)

10. Let $G$ be a finite graph with eigenvalues $\lambda_1, \ldots, \lambda_p$. Let $G^\star$ be the graph with the same vertex set as $G$ and with $\eta(u, v)$ edges between vertices $u$ and $v$ (including $u = v$), where $\eta(u, v)$ is the number of walks in $G$ of length two from $u$ to $v$. For example,

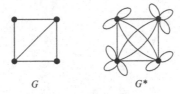

$$G \qquad\qquad\qquad G^\star$$

   Find the eigenvalues of $G^\star$ in terms of those of $G$.

11. (*) Let $K_n^o$ denote the complete graph with $n$ vertices, with one loop at each vertex. (Thus $A(K_n^o) = J_n$, the $n \times n$ all 1's matrix, and $K_n^o$ has $\binom{n+1}{2}$ edges.) Let $K_n^o - K_m^o$ denote $K_n^o$ with the edges of $K_m^o$ removed, i.e., choose $m$ vertices of $K_n^o$ and remove all edges between these vertices (including loops). (Thus $K_n^o - K_m^o$ has $\binom{n+1}{2} - \binom{m+1}{2}$ edges.) Find the number $C(\ell)$ of closed walks in $\Gamma = K_{21}^o - K_{18}^o$ of length $\ell \geq 1$.

12. (a) Let $G$ be a finite graph, and let $\Delta$ be the maximum degree of any vertex of $G$. Let $\lambda_1$ be the largest eigenvalue of the adjacency matrix $A(G)$. Show that $\lambda_1 \leq \Delta$.
   (b) (*) Suppose that $G$ is simple (no loops or multiple edges) and has a total of $q$ edges. Show that $\lambda_1 \leq \sqrt{2q}$.

13. Let $G$ be a finite simple graph with at least two vertices. Suppose that for some $\ell \geq 1$, the number of walks of length $\ell$ between any two vertices $u$, $v$ (including $u = v$) is odd. Show that there is a nonempty subset $S$ of the vertices such that $S$ has an even number of elements and such that every vertex $v$ of $G$ is adjacent to an even number of vertices in $S$. (In a simple graph, no vertex is adjacent to itself.)

# Chapter 2
# Cubes and the Radon Transform

Let us now consider a more interesting example of a graph $G$, one whose eigenvalues have come up in a variety of applications. Let $\mathbb{Z}_2$ denote the cyclic group of order 2, with elements 0 and 1 and group operation being addition modulo 2. Thus $0 + 0 = 0$, $0 + 1 = 1 + 0 = 1$, and $1 + 1 = 0$. Let $\mathbb{Z}_2^n$ denote the direct product of $\mathbb{Z}_2$ with itself $n$ times, so the elements of $\mathbb{Z}_2^n$ are $n$-tuples $(a_1, \ldots, a_n)$ of 0's and 1's, under the operation of component-wise addition. Define a graph $C_n$, called the *n-cube*, as follows: the vertex set of $C_n$ is given by $V(C_n) = \mathbb{Z}_2^n$, and two vertices $u$ and $v$ are connected by an edge if they differ in exactly one component. Equivalently, $u + v$ has exactly one nonzero component. If we regard $\mathbb{Z}_2^n$ as consisting of *real* vectors, then these vectors form the set of vertices of an $n$-dimensional cube. Moreover, two vertices of the cube lie on an edge (in the usual geometric sense) if and only if they form an edge of $C_n$. This explains why $C_n$ is called the $n$-cube. We also see that walks in $C_n$ have a nice geometric interpretation—they are simply walks along the edges of an $n$-dimensional cube.

We want to determine explicitly the eigenvalues and eigenvectors of $C_n$. We will do this by a somewhat indirect but extremely useful and powerful technique, the finite Radon transform. Let $\mathcal{V}$ denote the set of all functions $f \colon \mathbb{Z}_2^n \to \mathbb{R}$, where $\mathbb{R}$ denotes the field of real numbers.[1] Note that $\mathcal{V}$ is a vector space over $\mathbb{R}$ of dimension $2^n$ [why?]. If $u = (u_1, \ldots, u_n)$ and $v = (v_1, \ldots, v_n)$ are elements of $\mathbb{Z}_2^n$, then define their *dot product* by

$$u \cdot v = u_1 v_1 + \cdots + u_n v_n, \tag{2.1}$$

where the computation is performed modulo 2. Thus we regard $u \cdot v$ as an element of $\mathbb{Z}_2$. The expression $(-1)^{u \cdot v}$ is defined to be the *real number* $+1$ or $-1$, depending on whether $u \cdot v = 0$ or 1, respectively. Since for integers $k$ the value of $(-1)^k$

---

[1] For abelian groups other than $\mathbb{Z}_2^n$ it is necessary to use complex numbers rather than real numbers. We could use complex numbers here, but there is no need to do so.

© Springer International Publishing AG, part of Springer Nature 2018
R. P. Stanley, *Algebraic Combinatorics*, Undergraduate Texts in Mathematics,
https://doi.org/10.1007/978-3-319-77173-1_2

depends only on $k \pmod 2$, it follows that we can treat $u$ and $v$ as integer vectors without affecting the value of $(-1)^{u \cdot v}$. Thus, for instance, formulas such as

$$(-1)^{u \cdot (v+w)} = (-1)^{u \cdot v + u \cdot w} = (-1)^{u \cdot v}(-1)^{u \cdot w}$$

are well defined and valid. From a more algebraic viewpoint, the map $\mathbb{Z} \to \{-1, 1\}$ sending $n$ to $(-1)^n$ is a group homomorphism, where of course the product on $\{-1, 1\}$ is multiplication.

We now define two important bases of the vector space $\mathcal{V}$. There will be one basis element of each basis for each $u \in \mathbb{Z}_2^n$. The first basis, denoted $B_1$, has elements $f_u$ defined as follows:

$$f_u(v) = \delta_{uv}, \tag{2.2}$$

the Kronecker delta. It is easy to see that $B_1$ is a basis, since any $g \in \mathcal{V}$ satisfies

$$g = \sum_{u \in \mathbb{Z}_2^n} g(u) f_u \tag{2.3}$$

[why?]. Hence $B_1$ spans $\mathcal{V}$, so since $\#B_1 = \dim \mathcal{V} = 2^n$, it follows that $B_1$ is a basis. The second basis, denoted $B_2$, has elements $\chi_u$ defined as follows:

$$\chi_u(v) = (-1)^{u \cdot v}.$$

In order to show that $B_2$ is a basis, we will use an inner product on $\mathcal{V}$ (denoted $\langle \cdot, \cdot \rangle$) defined by

$$\langle f, g \rangle = \sum_{u \in \mathbb{Z}_2^n} f(u) g(u).$$

Note that this inner product is just the usual dot product with respect to the basis $B_1$.

**2.1 Lemma.** *The set $B_2 = \{\chi_u : u \in \mathbb{Z}_2^n\}$ forms a basis for $\mathcal{V}$.*

*Proof.* Since $\#B_2 = \dim \mathcal{V}$ (= $2^n$), it suffices to show that $B_2$ is linearly independent. In fact, we will show that the elements of $B_2$ are orthogonal.[2] We have

$$\langle \chi_u, \chi_v \rangle = \sum_{w \in \mathbb{Z}_2^n} \chi_u(w) \chi_v(w)$$

$$= \sum_{w \in \mathbb{Z}_2^n} (-1)^{(u+v) \cdot w}.$$

---

[2]Recall from linear algebra that nonzero orthogonal vectors in a real vector space are linearly independent.

It is left as an easy exercise to the reader to show that for any $y \in \mathbb{Z}_2^n$, we have

$$\sum_{w \in \mathbb{Z}_2^n} (-1)^{y \cdot w} = \begin{cases} 2^n, & \text{if } y = \mathbf{0}, \\ 0, & \text{otherwise}, \end{cases}$$

where $\mathbf{0}$ denotes the identity element of $\mathbb{Z}_2^n$ (the vector $(0, 0, \ldots, 0)$). Thus $\langle \chi_u, \chi_v \rangle \neq 0$ if and only $u + v = \mathbf{0}$, i.e., $u = v$, so the elements of $B_2$ are orthogonal (and nonzero). Hence they are linearly independent as desired.                   $\square$

We now come to the key definition of the Radon transform.

Given a subset $\Gamma$ of $\mathbb{Z}_2^n$ and a function $f \in \mathcal{V}$, define a new function $\Phi_\Gamma f \in \mathcal{V}$ by

$$\Phi_\Gamma f(v) = \sum_{w \in \Gamma} f(v + w).$$

The function $\Phi_\Gamma f$ is called the (discrete or finite) Radon transform of $f$ (on the group $\mathbb{Z}_2^n$, with respect to the subset $\Gamma$).

We have defined a map $\Phi_\Gamma : \mathcal{V} \to \mathcal{V}$. It is easy to see that $\Phi_\Gamma$ is a linear transformation; we want to compute its eigenvalues and eigenvectors.

**2.2 Theorem.** *The eigenvectors of $\Phi_\Gamma$ are the functions $\chi_u$, where $u \in \mathbb{Z}_2^n$. The eigenvalue $\lambda_u$ corresponding to $\chi_u$ (i.e., $\Phi_\Gamma \chi_u = \lambda_u \chi_u$) is given by*

$$\lambda_u = \sum_{w \in \Gamma} (-1)^{u \cdot w}.$$

*Proof.* Let $v \in \mathbb{Z}_2^n$. Then

$$\Phi_\Gamma \chi_u(v) = \sum_{w \in \Gamma} \chi_u(v + w)$$

$$= \sum_{w \in \Gamma} (-1)^{u \cdot (v + w)}$$

$$= \left( \sum_{w \in \Gamma} (-1)^{u \cdot w} \right) (-1)^{u \cdot v}$$

$$= \left( \sum_{w \in \Gamma} (-1)^{u \cdot w} \right) \chi_u(v).$$

Hence

$$\Phi_\Gamma \chi_u = \left( \sum_{w \in \Gamma} (-1)^{u \cdot w} \right) \chi_u,$$

as desired.                                                          $\square$

Note that because the $\chi_u$'s form a basis for $\mathcal{V}$ by Lemma 2.1, it follows that Theorem 2.2 yields a complete set of eigenvalues and eigenvectors for $\Phi_\Gamma$. Note also that the eigenvectors $\chi_u$ of $\Phi_\Gamma$ are independent of $\Gamma$; only the eigenvalues depend on $\Gamma$.

Now we come to the payoff. Let $\Delta = \{\delta_1, \ldots, \delta_n\}$, where $\delta_i$ is the $i$th unit coordinate vector (i.e., $\delta_i$ has a 1 in position $i$ and 0's elsewhere). Note that the $j$th coordinate of $\delta_i$ is just $\delta_{ij}$ (the Kronecker delta), explaining our notation $\delta_i$. Let $[\Phi_\Delta]$ denote the matrix of the linear transformation $\Phi_\Delta \colon \mathcal{V} \to \mathcal{V}$ with respect to the basis $B_1$ of $\mathcal{V}$ given by (2.2).

**2.3 Lemma.** *We have* $[\Phi_\Delta] = A(C_n)$, *the adjacency matrix of the $n$-cube.*

*Proof.* Let $v \in \mathbb{Z}_2^n$. We have

$$\Phi_\Delta f_u(v) = \sum_{w \in \Delta} f_u(v + w)$$

$$= \sum_{w \in \Delta} f_{u+w}(v),$$

since $u = v + w$ if and only if $u + w = v$. There follows

$$\Phi_\Delta f_u = \sum_{w \in \Delta} f_{u+w}. \tag{2.4}$$

Equation (2.4) says that the $(u, v)$-entry (short for $(f_u, f_v)$-entry) of the matrix $[\Phi_\Delta]$ is given by

$$(\Phi_\Delta)_{uv} = \begin{cases} 1, & \text{if } u + v \in \Delta \\ 0, & \text{otherwise.} \end{cases}$$

Now $u + v \in \Delta$ if and only if $u$ and $v$ differ in exactly one coordinate. This is just the condition for $uv$ to be an edge of $C_n$, so the proof follows.                                      □

**2.4 Corollary.** *The eigenvectors $E_u$ ($u \in \mathbb{Z}_2^n$) of $A(C_n)$ (regarded as linear combinations of the vertices of $C_n$, i.e., of the elements of $\mathbb{Z}_2^n$) are given by*

$$E_u = \sum_{v \in \mathbb{Z}_2^n} (-1)^{u \cdot v} v. \tag{2.5}$$

*The eigenvalue $\lambda_u$ corresponding to the eigenvector $E_u$ is given by*

$$\lambda_u = n - 2\omega(u), \tag{2.6}$$

*where $\omega(u)$ is the number of 1's in u. (The integer $\omega(u)$ is called the* Hamming weight *or simply the* weight *of u.) Hence $A(C_n)$ has $\binom{n}{i}$ eigenvalues equal to $n - 2i$, for each $0 \leq i \leq n$.*

*Proof.* For any function $g \in \mathcal{V}$ we have by (2.3) that

$$g = \sum_v g(v) f_v.$$

Applying this equation to $g = \chi_u$ gives

$$\chi_u = \sum_v \chi_u(v) f_v = \sum_v (-1)^{u \cdot v} f_v. \tag{2.7}$$

Equation (2.7) expresses the eigenvector $\chi_u$ of $\Phi_\Delta$ (or even $\Phi_\Gamma$ for any $\Gamma \subseteq \mathbb{Z}_2^n$) as a linear combination of the functions $f_v$. But $\Phi_\Delta$ has the same matrix with respect to the basis of the $f_v$'s as $A(C_n)$ has with respect to the vertices $v$ of $C_n$. Hence the expansion of the eigenvectors of $\Phi_\Delta$ in terms of the $f_v$'s has the same coefficients as the expansion of the eigenvectors of $A(C_n)$ in terms of the $v$'s, so (2.5) follows.

According to Theorem 2.2 the eigenvalue $\lambda_u$ corresponding to the eigenvector $\chi_u$ of $\Phi_\Delta$ (or equivalently, the eigenvector $E_u$ of $A(C_n)$) is given by

$$\lambda_u = \sum_{w \in \Delta} (-1)^{u \cdot w}. \tag{2.8}$$

Now $\Delta = \{\delta_1, \ldots, \delta_n\}$ and $\delta_i \cdot u$ is 1 if $u$ has a one in its $i$th coordinate and is 0 otherwise. Hence the sum in (2.8) has $n - \omega(u)$ terms equal to $+1$ and $\omega(u)$ terms equal to $-1$, so $\lambda_u = (n - \omega(u)) - \omega(u) = n - 2\omega(u)$, as claimed. $\square$

We have all the information needed to count walks in $C_n$.

**2.5 Corollary.** *Let $u, v \in \mathbb{Z}_2^n$, and suppose that $\omega(u + v) = k$ (i.e., u and v disagree in exactly k coordinates). Then the number of walks of length $\ell$ in $C_n$ between u and v is given by*

$$(A^\ell)_{uv} = \frac{1}{2^n} \sum_{i=0}^{n} \sum_{j=0}^{k} (-1)^j \binom{k}{j} \binom{n-k}{i-j} (n - 2i)^\ell, \tag{2.9}$$

*where we set $\binom{n-k}{i-j} = 0$ if $j > i$. In particular,*

$$(A^\ell)_{uu} = \frac{1}{2^n} \sum_{i=0}^{n} \binom{n}{i} (n - 2i)^\ell. \tag{2.10}$$

*Proof.* Let $E_u$ and $\lambda_u$ be as in Corollary 2.4. In order to apply Corollary 1.2, we need the eigenvectors to be of *unit* length (where we regard the $f_v$'s as an orthonormal basis of $\mathcal{V}$). By (2.5), we have

$$|E_u|^2 = \sum_{v \in \mathbb{Z}_2^n} ((-1)^{u \cdot v})^2 = 2^n.$$

Hence we should replace $E_u$ by $E_u' = \frac{1}{2^{n/2}} E_u$ to get an orthonormal basis. According to Corollary 1.2, we thus have

$$(A^\ell)_{uv} = \frac{1}{2^n} \sum_{w \in \mathbb{Z}_2^n} E_{uw} E_{vw} \lambda_w^\ell.$$

Now $E_{uw}$ by definition is the coefficient of $f_w$ in the expansion (2.5), i.e., $E_{uw} = (-1)^{u \cdot w}$ (and similarly for $E_v$), while $\lambda_w = n - 2\omega(w)$. Hence

$$(A^\ell)_{uv} = \frac{1}{2^n} \sum_{w \in \mathbb{Z}_2^n} (-1)^{(u+v) \cdot w} (n - 2\omega(w))^\ell. \tag{2.11}$$

The number of vectors $w$ of Hamming weight $i$ which have $j$ 1's in common with $u + v$ is $\binom{k}{j}\binom{n-k}{i-j}$, since we can choose the $j$ 1's in $u + v$ which agree with $w$ in $\binom{k}{j}$ ways, while the remaining $i - j$ 1's of $w$ can be inserted in the $n - k$ remaining positions in $\binom{n-k}{i-j}$ ways. Since $(u + v) \cdot w \equiv j \pmod 2$, the sum (2.11) reduces to (2.9) as desired. Clearly setting $u = v$ in (2.9) yields (2.10), completing the proof.

$\square$

It is possible to give a direct proof of (2.10) avoiding linear algebra, though we do not do so here. Thus by Corollary 1.3 and Lemma 1.7 (exactly as was done for $K_n$) we have another determination of the eigenvalues of $C_n$. With a little more work one can also obtain a direct proof of (2.9). Later in Example 9.12, however, we will use the eigenvalues of $C_n$ to obtain a combinatorial result for which a nonalgebraic proof was found only recently and is by no means easy.

**2.6 Example.** Setting $k = 1$ in (2.9) yields

$$(A^\ell)_{uv} = \frac{1}{2^n} \sum_{i=0}^{n} \left[ \binom{n-1}{i} - \binom{n-1}{i-1} \right] (n - 2i)^\ell$$

$$= \frac{1}{2^n} \sum_{i=0}^{n-1} \binom{n-1}{i} \frac{(n-2i)^{\ell+1}}{n-i}.$$

NOTE (for those familiar with the representation theory of finite groups). The functions $\chi_u \colon \mathbb{Z}_2^n \to \mathbb{R}$ are just the irreducible (complex) characters of the group $\mathbb{Z}_2^n$, and the orthogonality of the $\chi_u$'s shown in the proof of Lemma 2.1 is the usual orthogonality relation for the irreducible characters of a finite group. The results of this chapter extend readily to any finite abelian group. Exercise 2.5 does the case $\mathbb{Z}_n$, the cyclic group of order $n$. For nonabelian finite groups the situation is much more complicated because not all irreducible representations have degree one (i.e., are homomorphisms $G \to \mathbb{C}^*$, the multiplicative group of $\mathbb{C}$), and there do not exist formulas as explicit as the ones for abelian groups.

We can give a little taste of the situation for arbitrary groups as follows. Let $G$ be a finite group, and let $M(G)$ be its multiplication table. Regard the entries of $M(G)$ as *commuting* indeterminates, so that $M(G)$ is simply a matrix with indeterminate entries. For instance, let $G = \mathbb{Z}_3$. Let the elements of $G$ be $a, b, c$, where say $a$ is the identity. Then

$$M(G) = \begin{bmatrix} a & b & c \\ b & c & a \\ c & a & b \end{bmatrix}.$$

We can compute that $\det M(G) = (a + b + c)(a + \omega b + \omega^2 c)(a + \omega^2 b + \omega c)$, where $\omega = e^{2\pi i/3}$. In general, when $G$ is abelian, Dedekind knew that $\det M(G)$ factors into certain explicit linear factors over $\mathbb{C}$. Theorem 2.2 is equivalent to this statement for the group $G = \mathbb{Z}_2^n$ [why?]. Equation (13.5) gives the factorization for $G = \mathbb{Z}_n$. (For each $w \in G$ one needs to interchange the row indexed by the group element $w$ with the row indexed by $w^{-1}$ in order to convert $M(\mathbb{Z}_n)$ to the circulant matrices of (13.5), but these operations only affect the sign of the determinant.) Dedekind asked Frobenius about the factorization of $\det M(G)$, known as the *group determinant*, for nonabelian finite $G$. For instance, let $G = \mathfrak{S}_3$ (the symmetric group of all permutations of $1, 2, 3$), with elements (in cycle notation) $a = (1)(2)(3)$, $b = (1, 2)(3)$, $c = (1, 3)(2)$, $d = (1)(2, 3)$, $e = (1, 2, 3)$, and $f = (1, 3, 2)$. Then $\det M(G) = f_1 f_2 f_3^2$, where

$$f_1 = a + b + c + d + e + f$$
$$f_2 = -a + b + c + d - e - f$$
$$f_3 = a^2 - b^2 - c^2 - d^2 + e^2 + f^2 - ae - af + bc + bd + cd - ef.$$

Frobenius showed that in general there is a set $\mathcal{P}$ of irreducible homogeneous polynomials $f$, of some degree $d_f$, where $\#\mathcal{P}$ is the number of conjugacy classes of $G$, for which

$$\det M(G) = \prod_{f \in \mathcal{P}} f^{d_f}.$$

Note that taking the degree of both sides gives $\#G = \sum_f d_f^2$. Frobenius' result was a highlight in his development of group representation theory. The numbers $d_f$ are just the degrees of the irreducible (complex) representations of $G$. For the symmetric group $\mathfrak{S}_n$, these degrees are the numbers $f^\lambda$ of Theorem 8.1, and Appendix 1 of Chapter 8 gives a bijective proof that $\sum_\lambda (f^\lambda)^2 = n!$.

## Notes for Chapter 2

The Radon transform first arose in a continuous setting in the paper [107] of Radon and has been applied to such areas as computerized tomography. The finite version was first defined by Bolker [10]. For some further applications to combinatorics see Kung [80]. For the Radon transform on the $n$-cube $\mathbb{Z}_2^n$, see Diaconis and Graham [34]. For the generalization to $\mathbb{Z}_k^n$, see DeDeo and Velasquez [33].

For an exposition of the development of group representation theory by Frobenius and other pioneers, see the survey articles of Hawkins [63–65].

## Exercises for Chapter 2

1. (a) Start with $n$ coins heads up. Choose a coin at random (each equally likely) and turn it over. Do this a total of $\ell$ times. What is the probability that all coins will have heads up? (Don't solve this from scratch; rather use some previous results.)
   (b) Same as (a), except now compute the probability that all coins have tails up.
   (c) Same as (a), but now we turn over two coins at a time.
2. (a) (difficult) (*) For $k < n/2$ let $\mathcal{C}_{n,k}$ be the subgraph of the cube $\mathcal{C}_n$ spanned by all vertices of $\mathcal{C}_n$ with $k - 1$ or $k$ 1's (so the edges of $\mathcal{C}_{n,k}$ consist of all edges of $\mathcal{C}_n$ that connect two vertices of $\mathcal{C}_{n,k}$; there are a total of $k\binom{n}{k}$ edges). Show that the characteristic polynomial of $A = A(\mathcal{C}_{n,k})$ is given by

$$\det(A - xI) = \pm x^{\binom{n}{k} - \binom{n}{k-1}} \prod_{i=1}^{k} (x^2 - i(n - 2k + i + 1))^{\binom{n}{k-i} - \binom{n}{k-i-1}},$$

   where we set $\binom{n}{-1} = 0$.
   (b) Find the number of closed walks in $\mathcal{C}_{n,k}$ of length $\ell$ beginning and ending with a fixed vertex $v$.
3. (very difficult, and unrelated to the text) (*) Let $n = 2k + 1$. Show that the graphs $\mathcal{C}_{n,k+1}$ of Problem 2 above have a Hamiltonian cycle, i.e., a closed path that contains every vertex exactly once. A *closed path* in a graph $G$ is a closed walk that does not repeat any vertices except at the last step.

4. Let $G$ be the graph with vertex set $\mathbb{Z}_2^n$ (the same as the $n$-cube) and with edge set defined as follows: $\{u, v\}$ is an edge of $G$ if $u$ and $v$ differ in exactly *two* coordinates (i.e., if $\omega(u, v) = 2$). What are the eigenvalues of $G$?

5. This problem is devoted to the graph $Z_n$ with vertex set $\mathbb{Z}_n$ (the cyclic group of order $n$, with elements $0, 1, \ldots, n - 1$ under the operation of addition modulo $n$) and edges consisting of all pairs $\{i, i + 1\}$ (with $i + 1$ computed in $\mathbb{Z}_n$, so $(n - 1) + 1 = 0$). The graph $Z_n$ is called an $n$-*cycle*. We will develop properties of its adjacency matrix analogously to what was done for the $n$-cube $C_n$. It will be necessary to work over the complex numbers $\mathbb{C}$. Recall that there are exactly $n$ complex numbers $z$ (called $n$th roots of unity) satisfying $z^n = 1$. They are given by $\zeta^0 = 1, \zeta^1 = \zeta, \zeta^2, \ldots, \zeta^{n-1}$, where $\zeta = e^{2\pi i/n}$.

   (a) Draw the graphs $Z_3$, $Z_4$, and $Z_5$.
   (b) Let $\mathcal{V}$ be the complex vector space of all functions $f : \mathbb{Z}_n \to \mathbb{C}$. What is the dimension of $\mathcal{V}$?
   (c) (*) If $k \in \mathbb{Z}$, then note that $\zeta^k$ depends only on the value of $k$ modulo $n$. Hence if $u \in \mathbb{Z}_n$ then we can define $\zeta^u$ by regarding $u$ as an ordinary integer, and the usual laws of exponents such as $\zeta^{u+v} = \zeta^u \zeta^v$ (where $u, v \in \mathbb{Z}_n$) still hold. For $u \in \mathbb{Z}_n$ define $\chi_u \in \mathcal{V}$ by $\chi_u(v) = \zeta^{uv}$. Let $B = \{\chi_u : u \in \mathbb{Z}_n\}$. Show that $B$ is a basis for $\mathcal{V}$.
   (d) Given $\Gamma \subseteq \mathbb{Z}_n$ and $f \in \mathcal{V}$, define $\Phi_\Gamma f \in \mathcal{V}$ by

$$\Phi_\Gamma f(v) = \sum_{w \in \Gamma} f(v + w).$$

   Show that the eigenvectors of $\Phi_\Gamma$ are the functions $\chi_u$, with corresponding eigenvalue $\lambda_u = \sum_{w \in \Gamma} \zeta^{uw}$.
   (e) Let $\Delta = \{1, n - 1\} \subseteq \mathbb{Z}_n$. Define $f_u \in \mathcal{V}$ by $f_u(v) = \delta_{uv}$. Let $F = \{f_u : u \in \mathbb{Z}_n\}$. It is clear that $F$ is a basis for $\mathcal{V}$ (just as for $C_n$). Show that the matrix $[\Phi_\Delta]$ of $\Phi_\Delta$ with respect to the basis $F$ is just $A(Z_n)$, the adjacency matrix of $Z_n$.
   (f) Show that the eigenvalues of $A(Z_n)$ are the numbers $2\cos(\frac{2\pi j}{n})$, where $0 \leq j \leq n - 1$. What are the corresponding eigenvectors?
   (g) How many closed walks in $Z_n$ are of length $\ell$ and start at $0$? Give the answers in the cases $n = 4$ and $n = 6$ without using trigonometric functions, complex exponentials, etc.
   (h) Let $Z_n^{(2)}$ be the graph with vertex set $\mathbb{Z}_n$ and edges $\{i, j\}$ for $j - i = 1$ or $j - i = 2$. How many closed walks in $Z_n^{(2)}$ are of length $\ell$ and start at $0$? Try to express your answer in terms of trigonometric functions, and not involving complex numbers.

6. Let $\widetilde{C}_n$ be the graph obtained from the $n$-cube graph $C_n$ by adding an edge between every vertex $v$ and its antipode (the vertex which differs from $v$ in all $n$ coordinates). Find the number of closed walks in $\widetilde{C}_n$ of length $\ell$ which begin (and hence end) at the origin $\mathbf{0} = (0, 0, \ldots, 0)$.

# Chapter 3
# Random Walks

Let $G$ be a finite graph. We assume throughout this chapter that $G$ has at least two vertices and is *connected*, i.e., there exists a walk between any two vertices of $G$. We consider a random walk on the vertices of $G$ of the following type. Start at a vertex $u$. (The vertex $u$ could be chosen randomly according to some probability distribution or could be specified in advance.) Among all the edges incident to $u$, choose one uniformly at random (i.e., if there are $k$ edges incident to $u$, then each of these edges is chosen with probability $1/k$). Travel to the vertex $v$ at the other end of the chosen edge and continue as before from $v$. Readers with some familiarity with probability theory will recognize this random walk as a special case of a finite-state Markov chain. Many interesting questions may be asked about such walks; the basic one is to determine the probability of being at a given vertex after a given number $\ell$ of steps.

Suppose vertex $u$ has *degree* $d_u$, i.e., there are $d_u$ edges incident to $u$ (counting loops at $u$ once only). Let $M = M(G)$ be the matrix whose rows and columns are indexed by the vertex set $\{v_1, \ldots, v_p\}$ of $G$ and whose $(u, v)$-entry is given by

$$M_{uv} = \frac{\mu_{uv}}{d_u}, \tag{3.1}$$

where $\mu_{uv}$ is the number of edges between $u$ and $v$ (which for simple graphs will be 0 or 1). Thus $M_{uv}$ is just the probability that if one starts at $u$, then the next step will be to $v$. We call $M$ the *probability matrix* associated with $G$. An elementary probability theory argument (equivalent to Theorem 1.1) shows that if $\ell$ is a positive integer, then $(M^\ell)_{uv}$ is equal to the probability that one ends up at vertex $v$ in $\ell$ steps given that one has started at $u$. Suppose now that the starting vertex is not specified, but rather we are given probabilities $\rho_u$ summing to 1 and that we start at vertex $u$ with probability $\rho_u$. Let $P$ be the row vector $P = [\rho_{v_1}, \ldots, \rho_{v_p}]$. Then again an elementary argument shows that if $PM^\ell = [\sigma_{v_1}, \ldots, \sigma_{v_p}]$, then $\sigma_v$ is the probability of ending up at $v$ in $\ell$ steps (with the given starting distribution). By reasoning as in

© Springer International Publishing AG, part of Springer Nature 2018

R. P. Stanley, *Algebraic Combinatorics*, Undergraduate Texts in Mathematics, https://doi.org/10.1007/978-3-319-77173-1_3

Chapter 1, we see that if we know the eigenvalues and eigenvectors of $M$, then we can compute the crucial probabilities $(M^\ell)_{uv}$ and $\sigma_u$.

Since the matrix $M$ is not the same as the adjacency matrix $A$, what does all this have to do with adjacency matrices? The answer is that in one important case $M$ is just a scalar multiple of $A$. We say that the graph $G$ is *regular of degree d* if each $d_u = d$, i.e., each vertex is incident to $d$ edges. In this case it's easy to see that $M(G) = \frac{1}{d}A(G)$. Hence the eigenvectors $E_u$ of $M(G)$ and $A(G)$ are the same, and the eigenvalues are related by $\lambda_u(M) = \frac{1}{d}\lambda_u(A)$. Thus random walks on a regular graph are closely related to the adjacency matrix of the graph.

**3.1 Example.** Consider a random walk on the $n$-cube $C_n$ which begins at the "origin" (the vector $(0, \ldots, 0)$). What is the probability $p_\ell$ that after $\ell$ steps one is again at the origin? Before applying any formulas, note that after an even (respectively, odd) number of steps, one must be at a vertex with an even (respectively, odd) number of 1's. Hence $p_\ell = 0$ if $\ell$ is odd. Now note that $C_n$ is regular of degree $n$. Thus by (2.6), we have

$$\lambda_u(M(C_n)) = \frac{1}{n}(n - 2\omega(u)).$$

By (2.10) we conclude that

$$p_\ell = \frac{1}{2^n n^\ell} \sum_{i=0}^{n} \binom{n}{i} (n - 2i)^\ell.$$

Note that the above expression for $p_\ell$ does indeed reduce to 0 when $\ell$ is odd.

It is worth noting that even though the probability matrix $M$ need not be a symmetric matrix, nonetheless it has only real eigenvalues.

**3.2 Theorem.** *Let G be a finite graph. Then the probability matrix $M = M(G)$ is diagonalizable and has only real eigenvalues.*

*Proof.* Since we are assuming that $G$ is connected and has at least two vertices, it follows that $d_v > 0$ for every vertex $v$ of $G$. Let $D$ be the diagonal matrix whose rows and columns are indexed by the vertices of $G$, with $D_{vv} = \sqrt{d_v}$. Then

$$(DMD^{-1})_{uv} = \sqrt{d_u} \cdot \frac{\mu_{uv}}{d_u} \cdot \frac{1}{\sqrt{d_v}}$$

$$= \frac{\mu_{uv}}{\sqrt{d_u d_v}}.$$

Hence $DMD^{-1}$ is a symmetric matrix and thus has only real eigenvalues. But if $B$ and $C$ are any $p \times p$ matrices with $C$ invertible, then $B$ and $CBC^{-1}$ have the same characteristic polynomial and hence the same eigenvalues. Therefore all the eigenvalues of $M$ are real. Moreover, $B$ is diagonalizable if and only if $CBC^{-1}$

is diagonalizable. (In fact, $B$ and $CBC^{-1}$ have the same Jordan canonical form.) Since a symmetric matrix is diagonalizable, it follows that $M$ is also diagonalizable.  □

Let us give one further example of the connection between linear algebra and random walks on graphs. Let $u$ and $v$ be vertices of a connected graph $G$. Define the *access time* or *hitting time* $H(u, v)$ to be the expected number of steps that a random walk (as defined above) starting at $u$ takes to reach $v$ for the first time. Thus if the probability is $p_n$ that we reach $v$ for the first time in $n$ steps, then by definition of expectation we have

$$H(u, v) = \sum_{n \geq 1} np_n. \tag{3.2}$$

Conceivably this sum could be infinite, though we will see below that this is not the case. Note that $H(v, v) = 0$.

As an example, suppose that $G$ has three vertices $u, v, w$ with an edge between $u$ and $w$ and another edge between $w$ and $v$. We can compute $H(u, v)$ as follows. After one step we will be at $w$. Then with probability $\frac{1}{2}$ we will step to $v$ and with probability $\frac{1}{2}$ back to $u$. Hence [why?]

$$H(u, v) = \frac{1}{2} \cdot 2 + \frac{1}{2}(2 + H(u, v)). \tag{3.3}$$

Solving this linear equation gives $H(u, v) = 4$.

We want to give a formula for the access time $H(u, v)$ in terms of linear algebra. The proof requires some basic results on eigenvalues and eigenvectors of nonnegative matrices, which we will explain and then state without proof. An $r \times r$ real matrix $B$ is called *nonnegative* if every entry is nonnegative. We say that $B$ is *irreducible* if it is not the $1 \times 1$ matrix $[0]$ and if there does not exist a permutation matrix $P$ (a matrix with one 1 in every row and column, and all other entries 0) such that

$$PBP^{-1} = \begin{bmatrix} C & D \\ 0 & E \end{bmatrix},$$

where $C$ and $E$ are square matrices of size greater than zero. For instance, the adjacency matrix $A$ and probability matrix $M$ of a graph $G$ are irreducible if and only if $G$ is connected and is not an *isolated vertex* (i.e., a vertex $v$ incident to no edges, not even a loop from $v$ to itself). We now state without proof a version of the *Perron–Frobenius theorem*. There are some other parts of the Perron–Frobenius theorem that we don't need here and are omitted.

**3.3 Theorem.** *Let $B$ be a nonnegative irreducible square matrix. If $\rho$ is the maximum absolute value of the eigenvalues of $B$, then $\rho > 0$, and there is an* *eigenvalue equal to ρ. Moreover, there is an eigenvector for ρ (unique up to multiplication by a positive real number) all of whose entries are positive.*

Now let $M$ be the probability matrix defined by (3.1). Let $M[v]$ denote $M$ with the row and column indexed by $v$ deleted. Thus if $G$ has $p$ vertices, then $M[v]$ is a $(p-1) \times (p-1)$ matrix. Let $T[v]$ be the column vector of length $p-1$ whose rows are indexed by the vertices $w \neq v$, with $T[v]_w = \mu_{wv}/d_w$. Write $I_{p-1}$ for the identity matrix of size $p-1$.

**3.4 Theorem.** *The matrix $I_{p-1} - M[v]$ is invertible, and*

$$H(u, v) = ((I_{p-1} - M[v])^{-2}T[v])_u, \qquad (3.4)$$

*the u-entry of the column vector $(I_{p-1} - M[v])^{-2}T[v]$.*

*Proof.* We first give a "formal" argument and then justify its validity. The probability that when we take $n$ steps from $u$, we never reach $v$ and end up at some vertex $w$ is $(M[v]^n)_{uw}$ [why?]. The probability that once we reach $w$ the next step is to $v$ is $\mu(w, v)/d_w$. Hence by definition of expectation we have

$$H(u, v) = \sum_{w \neq v} \sum_{n \geq 0} (n + 1)\frac{\mu_{wv}}{d_w}(M[v]^n)_{uw}. \qquad (3.5)$$

We claim that if $x$ is a complex number satisfying $|x| < 1$, then

$$\sum_{n \geq 0}(n + 1)x^n = (1 - x)^{-2}. \qquad (3.6)$$

This identity is a simple exercise in calculus. For instance, we can compute the coefficient of $x^n$ in the product $(1 - x)^2 \sum_{n \geq 0}(n + 1)x^n$. We can also differentiate the familiar identity

$$\sum_{n \geq 0}x^n = \frac{1}{1 - x}. \qquad (3.7)$$

Another proof is obtained by expanding $(1 - x)^{-2}$ by the binomial theorem for the exponent $-2$. Convergence for $|x| < 1$ follows for example from the corresponding result for (3.7).

Let us "blindly" apply (3.6) to (3.5). We obtain

$$H(u, v) = \sum_{w \neq v}((I_{p-1} - M[v])^{-2})_{uw}\frac{\mu_{wv}}{d_w}$$

$$= ((I_{p-1} - M[v])^{-2}T[v])_u, \qquad (3.8)$$

as claimed.

It remains to justify our derivation of (3.8). For an arbitrary real (or complex) $r \times r$ matrix $B$, we can define $\sum_{n \geq 0}(n+1)B^n$ entry-wise, that is, we set $\sum_{n \geq 0}(n+1)B^n = C$ if

$$\sum_{n \geq 0}(n+1)(B^n)_{ij} = C_{ij}$$

for all $i$ and $j$ indexing the rows and columns of $B$ and $C$.

It is straightforward to verify by induction on $m$ the identity

$$(I_r - B)^2 \left(I_r + 2B + 3B^2 + \cdots + mB^{m-1}\right) = I_r - (m+1)B^m + mB^{m+1}.$$

$$(3.9)$$

Suppose that $B$ is diagonalizable and that all eigenvalues $\lambda_1, \ldots, \lambda_r$ of $B$ satisfy $|\lambda_j| < 1$. Note that our proof of (1.1) extends to any diagonalizable matrix. (The matrix $U$ need not be orthogonal, but this is irrelevant to the proof.) Hence

$$(B^n)_{ij} = c_1\lambda_1^n + \cdots + c_r\lambda_r^n,$$

where $c_1, \ldots, c_r$ are complex numbers (independent of $n$). Hence from (3.9) we see that the limit as $m \to \infty$ of the right-hand side approaches $I_r$. It follows [why?] that $\sum_{n \geq 0}(n+1)B^n$ converges to $(I_r - B)^{-2}$.

NOTE. The above argument shows that $I_r - B$ is indeed invertible. This fact is also an immediate consequence of the hypothesis that all eigenvalues of $B$ have absolute value less than one, since in particular there is no eigenvalue $\lambda = 1$.

From the discussion above, it remains to show that $M[v]$ is diagonalizable, with all eigenvalues of absolute value less than one. The diagonalizability of $M[v]$ is shown in exactly the same way as for $M$ in Theorem 3.2. (Thus we see also that $M[v]$ has real eigenvalues, though we don't need this fact here.) It remains to show that the eigenvalues $\theta_1, \ldots, \theta_{p-1}$ of $M[v]$ satisfy $|\theta_j| < 1$. We would like to apply Theorem 3.3 to the matrix $M[v]$, but this matrix might not be irreducible since the graph $G-v$ (defined by deleting from $G$ the vertex $v$ and all incident edges) need not be connected or may be just an isolated vertex. If $G - v$ has connected components $H_1, \ldots, H_m$, then we can order the vertices of $G - v$ so that $M[v]$ has the block structure

$$M[v] = \begin{bmatrix} N_1 & 0 & \cdots & 0 \\ 0 & N_2 & \cdots & 0 \\ & & \vdots & \\ 0 & 0 & \cdots & N_m \end{bmatrix},$$

where each $N_i$ is irreducible or is the $1 \times 1$ matrix $[0]$ (corresponding to $H_i$ being an isolated vertex). The eigenvalues of $M[v]$ are the eigenvalues of the $N_i$'s.

We need to show that each eigenvalue of $N_i$ has absolute value less than one. If $N_i = [0]$ then the only eigenvalue is 0, so we may assume that $H_i$ is not an isolated vertex. Suppose that $H_i$ has $k$ vertices, so $N_i$ is a $k \times k$ matrix. Let $\rho_i$ be the largest real eigenvalue of $N_i$, so by Theorem 3.3 all eigenvalues $\lambda$ of $N_i$ satisfy $|\lambda| \leq \rho_i$. Let $U = [u_1, \ldots, u_k]$ be a left eigenvector for $\rho_i$ with positive entries (which exists by Theorem 3.3), so $U N_i = \rho_i U$. Let $V$ be the column vector of length $k$ of all 1's. Consider the matrix product $U N_i V$. On the one hand we have

$$U N_i V = (\rho_i U)V = \rho_i(u_1 + \cdots + u_k). \tag{3.10}$$

On the other hand, if $\sigma_j$ denotes the $j$th row sum of $N_i$, then

$$U N_i V = U[\sigma_1, \ldots, \sigma_k]^t = \sigma_1 u_1 + \cdots + \sigma_k u_k, \tag{3.11}$$

where $^t$ denotes transpose. Now every $\sigma_j$ satisfies $0 \leq \sigma_j \leq 1$, and at least one $\sigma_h$ satisfies $\sigma_h < 1$ [why?]. Since each $u_j > 0$, it follows from (3.11) that $U N_i V < u_1 + \cdots + u_k$. Comparing with (3.10) gives $\rho_i < 1$.

Since the eigenvalues of $M[v]$ are just the eigenvalues of the $N_i$'s, we see that all eigenvalues $\theta$ of $M[v]$ satisfy $|\theta| < 1$. This completes the proof of Theorem 3.4.

$\square$

**3.5 Example.** Let $G$ be the graph of Figure 3.1 with $v = v_4$. Then

$$M = \begin{bmatrix} \dfrac{1}{3} & \dfrac{1}{3} & 0 & \dfrac{1}{3} \\[2mm] \dfrac{1}{4} & 0 & \dfrac{1}{4} & \dfrac{1}{2} \\[2mm] 0 & \dfrac{1}{2} & 0 & \dfrac{1}{2} \\[2mm] \dfrac{1}{4} & \dfrac{1}{2} & \dfrac{1}{4} & 0 \end{bmatrix}$$

$$I_3 - M[v] = \begin{bmatrix} \dfrac{2}{3} & -\dfrac{1}{3} & 0 \\[2mm] -\dfrac{1}{4} & 1 & -\dfrac{1}{4} \\[2mm] 0 & -\dfrac{1}{2} & 1 \end{bmatrix}$$

$$(I_3 - M[v])^{-2} = \begin{bmatrix} \dfrac{55}{16} & \dfrac{13}{6} & \dfrac{17}{24} \\[2mm] \dfrac{13}{8} & \dfrac{7}{3} & \dfrac{11}{12} \\[2mm] \dfrac{17}{16} & \dfrac{11}{6} & \dfrac{13}{8} \end{bmatrix}$$

**Fig. 3.1** A graph for
Example 3.5

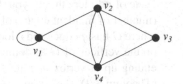

$$(I_3 - M[v])^{-2} \begin{bmatrix} \dfrac{1}{3} \\[2mm] \dfrac{1}{2} \\[2mm] \dfrac{1}{2} \end{bmatrix} = \begin{bmatrix} \dfrac{31}{12} \\[2mm] \dfrac{13}{6} \\[2mm] \dfrac{25}{12} \end{bmatrix}.$$

Thus $H(v_1, v) = 31/12$, $H(v_2, v) = 13/6$, and $H(v_3, v) = 25/12$.

NOTE. The method used to prove that $\sum_{n \geq 0}(n + 1)B^n$ converges when all eigenvalues of $B$ have absolute value less than one can be extended, with a little more work (mostly concerned with non-diagonalizability), to show the following. Let $F(x) = \sum_{n \geq 0} a_n x^n$ be a power series with complex coefficients $a_n$. Let $\alpha > 0$ be such that $F(x)$ converges whenever $|x| < \alpha$. Let $B$ be a square matrix (over the complex numbers) whose eigenvalues $\lambda$ all satisfy $|\lambda| < \alpha$. Then the matrix power series $\sum_{n \geq 0} a_n B^n$ converges in the entry-wise sense described above.

## Notes for Chapter 3

Random walks on graphs is a vast subject, of which we have barely scratched the surface. Two typical questions considerably deeper than what we have considered are the following: how rapidly does a random walk approach the stationary distribution of Exercise 3.1? Assuming $G$ is connected, what is the expected number of steps needed to visit every vertex? For a nice survey of random walks in graphs, see Lovász [84]. The topic of matrix power series is part of the subject of *matrix analysis*. For further information, see for instance Chapter 5 of the text by Horn and Johnson [68]. Our proof of Theorem 3.4 is somewhat "naive," avoiding the development of the theory of matrix norms.

## Exercises for Chapter 3

1. Let $G$ be a (finite) graph with vertices $v_1, \ldots, v_p$. Assume that some power of the probability matrix $M(G)$ defined by (3.1) has positive entries. (It's not hard to see that this is equivalent to $G$ being connected and containing at least one

cycle of odd length, but you don't have to show this.) Let $d_k$ denote the degree (number of incident edges) of vertex $v_k$. Let $D = d_1 + d_2 + \cdots + d_p = 2q - r$, where $G$ has $q$ edges and $r$ loops. Start at any vertex of $G$ and do a random walk on the vertices of $G$ as defined in the text. Let $p_k(\ell)$ denote the probability of ending up at vertex $v_k$ after $\ell$ steps. Assuming the Perron–Frobenius theorem (Theorem 3.3), show that

$$\lim_{\ell \to \infty} p_k(\ell) = d_k/D.$$

This limiting probability distribution on the set of vertices of $G$ is called the *stationary distribution* of the random walk.

2. (a) Let $G$ be a finite graph (allowing loops and multiple edges). Suppose that there is some integer $\ell > 0$ such that the number of walks of length $\ell$ from any fixed vertex $u$ to any fixed vertex $v$ is independent of $u$ and $v$. Show that $G$ has the same number $k$ of edges between any two vertices (including $k$ loops at each vertex).

   (b) Let $G$ be a finite graph (allowing loops and multiple edges) with the following property. There is some integer $\ell > 0$ such that if we start at any vertex of $G$ and do a random walk (in the sense of the text) for $\ell$ steps, then we are equally likely to be at any vertex. In other words, if $G$ has $p$ vertices then the probability that the walk ends at vertex $v$ is exactly $1/p$ for any $v$. Show that we have the same conclusion as (a), i.e., $G$ has the same number $k$ of edges between any two vertices.

3. (a) Let $P(x)$ be a nonzero polynomial with real coefficients. Show that the following two conditions are equivalent.

   - There exists a nonzero polynomial $Q(x)$ with real coefficients such that all coefficients of $P(x)Q(x)$ are nonnegative.
   - There does not exist a real number $a > 0$ such that $P(a) = 0$.

   (b) (difficult) Let $G$ be a *connected* finite graph, and let $M$ be the probability matrix defined by (3.1). Show that the following two conditions are equivalent.

   - There exists a probability distribution $P$ on $\mathbb{P}$ (so $P(k)$ is the probability of choosing $k \in \mathbb{P}$) such that if we first choose $k$ from the distribution $P$ and then start at any vertex of $G$ and walk exactly $k$ steps according to the random walk described in the text, then we are equally likely to be at any vertex of $G$.
   - The graph $G$ is regular, and no positive real number except 1 is an eigenvalue of $M$.

4. (*) Fix $0 \le p \le 1$. Start at the vertex $(0, 0, \ldots, 0)$ of the $n$-cube $C_n$. Walk along the edges of the cube according to the following rule: after each unit of time, either stay where you are with probability $p$ or step to a neighboring vertex randomly (uniformly). Thus the probability of stepping to a particular

neighboring vertex is $(1 - p)/n$. Find a formula for the probability $P(\ell)$ that after $\ell$ units of time you are again at $(0, 0, \ldots, 0)$. For instance, $P(0) = 1$ and $P(1) = p$. Express your formula as a finite sum.

5. This problem is not directly related to the text but is a classic problem with a very clever elegant solution. Let $G$ be the graph with vertex set $\mathbb{Z}_n$ (the integers modulo $n$), with an edge between $i$ and $i + 1$ for all $i \in \mathbb{Z}_n$. Hence $G$ is just an $n$-cycle. Start at vertex $0$ and do a random walk as in the text, so from vertex $i$ walk to $i - 1$ or $i + 1$ with probability $1/2$ each. For each $i \in \mathbb{Z}_n$, find the probability that vertex $i$ is the last vertex to be visited for the first time. In other words, at the first time we arrive at vertex $i$, we have visited all the other vertices at least once each. For instance, $p_0 = 0$ (if $n > 1$), since vertex $0$ is the *first* vertex to be visited.

6. Let $G$ be the $3 \times 3$ "grid graph" of Example 10.7. Using your favorite software for linear algebra, compute the access times $H(u, v)$ for every pair $u, v$ of vertices of $G$.

7. (a) Show that if $u$ and $v$ are two vertices of a connected graph $G$, then we need not have $H(u, v) = H(v, u)$, where $H$ denotes access time. What if $G$ is also assumed to be regular?

   (b) (difficult) For each $n \geq 1$, what is the maximum possible value of $H(u, v) - H(v, u)$ for two vertices $u, v$ of a connected simple graph with $n$ vertices?

8. (*) Let $u$ and $v$ be distinct vertices of the complete graph $K_n$. Show that $H(u, v) = n - 1$.

9. (*) Let $P_n$ be the graph with vertices $v_1, \ldots, v_n$ and an edge between $v_i$ and $v_{i+1}$ for all $1 \leq i \leq n - 1$. Show that $H(v_1, v_n) = (n - 1)^2$. What about $H(v_i, v_j)$ for any $i \neq j$? What if we also have an edge between $v_1$ and $v_n$?

10. Let $K_{mn}$ be a complete bipartite graph with vertex bipartition $(A_1, A_2)$, where $\#A_1 = m$ and $\#A_2 = n$. Find the access time $H(u, v)$ between every pair of distinct vertices. There will be two inequivalent cases: both $u$ and $v$ lie in the same $A_i$, or they lie in different $A_i$'s.

11. (*) For any three vertices $u, v, w$ of a graph $G$, show that

$$H(u, v) + H(v, w) + H(w, u) = H(u, w) + H(w, v) + H(v, u).$$

12. Let $k \geq 0$, and let $u$ and $v$ be vertices of a graph $G$. Define the *kth binomial moment* $H_k(u, v)$ of the access time to be the average value (expectation) of $\binom{n}{k}$, where $n$ is the number of steps that a random walk starting at $u$ takes to reach $v$ for the first time. Thus in the notation of (3.2) we have

$$H_k(u, v) = \sum_{n \geq 1} \binom{n}{k} p_n.$$

Let $x$ be an indeterminate. Following the notation of (3.4), show that

$$\sum_{k\geq 0} H_k(u, v)x^k = ((I_{p-1} - (x+1)M[v])^{-1}T[v])_u.$$

13. (*) Generalizing Exercise 3.8 above, show that for any two distinct vertices $u$, $v$ of the complete graph $K_n$, the $k$th binomial moment of the access time is given by $H_k(u, v) = (n-1)(n-2)^{k-1}$, $k \geq 1$. (When $n = 2$ and $k = 1$, we should set $0^0 = 1$.)

# Chapter 4
# The Sperner Property

In this chapter we consider a surprising application of certain adjacency matrices to some problems in extremal set theory. An important role will also be played by finite groups in Chapter 5, which is a continuation of the present chapter. In general, extremal set theory is concerned with finding (or estimating) the most or least number of sets satisfying given set-theoretic or combinatorial conditions. For example, a typical easy problem in extremal set theory is the following: what is the most number of subsets of an $n$-element set with the property that any two of them intersect? (Can you solve this problem?) The problems to be considered here are most conveniently formulated in terms of *partially ordered sets* or posets for short. Thus we begin with discussing some basic notions concerning posets.

**4.1 Definition.** A *poset* $P$ is a finite set, also denoted $P$, together with a binary relation denoted $\leq$ satisfying the following axioms:

(P1) (reflexivity) $x \leq x$ for all $x \in P$
(P2) (antisymmetry) If $x \leq y$ and $y \leq x$, then $x = y$
(P3) (transitivity) If $x \leq y$ and $y \leq z$, then $x \leq z$

One easy way to obtain a poset is the following. Let $P$ be any collection of sets. If $x, y \in P$, then define $x \leq y$ in $P$ if $x \subseteq y$ as sets. It is easy to see that this definition of $\leq$ makes $P$ into a poset. If $P$ consists of *all* subsets of an $n$-element set $S$, then $P$ is called a (finite) *Boolean algebra* of *rank* $n$ and is denoted by $B_S$. If $S = \{1, 2, \ldots, n\}$, then we denote $B_S$ simply by $B_n$. Boolean algebras will play an important role throughout this chapter and the next.

There is a simple way to represent small posets pictorially. The *Hasse diagram* of a poset $P$ is a planar drawing, with elements of $P$ drawn as dots. If $x < y$ in $P$ (i.e., $x \leq y$ and $x \neq y$), then $y$ is drawn "above" $x$ (i.e., with a larger vertical coordinate). An edge is drawn between $x$ and $y$ if $y$ *covers* $x$, i.e., $x < y$ and no element $z$ satisfies $x < z < y$. We then write $x \lessdot y$ or $y \gtrdot x$. By the transitivity property (P3), all the relations of a finite poset are determined by the cover relations,

© Springer International Publishing AG, part of Springer Nature 2018
R. P. Stanley, *Algebraic Combinatorics*, Undergraduate Texts in Mathematics,
https://doi.org/10.1007/978-3-319-77173-1_4

so the Hasse diagram determines $P$. (This is not true for infinite posets; for instance, the real numbers $\mathbb{R}$ with their usual order is a poset with no cover relations.) The Hasse diagram of the Boolean algebra $B_3$ looks like

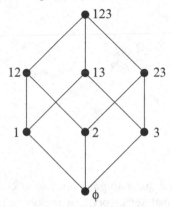

We say that two posets $P$ and $Q$ are *isomorphic* if there is a bijection (one-to-one and onto function) $\varphi \colon P \to Q$ such that $x \leq y$ in $P$ if and only if $\varphi(x) \leq \varphi(y)$ in $Q$. Thus one can think that two posets are isomorphic if they differ only in the names of their elements. This is exactly analogous to the notion of isomorphism of groups, rings, etc. It is an instructive exercise (see Exercise 4.1) to draw Hasse diagrams of the one poset of *order* (number of elements) one (up to isomorphism), the two posets of order two, the five posets of order three, and the sixteen posets of order four. More ambitious readers can try the 63 posets of order five, the 318 of order six, the 2,045 of order seven, the 16,999 of order eight, the 183,231 of order nine, the 2,567,284 of order ten, the 46,749,427 of order eleven, the 1,104,891,746 of order twelve, the 33,823,827,452 of order thirteen, the 1,338,193,159,771 of order fourteen, the 68,275,077,901,156 of order fifteen, and the 4,483,130,665,195,087 of order sixteen. Beyond this the number is not currently known.

A *chain* $C$ in a poset is a totally ordered subset of $P$, i.e., if $x, y \in C$ then either $x \leq y$ or $y \leq x$ in $P$. A finite chain is said to have *length* $n$ if it has $n + 1$ elements. Such a chain thus has the form $x_0 < x_1 < \cdots < x_n$. We say that a finite poset is *graded of rank* $n$ if every maximal chain has length $n$. (A chain is *maximal* if it's contained in no larger chain.) For instance, the Boolean algebra $B_n$ is graded of rank $n$ [why?]. A chain $y_0 < y_1 < \cdots < y_j$ is said to be *saturated* if each $y_{i+1}$ covers $y_i$. Such a chain need not be maximal since there can be elements of $P$ less than $y_0$ or greater than $y_j$. If $P$ is graded of rank $n$ and $x \in P$, then we say that $x$ has *rank* $j$, denoted $\rho(x) = j$, if the largest saturated chain of $P$ with top element $x$ has length $j$. Thus [why?] if we let $P_j = \{x \in P : \rho(x) = j\}$, then $P$ is a *disjoint* union $P = P_0 \uplus P_1 \uplus \cdots \uplus P_n$, and every maximal chain of $P$ has the form $x_0 < x_1 < \cdots < x_n$ where $\rho(x_j) = j$. We call $P_j$ the $j$th *level* of $P$. We write $p_j = \#P_j$, the number of elements of $P$ of rank $j$. For example, if $P = B_n$ then $\rho(x) = |x|$ (the cardinality of $x$ as a set) and

$$p_j = \#\{x \subseteq \{1, 2, \dots, n\} : |x| = j\} = \binom{n}{j}.$$

(Note that we use both $|S|$ and $\#S$ for the cardinality of a finite set $S$.) If a graded poset $P$ of rank $n$ has $p_i$ elements of rank $i$, then define the *rank-generating function*

$$F(P, q) = \sum_{i=0}^{n} p_i q^i = \sum_{x \in P} q^{\rho(x)}.$$

For instance, $F(B_n, q) = (1 + q)^n$ [why?].

We say that a graded poset $P$ of rank $n$ (always assumed to be finite) is *rank-symmetric* if $p_i = p_{n-i}$ for $0 \le i \le n$ and *rank-unimodal* if $p_0 \le p_1 \le \cdots \le p_j \ge p_{j+1} \ge p_{j+2} \ge \cdots \ge p_n$ for some $0 \le j \le n$. If $P$ is both rank-symmetric and rank-unimodal, then we clearly have

$$p_0 \le p_1 \le \cdots \le p_m \ge p_{m+1} \ge \cdots \ge p_n, \text{ if } n = 2m$$

$$p_0 \le p_1 \le \cdots \le p_m = p_{m+1} \ge p_{m+2} \ge \cdots \ge p_n, \text{ if } n = 2m + 1.$$

We also say that the sequence $p_0, p_1, \dots, p_n$ itself or the polynomial $F(q) = p_0 + p_1 q + \cdots + p_n q^n$ is *symmetric* or *unimodal*, as the case may be. For instance, $B_n$ is rank-symmetric and rank-unimodal, since it is well-known (and easy to prove) that the sequence $\binom{n}{0}, \binom{n}{1}, \dots, \binom{n}{n}$ (the $n$th row of Pascal's triangle) is symmetric and unimodal. Thus the polynomial $(1 + q)^n$ is symmetric and unimodal.

A few more definitions, and then finally some results! An *antichain* in a poset $P$ is a subset $A$ of $P$ for which no two elements are comparable, i.e., we can never have $x, y \in A$ and $x < y$. For instance, in a graded poset $P$ the "levels" $P_j$ are antichains [why?]. We will be concerned with the problem of finding the largest antichain in a poset. Consider for instance the Boolean algebra $B_n$. The problem of finding the largest antichain in $B_n$ is clearly equivalent to the following problem in extremal set theory: find the largest collection of subsets of an $n$-element set such that no element of the collection contains another (as a subset). A good guess would be to take all the subsets of cardinality $\lfloor n/2 \rfloor$ (where $\lfloor t \rfloor$ denotes the greatest integer $\le t$), giving a total of $\binom{n}{\lfloor n/2 \rfloor}$ sets in all. But how can we actually prove there is no larger collection? Such a proof was first given by Emanuel Sperner in 1927 and is known as *Sperner's theorem*. We will give three proofs of Sperner's theorem in this chapter: one proof uses linear algebra and will be applied to certain other situations; the second proof is an elegant combinatorial argument due to David Lubell in 1966; while the third proof is another combinatorial argument closely related to the linear algebra proof. We present the last two proofs for their "cultural value." Our extension of Sperner's theorem to certain other situations will involve the following crucial definition.

**4.2 Definition.** Let $P$ be a graded poset of rank $n$. We say that $P$ has the *Sperner property* or is a *Sperner poset* if

$$\max\{\#A : A \text{ is an antichain of } P\} = \max\{\#P_i : 0 \le i \le n\}.$$

In other words, no antichain is larger than the largest level $P_i$.

Thus Sperner's theorem is equivalent to saying that $B_n$ has the Sperner property. Note that if $P$ has the Sperner property then there may still be antichains of maximum cardinality other than the biggest $P_i$; there just can't be any bigger antichains.

**4.3 Example.** A simple example of a graded poset that fails to satisfy the Sperner property is the following:

We now will discuss a simple combinatorial condition which guarantees that certain graded posets $P$ are Sperner. We define an *order-matching* from $P_i$ to $P_{i+1}$ to be a *one-to-one* function $\mu : P_i \rightarrow P_{i+1}$ satisfying $x < \mu(x)$ for all $x \in P_i$. Clearly if such an order-matching exists then $p_i \le p_{i+1}$ (since $\mu$ is one-to-one). Easy examples (such as the diagram above) show that the converse is false, i.e., if $p_i \le p_{i+1}$ then there need not exist an order-matching from $P_i$ to $P_{i+1}$. We similarly define an order-matching from $P_i$ to $P_{i-1}$ to be a one-to-one function $\mu : P_i \rightarrow P_{i-1}$ satisfying $\mu(x) < x$ for all $x \in P_i$.

**4.4 Proposition.** *Let $P$ be a graded poset of rank n. Suppose there exists an integer $0 \le j \le n$ and order-matchings*

$$P_0 \rightarrow P_1 \rightarrow P_2 \rightarrow \cdots \rightarrow P_j \leftarrow P_{j+1} \leftarrow P_{j+2} \leftarrow \cdots \leftarrow P_n. \qquad (4.1)$$

*Then $P$ is rank-unimodal and Sperner.*

*Proof.* Since order-matchings are one-to-one it is clear that

$$p_0 \le p_1 \le \cdots \le p_j \ge p_{j+1} \ge p_{j+2} \ge \cdots \ge p_n.$$

Hence $P$ is rank-unimodal.

Define a graph $G$ as follows. The vertices of $G$ are the elements of $P$. Two vertices $x, y$ are connected by an edge if one of the order-matchings $\mu$ in the statement of the proposition satisfies $\mu(x) = y$. (Thus $G$ is a subgraph of the Hasse diagram of $P$.) Drawing a picture will convince you that $G$ consists of a disjoint union of paths, including single-vertex paths not involved in any of the order-matchings. The vertices of each of these paths form a chain in $P$. Thus we have partitioned the elements of $P$ into disjoint chains. Since $P$ is rank-unimodal with biggest level $P_j$, all of these chains must pass through $P_j$ [why?]. Thus the number of chains is exactly $p_j$. Any antichain $A$ can intersect each of these chains at most once, so the cardinality $|A|$ of $A$ cannot exceed the number of chains, i.e., $|A| \le p_j$. Hence by definition $P$ is Sperner. □

It is now finally time to bring some linear algebra into the picture. For any (finite) set $S$, we let $\mathbb{R}S$ denote the real vector space consisting of all formal linear combinations (with real coefficients) of elements of $S$. Thus $S$ is a basis for $\mathbb{R}S$, and in fact we could have simply defined $\mathbb{R}S$ to be the real vector space with basis $S$. The next lemma relates the combinatorics we have just discussed to linear algebra and will allow us to prove that certain posets are Sperner by the use of linear algebra (combined with some finite group theory).

**4.5 Lemma.** *Suppose there exists a linear transformation $U : \mathbb{R}P_i \rightarrow \mathbb{R}P_{i+1}$ ($U$ stands for "up") satisfying:*

- *$U$ is one-to-one.*
- *For all $x \in P_i$, $U(x)$ is a linear combination of elements $y \in P_{i+1}$ satisfying $x < y$. (We then call $U$ an order-raising operator.)*

*Then there exists an order-matching $\mu : P_i \rightarrow P_{i+1}$.*

*Similarly, suppose there exists a linear transformation $U : \mathbb{R}P_i \rightarrow \mathbb{R}P_{i+1}$ satisfying:*

- *$U$ is onto.*
- *$U$ is an order-raising operator.*

*Then there exists an order-matching $\mu : P_{i+1} \rightarrow P_i$.*

*Proof.* Suppose $U : \mathbb{R}P_i \rightarrow \mathbb{R}P_{i+1}$ is a one-to-one order-raising operator. Let $[U]$ denote the matrix of $U$ with respect to the bases $P_i$ of $\mathbb{R}P_i$ and $P_{i+1}$ of $\mathbb{R}P_{i+1}$. Thus the rows of $[U]$ are indexed by the elements $y_1, \dots, y_{p_{i+1}}$ of $P_{i+1}$ (in some order) and the columns by the elements $x_1, \dots, x_{p_i}$ of $P_i$. Since $U$ is one-to-one, the rank of $[U]$ is equal to $p_i$ (the number of columns). Since the row rank of a matrix equals its column rank, $[U]$ must have $p_i$ linearly independent rows. Say we have labelled the elements of $P_{i+1}$ so that the first $p_i$ rows of $[U]$ are linearly independent.

Let $A = (a_{ij})$ be the $p_i \times p_i$ matrix whose rows are the first $p_i$ rows of $[U]$. (Thus $A$ is a square submatrix of $[U]$.) Since the rows of $A$ are linearly independent, we have

$$\det(A) = \sum \pm a_{1\pi(1)} \cdots a_{p_i \pi(p_i)} \neq 0,$$

where the sum is over all permutations $\pi$ of $1, \dots, p_i$. Thus some term $\pm a_{1\pi(1)} \cdots a_{p_i \pi(p_i)}$ of the above sum is nonzero. Since $U$ is order-raising, this means that [why?] $y_k > x_{\pi(k)}$ for $1 \le k \le p_i$. Hence the map $\mu : P_i \rightarrow P_{i+1}$ defined by $\mu(x_k) = y_{\pi^{-1}(k)}$ is an order-matching, as desired.

The case when $U$ is onto rather than one-to-one is proved by a completely analogous argument. It can also be deduced from the one-to-one case by considering the transpose of the matrix $[U]$. □

NOTE. Although it does not really help in understanding the theory, it is interesting to regard a one-to-one order-raising operator as a "quantum order-matching." Rather than choosing a single element $y = \mu(x)$ that is matched with $x \in P_i$, we choose all possible elements $y \in P_{i+1}$ satisfying $y > x$ at the same time. If $U(x) = \sum_{y>x} c_y y$ (where $c_y \in \mathbb{R}$), then we are choosing $y$ with "weight" $c_y$. As explained in the proof of Lemma 4.5 above, we "break the symmetry" and obtain a single matched element $\mu(x)$ by choosing some nonvanishing term in the expansion of a determinant.

We now want to apply Proposition 4.4 and Lemma 4.5 to the Boolean algebra $B_n$. For each $0 \le i < n$, we need to define a linear transformation $U_i \colon \mathbb{R}(B_n)_i \to \mathbb{R}(B_n)_{i+1}$, and then prove it has the desired properties. We simply define $U_i$ to be the simplest possible order-raising operator, namely, for $x \in (B_n)_i$, let

$$U_i(x) = \sum_{\substack{y \in (B_n)_{i+1} \\ y > x}} y. \tag{4.2}$$

Note that since $(B_n)_i$ is a basis for $\mathbb{R}(B_n)_i$, (4.2) does indeed define a unique linear transformation $U_i \colon \mathbb{R}(B_n)_i \to \mathbb{R}(B_n)_{i+1}$. By definition $U_i$ is order-raising; we want to show that $U_i$ is one-to-one for $i < n/2$ and onto for $i \ge n/2$. There are several ways to show this using only elementary linear algebra; we will give what is perhaps the simplest proof, though it is quite tricky. The idea is to introduce "dual" or "adjoint" operators $D_i \colon \mathbb{R}(B_n)_i \to \mathbb{R}(B_n)_{i-1}$ to the $U_i$'s ($D$ stands for "down"), defined by

$$D_i(y) = \sum_{\substack{x \in (B_n)_{i-1} \\ x < y}} x, \tag{4.3}$$

for all $y \in (B_n)_i$. Let $[U_i]$ denote the matrix of $U_i$ with respect to the bases $(B_n)_i$ and $(B_n)_{i+1}$, and similarly let $[D_i]$ denote the matrix of $D_i$ with respect to the bases $(B_n)_i$ and $(B_n)_{i-1}$. A key observation which we will use later is that

$$[D_i] = [U_{i-1}]^t, \tag{4.4}$$

i.e., the matrix $[D_i]$ is the transpose of the matrix $[U_{i-1}]$ [why?]. Now let $I_i \colon \mathbb{R}(B_n)_i \to \mathbb{R}(B_n)_i$ denote the identity transformation on $\mathbb{R}(B_n)_i$, i.e., $I_i(u) = u$ for all $u \in \mathbb{R}(B_n)_i$. The next lemma states (in linear algebraic terms) the fundamental combinatorial property of $B_n$ which we need. For this lemma set $U_n = 0$ and $D_0 = 0$ (the 0 linear transformation between the appropriate vector spaces).

**4.6 Lemma.** *Let $0 \le i \le n$. Then*

$$D_{i+1}U_i - U_{i-1}D_i = (n - 2i)I_i. \tag{4.5}$$

*(Linear transformations are multiplied right-to-left, so $AB(u) = A(B(u))$.)*

*Proof.* Let $x \in (B_n)_i$. We need to show that if we apply the left-hand side of (4.5) to $x$, then we obtain $(n - 2i)x$. We have

$$D_{i+1}U_i(x) = D_{i+1}\left(\sum_{\substack{|y|=i+1 \\ x \subset y}} y\right)$$

$$= \sum_{\substack{|y|=i+1 \\ x \subset y}} \sum_{\substack{|z|=i \\ z \subset y}} z.$$

If $x, z \in (B_n)_i$ satisfy $|x \cap z| < i - 1$, then there is no $y \in (B_n)_{i+1}$ such that $x \subset y$ and $z \subset y$. Hence the coefficient of $z$ in $D_{i+1}U_i(x)$ when it is expanded in terms of the basis $(B_n)_i$ is 0. If $|x \cap z| = i - 1$, then there is one such $y$, namely, $y = x \cup z$. Finally if $x = z$ then $y$ can be any element of $(B_n)_{i+1}$ containing $x$, and there are $n - i$ such $y$ in all. It follows that

$$D_{i+1}U_i(x) = (n - i)x + \sum_{\substack{|z|=i \\ |x \cap z|=i-1}} z. \tag{4.6}$$

By exactly analogous reasoning (which the reader should check), we have for $x \in (B_n)_i$ that

$$U_{i-1}D_i(x) = ix + \sum_{\substack{|z|=i \\ |x \cap z|=i-1}} z. \tag{4.7}$$

Subtracting (4.7) from (4.6) yields $(D_{i+1}U_i - U_{i-1}D_i)(x) = (n - 2i)x$, as desired. □

**4.7 Theorem.** *The operator $U_i$ defined above is one-to-one if $i < n/2$ and is onto if $i \geq n/2$.*

*Proof.* Recall that $[D_i] = [U_{i-1}]^t$. From linear algebra we know that a (rectangular) matrix times its transpose is *positive semidefinite* (or just *semidefinite* for short) and hence has nonnegative (real) eigenvalues. In particular, by (4.4) the matrix $[U_{i-1}][D_i]$ is semidefinite. By Lemma 4.6 we have

$$D_{i+1}U_i = U_{i-1}D_i + (n - 2i)I_i.$$

Thus the eigenvalues of $D_{i+1}U_i$ are obtained from the eigenvalues of $U_{i-1}D_i$ by adding $n - 2i$. Since the eigenvalues of $U_{i-1}D_i$ are nonnegative (by the semidefinite property) and we are assuming that $n - 2i > 0$, it follows that the eigenvalues of $D_{i+1}U_i$ are strictly positive. Hence $D_{i+1}U_i$ is invertible (since it has no 0 eigenvalues). But this implies that $U_i$ is one-to-one [why?], as desired.

The case $i \geq n/2$ is done by a "dual" argument (or in fact can be deduced directly from the $i < n/2$ case by using the fact that the poset $B_n$ is "self-dual," though we will not go into this). Namely, from the fact that

$$U_i D_{i+1} = D_{i+2} U_{i+1} + (2i + 2 - n) I_{i+1}$$

we get that $U_i D_{i+1}$ is invertible, so now $U_i$ is onto, completing the proof.                □

Combining Proposition 4.4, Lemma 4.5, and Theorem 4.7, we obtain the famous theorem of Sperner.

**4.8 Corollary.**   *The Boolean algebra $B_n$ has the Sperner property.*

It is natural to ask whether there is a less indirect proof of Corollary 4.8. In fact, several nice proofs are known; we first give one due to David Lubell, mentioned before Definition 4.2.

**Lubell's Proof of Sperner's Theorem.** First we count the total number of maximal chains $\emptyset = x_0 < x_1 < \cdots < x_n = \{1, \ldots, n\}$ in $B_n$. There are $n$ choices for $x_1$, then $n - 1$ choices for $x_2$, etc., so there are $n!$ maximal chains in all. Next we count the number of maximal chains $x_0 < x_1 < \cdots < x_i = x < \cdots < x_n$ which contain a given element $x$ of rank $i$. There are $i$ choices for $x_1$, then $i - 1$ choices for $x_2$, up to one choice for $x_i$. Similarly there are $n - i$ choices for $x_{i+1}$, then $n - i - 1$ choices for $x_{i+2}$, etc., up to one choice for $x_n$. Hence the number of maximal chains containing $x$ is $i!(n - i)!$.

Now let $A$ be an antichain. If $x \in A$, then let $C_x$ be the set of maximal chains of $B_n$ which contain $x$. Since $A$ is an antichain, the sets $C_x$, $x \in A$ are pairwise disjoint. Hence

$$\left| \bigcup_{x \in A} C_x \right| = \sum_{x \in A} |C_x|$$

$$= \sum_{x \in A} (\rho(x))!(n - \rho(x))!.$$

Since the total number of maximal chains in the $C_x$'s cannot exceed the total number $n!$ of maximal chains in $B_n$, we have

$$\sum_{x \in A} (\rho(x))!(n - \rho(x))! \leq n!.$$

Divide both sides by $n!$ to obtain

$$\sum_{x \in A} \frac{1}{\binom{n}{\rho(x)}} \leq 1.$$

Since $\binom{n}{i}$ is maximized when $i = \lfloor n/2 \rfloor$, we have

$$\frac{1}{\binom{n}{\lfloor n/2 \rfloor}} \leq \frac{1}{\binom{n}{\rho(x)}},$$

for all $x \in A$ (or all $x \in B_n$). Thus

$$\sum_{x \in A} \frac{1}{\binom{n}{\lfloor n/2 \rfloor}} \leq 1,$$

or equivalently,

$$|A| \leq \binom{n}{\lfloor n/2 \rfloor}.$$

Since $\binom{n}{\lfloor n/2 \rfloor}$ is the size of the largest level of $B_n$, it follows that $B_n$ is Sperner.  $\square$

There is another nice way to show directly that $B_n$ is Sperner, namely, by constructing an explicit order-matching $\mu \colon (B_n)_i \to (B_n)_{i+1}$ when $i < n/2$. We will define $\mu$ by giving an example. Let $n = 21$, $i = 9$, and $S = \{3, 4, 5, 8, 12, 13, 17, 19, 20\}$. We want to define $\mu(S)$. Let $(a_1, a_2, \ldots, a_{21})$ be a sequence of $\pm 1$'s, where $a_i = 1$ if $i \in S$, and $a_i = -1$ if $i \notin S$. For the set $S$ above we get the sequence (writing $-$ for $-1$)

$$- - 1\,1\,1 - -1 - - - -1\,1 - - - -1 - 1\,1 -.$$

Replace any two consecutive terms $1 -$ with $0\,0$:

$$- - 1\,1\,0\,0 - 0\,0 - -1\,0\,0 - -0\,0\,1\,0\,0.$$

Ignore the $0$'s and replace any two consecutive terms $1 -$ with $0\,0$:

$$- - 1\,0\,0\,0\,0\,0\,0 - -0\,0\,0\,0 - 0\,0\,1\,0\,0.$$

Continue:

$$- - 0\,0\,0\,0\,0\,0\,0\,0 - 0\,0\,0\,0 - 0\,0\,1\,0\,0.$$

At this stage no further replacement is possible. The nonzero terms consist of a sequence of $-$'s followed by a sequence of 1's. There is at least one $-$ since $i < n/2$. Let $k$ be the position (coordinate) of the last $-$; here $k = 16$. Define $\mu(S) = S \cup \{k\} = S\cup\{16\}$. The reader can check that this procedure gives an order-matching. In particular, why is $\mu$ injective (one-to-one), i.e., why can we recover $S$ from $\mu(S)$?

It can be checked that if we glue together the order-matchings $(B_n)_i \to (B_n)_{i+1}$ for $i < n/2$ just defined, along with an obvious dual construction $(B_n)_i \to (B_n)_{i-1}$ for $i > n/2$ then we obtain more than just a partition of $B_n$ into saturated chains passing through the middle level ($n$ even) or middle two levels ($n$ odd), as in the proof of Proposition 4.4. We in fact have the additional property that these chains are all *symmetric*, i.e., they begin at some level $i \leq n/2$ and end at level $n - i$. Such a decomposition of a rank-symmetric, rank-unimodal graded poset $P$ into saturated chains is called a *symmetric chain decomposition*. A symmetric chain decomposition implies that for any $j \geq 1$, the largest size of a union of $j$ antichains is equal to the largest size of a union of $j$ levels of $P$ (Exercise 4.6). (The Sperner property corresponds to the case $j = 1$). It can be a challenging problem to decide whether certain posets have a symmetric chain decomposition (e.g., Exercises 5.5(b), 5.6 and 6.6), though we will not discuss this topic further here.

In view of the above elegant proof of Lubell and the explicit description of an order-matching $\mu : (B_n)_i \to (B_n)_{i+1}$, the reader may be wondering what was the point of giving a rather complicated and indirect proof using linear algebra. Admittedly, if all we could obtain from the linear algebra machinery we have developed was just another proof of Sperner's theorem, then it would have been hardly worth the effort. But in the next chapter we will show how Theorem 4.7, when combined with a little finite group theory, can be used to obtain many interesting combinatorial results for which simple, direct proofs are not known.

# Notes for Chapter 4

For further information on combinatorial aspects of partially ordered sets in general, see Caspard–Leclerc–Monjardet [21], Fishburn [40], Stanley [130, Ch. 3], and Trotter [136]. Sperner's theorem (Corollary 4.8) was first proved by Sperner [119]. The elegant proof of Lubell appears in [85]. A general reference on the Sperner property is the book by Engel [39]. For more general results on the combinatorics of finite sets, see Anderson [1]. The linear algebraic approach to the Sperner property discussed here is due independently to Pouzet [103] (further developed by Pouzet and Rosenberg [104]) and Stanley [123, 126]. For further information on explicit order matchings, symmetric chain decompositions, etc., see the text [1] of Anderson mentioned above.

# Exercises for Chapter 4

1. Draw Hasse diagrams of the 16 nonisomorphic four-element posets. For a more
   interesting challenge, draw also the 63 five-element posets. For those with lots of
   time to kill, draw the 318 six-element posets, the 2,045 seven-element posets, the
   16,999 eight-element posets, up to the 4,483,130,665,195,087 sixteen-element
   posets.

2. (a) Let $P$ be a finite poset and $f : P \to P$ an order-preserving bijection. That is,
       $f$ is a bijection (one-to-one and onto), and if $x \leq y$ in $P$ then $f(x) \leq f(y)$.
       Show that $f$ is an automorphism of $P$, i.e., $f^{-1}$ is order-preserving. (Try to
       use simple algebraic reasoning, though it's not necessary to do so.)
   (b) Show that the result of (a) need not be true if $P$ is infinite.

3. Let $F(q)$ and $G(q)$ be symmetric unimodal polynomials with nonnegative
   real coefficients. Show that $F(q)G(q)$ is also symmetric (easy) and unimodal
   (harder).

4. Let $q$ be a prime power, and let $\mathbb{F}_q$ denote the finite field with $q$ elements.
   Let $V = V_n(q) = \mathbb{F}_q^n$, the $n$-dimensional vector space over $\mathbb{F}_q$ of $n$-tuples of
   elements of $\mathbb{F}_q$. Let $B_n(q)$ denote the poset of all subspaces of $V$, ordered by
   inclusion. It's easy to see that $B_n(q)$ is graded of rank $n$, the rank of a subspace
   of $V$ being its dimension.

   (a) Draw the Hasse diagram of $B_3(2)$. (It has 16 elements.)
   (b) (*) Show that the number of elements of $B_n(q)$ of rank $k$ is given by the
       $q$-binomial coefficient

   $$\binom{n}{k} = \frac{(q^n - 1)(q^{n-1} - 1) \cdots (q^{n-k+1} - 1)}{(q^k - 1)(q^{k-1} - 1) \cdots (q - 1)}.$$

   (c) (*) Show that $B_n(q)$ is rank-symmetric.
   (d) Show that every element $x \in B_n(q)_k$ covers $[k] = 1 + q + \cdots + q^{k-1}$
       elements and is covered by $[n - k] = 1 + q + \cdots + q^{n-k-1}$ elements.
   (e) Define operators $U_i : \mathbb{R}B_n(q)_i \to \mathbb{R}B_n(q)_{i+1}$ and $D_i : \mathbb{R}B_n(q)_i \to \mathbb{R}B_n(q)_{i-1}$ by

   $$U_i(x) = \sum_{\substack{y \in B_n(q)_{i+1} \\ y > x}} y$$

   $$D_i(x) = \sum_{\substack{z \in B_n(q)_{i-1} \\ z < x}} z.$$

Show that

$$D_{i+1}U_i - U_{i-1}D_i = ([n-i] - [i])I_i.$$

(f)  Deduce that $B_n(q)$ is rank-unimodal and Sperner.

5.  (difficult) Let $S_1, S_2, \ldots, S_k$ be finite sets with $\#S_1 = \#S_2 = \cdots = \#S_k$. Let $P$ be the poset of all sets $T$ contained in some $S_i$, ordered by inclusion. In symbols,

$$P = 2^{S_1} \cup 2^{S_2} \cup \cdots \cup 2^{S_k},$$

where $2^S$ denotes the set of subsets of $S$. Is $P$ always rank-unimodal?

6.  Let $P$ be a rank-symmetric, rank-unimodal poset. Show that if $P$ has a symmetric chain decomposition, then for any $j \geq 1$ the largest size of a union of $j$ antichains is equal to the largest size of a union of $j$ levels of $P$.

7.  (a)  Let $P$ be a finite poset whose largest chain has $m$ elements. Show that $P$ is a union of $m$ antichains.

    (b)  (difficult) (*) Let $P$ be a finite poset whose largest antichain has $m$ elements. Show that $P$ is a union of $m$ chains.

# Chapter 5
# Group Actions on Boolean Algebras

Let us begin by reviewing some facts from group theory. Suppose that $X$ is an $n$-element set and that $G$ is a group. We say that $G$ *acts on* the set $X$ if for every element $\pi$ of $G$ we associate a permutation (also denoted $\pi$) of $X$, such that for all $x \in X$ and $\pi, \sigma \in G$ we have

$$\pi(\sigma(x)) = (\pi\sigma)(x).$$

Thus [why?] an action of $G$ on $X$ is the same as a homomorphism $\varphi : G \to \mathfrak{S}_X$, where $\mathfrak{S}_X$ denotes the symmetric group of all permutations of $X$. We sometimes write $\pi \cdot x$ instead of $\pi(x)$.

**5.1 Example.** (a) Let the real number $\alpha$ act on the $xy$-plane by rotation counterclockwise around the origin by an angle of $\alpha$ radians. It is easy to check that this defines an action of the group $\mathbb{R}$ of real numbers (under addition) on the $xy$-plane. The kernel of this action, i.e., the kernel of the homomorphism $\varphi : \mathbb{R} \to \mathfrak{S}_{\mathbb{R}^2}$, is the cyclic subgroup of $\mathbb{R}$ generated by $2\pi$.

(b) Now let $\alpha \in \mathbb{R}$ act by translation by a distance $\alpha$ to the right, i.e., adding $(\alpha, 0)$. This yields a completely different action of $\mathbb{R}$ on the $xy$-plane. This time the action is *faithful*, i.e., the kernel is the trivial subgroup $\{0\}$.

(c) Let $X = \{a, b, c, d\}$ and $G = \mathbb{Z}_2 \times \mathbb{Z}_2 = \{(0, 0), (0, 1), (1, 0), (1, 1)\}$. Let $G$ act as follows:

$$(0, 1) \cdot a = b, \quad (0, 1) \cdot b = a, \quad (0, 1) \cdot c = c, \quad (0, 1) \cdot d = d$$

$$(1, 0) \cdot a = a, \quad (1, 0) \cdot b = b, \quad (1, 0) \cdot c = d, \quad (1, 0) \cdot d = c.$$

The reader should check that this does indeed define an action. In particular, since $(1, 0)$ and $(0, 1)$ generate $G$, we don't need to define the action of $(0, 0)$ and $(1, 1)$—they are uniquely determined.

(d) Let $X$ and $G$ be as in (c), but now define the action by

© Springer International Publishing AG, part of Springer Nature 2018
R. P. Stanley, *Algebraic Combinatorics*, Undergraduate Texts in Mathematics,
https://doi.org/10.1007/978-3-319-77173-1_5

$$(0, 1) \cdot a = b, \quad (0, 1) \cdot b = a, \quad (0, 1) \cdot c = d, \quad (0, 1) \cdot d = c$$
$$(1, 0) \cdot a = c, \quad (1, 0) \cdot b = d, \quad (1, 0) \cdot c = a, \quad (1, 0) \cdot d = b.$$

Again one can check that we have an action of $\mathbb{Z}_2 \times \mathbb{Z}_2$ on $\{a, b, c, d\}$. The two actions of $G = \mathbb{Z}_2 \times \mathbb{Z}_2$ that we have just defined are quite different; for instance, in the first action we have some elements of $X$ fixed by some nonidentity element of $G$ (such as $(0, 1) \cdot c = c$), while the second action fails to have this property. See also Example 5.2(c, d) below for another fundamental way in which the two actions differ.

Recall what is meant by an *orbit* of the action of a group $G$ on a set $X$. Namely, we say that two elements $x, y$ of $X$ are *$G$-equivalent* if $\pi(x) = y$ for some $\pi \in G$. The relation of $G$-equivalence is an equivalence relation, and the equivalence classes are called orbits. Thus $x$ and $y$ are in the same orbit if $\pi(x) = y$ for some $\pi \in G$. The orbits form a *partition* of $X$, i.e., they are pairwise-disjoint, nonempty subsets of $X$ whose union is $X$. The orbit containing $x$ is denoted $Gx$; this is sensible notation since $Gx$ consists of all elements $\pi(x)$ where $\pi \in G$. Thus $Gx = Gy$ if and only if $x$ and $y$ are $G$-equivalent (i.e., in the same $G$-orbit). The set of all $G$-orbits is denoted $X/G$.

**5.2 Example.** (a) In Example 5.1(a), the orbits are circles with center $(0, 0)$, including the degenerate circle whose only point is $(0, 0)$.
  (b) In Example 5.1(b), the orbits are horizontal lines. Note that although in (a) and (b) the same group $G$ acts on the same set $X$, the orbits are different.
  (c) In Example 5.1(c), the orbits are $\{a, b\}$ and $\{c, d\}$.
  (d) In Example 5.1(d), there is only one orbit $\{a, b, c, d\}$. Again we have a situation in which a group $G$ acts on a set $X$ in two different ways, with different orbits.

We wish to consider the situation where $X = B_n$, the Boolean algebra of rank $n$ (so $|B_n| = 2^n$). We begin by defining an *automorphism* of a poset $P$ to be an isomorphism $\varphi \colon P \to P$. (This definition is exactly analogous to the definition of an automorphism of a group, ring, etc.) The set of all automorphisms of $P$ forms a group, denoted $\mathrm{Aut}(P)$ and called the *automorphism group* of $P$, under the operation of composition of functions (just as is the case for groups, rings, etc.)

Now consider the case $P = B_n$. Any permutation $\pi$ of $\{1, \ldots, n\}$ acts on $B_n$ as follows: if $x = \{i_1, i_2, \ldots, i_k\} \in B_n$, then

$$\pi(x) = \{\pi(i_1), \pi(i_2), \ldots, \pi(i_k)\}. \tag{5.1}$$

This action of $\pi$ on $B_n$ is an automorphism [why?]; in particular, if $|x| = i$, then also $|\pi(x)| = i$. Equation (5.1) defines an action of the symmetric group $\mathfrak{S}_n$ of all permutations of $\{1, \ldots, n\}$ on $B_n$ [why?]. (In fact, it is not hard to show that *every* automorphism of $B_n$ is of the form (5.1) for $\pi \in \mathfrak{S}_n$.) In particular, any subgroup $G$ of $\mathfrak{S}_n$ acts on $B_n$ *via* (5.1) (where we restrict $\pi$ to belong to $G$). In what follows this action is always meant.

**5.3 Example.** Let $n = 3$, and let $G$ be the subgroup of $\mathfrak{S}_3$ with elements $\iota$ and $(1, 2)$. Here $\iota$ denotes the identity permutation, and (using disjoint cycle notation) $(1, 2)$ denotes the permutation which interchanges 1 and 2 and fixes 3. There are six orbits of $G$ (acting on $B_3$). Writing, e.g., 13 as short for $\{1, 3\}$, the six orbits are $\{\emptyset\}$, $\{1, 2\}$, $\{3\}$, $\{12\}$, $\{13, 23\}$, and $\{123\}$.

We now define the class of posets which will be of interest to us here. Later we will give some special cases of particular interest.

Let $G$ be a subgroup of $\mathfrak{S}_n$. Define the *quotient poset* $B_n/G$ as follows. The elements of $B_n/G$ are the orbits of $G$. If $\mathfrak{o}$ and $\mathfrak{o}'$ are two orbits, then define $\mathfrak{o} \le \mathfrak{o}'$ in $B_n/G$ if there exist $x \in \mathfrak{o}$ and $y \in \mathfrak{o}'$ such that $x \le y$ in $B_n$. It's easy to check that this relation $\le$ is indeed a partial order.

**5.4 Example.** (a) Let $n = 3$ and $G$ be the group of order two generated by the cycle $(1, 2)$, as in Example 5.3. Then the Hasse diagram of $B_3/G$ is shown below, where each element (orbit) is labelled by one of its elements.

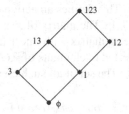

(b) Let $n = 5$ and $G$ be the group of order five generated by the cycle $(1, 2, 3, 4, 5)$. Then $B_5/G$ has Hasse diagram

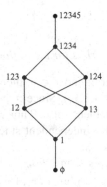

One simple property of a quotient poset $B_n/G$ is the following.

**5.5 Proposition.** *The quotient poset $B_n/G$ defined above is graded of rank n and rank-symmetric.*

*Proof.* We leave as an exercise the easy proof that $B_n/G$ is graded of rank $n$, and that the rank of an element $\mathfrak{o}$ of $B_n/G$ is just the rank in $B_n$ of any of the elements $x \in \mathfrak{o}$. Thus the number $p_i(B_n/G)$ of elements of rank $i$ is equal to the number of orbits $\mathfrak{o} \in (B_n)_i/G$. If $x \in B_n$, then let $\bar{x}$ denote the set-theoretic complement of $x$, i.e.,

$$\bar{x} = \{1, \ldots, n\} - x = \{1 \leq i \leq n : i \notin x\}.$$

Then $\{x_1, \ldots, x_j\}$ is an orbit of $i$-element subsets of $\{1, \ldots, n\}$ if and only if $\{\bar{x}_1, \ldots, \bar{x}_j\}$ is an orbit of $(n-i)$-element subsets [why?]. Hence $|(B_n)_i/G| = |(B_n)_{n-i}/G|$, so $B_n/G$ is rank-symmetric.                                    $\square$

Let $\pi \in \mathfrak{S}_n$. We associate with $\pi$ a linear transformation (still denoted $\pi$) $\pi : \mathbb{R}(B_n)_i \to \mathbb{R}(B_n)_i$ by the rule

$$\pi\left(\sum_{x \in (B_n)_i} c_x x\right) = \sum_{x \in (B_n)_i} c_x \pi(x),$$

where each $c_x$ is a real number. This defines an action of $\mathfrak{S}_n$, or of any subgroup $G$ of $\mathfrak{S}_n$, on the vector space $\mathbb{R}(B_n)_i$. The matrix of $\pi$ with respect to the basis $(B_n)_i$ is just a *permutation matrix*, i.e., a matrix with one 1 in every row and column, and 0's elsewhere. We will be interested in elements of $\mathbb{R}(B_n)_i$ which are fixed by every element of a subgroup $G$ of $\mathfrak{S}_n$. The set of all such elements is denoted $\mathbb{R}(B_n)_i^G$, so

$$\mathbb{R}(B_n)_i^G = \{v \in \mathbb{R}(B_n)_i : \pi(v) = v \text{ for all } \pi \in G\}.$$

**5.6 Lemma.** *A basis for $\mathbb{R}(B_n)_i^G$ consists of the elements*

$$v_{\mathfrak{o}} := \sum_{x \in \mathfrak{o}} x,$$

*where $\mathfrak{o} \in (B_n)_i/G$, the set of $G$-orbits for the action of $G$ on $(B_n)_i$.*

*Proof.* First note that if $\mathfrak{o}$ is an orbit and $x \in \mathfrak{o}$, then by definition of orbit we have $\pi(x) \in \mathfrak{o}$ for all $\pi \in G$ (or all $\pi \in \mathfrak{S}_n$). Since $\pi$ permutes the elements of $(B_n)_i$, it follows that $\pi$ permutes the elements of $\mathfrak{o}$. Thus $\pi(v_{\mathfrak{o}}) = v_{\mathfrak{o}}$, so $v_{\mathfrak{o}} \in \mathbb{R}(B_n)_i^G$. It is clear that the $v_{\mathfrak{o}}$'s are linearly independent since any $x \in (B_n)_i$ appears with nonzero coefficient in exactly one $v_{\mathfrak{o}}$.

It remains to show that the $v_{\mathfrak{o}}$'s span $\mathbb{R}(B_n)_i^G$, i.e., any $v = \sum_{x \in (B_n)_i} c_x x \in \mathbb{R}(B_n)_i^G$ can be written as a linear combination of $v_{\mathfrak{o}}$'s. Given $x \in (B_n)_i$, let $G_x = \{\pi \in G : \pi(x) = x\}$, the *stabilizer* of $x$. We leave as an easy exercise the standard fact that $\pi(x) = \sigma(x)$ (where $\pi, \sigma \in G$) if and only if $\pi$ and $\sigma$ belong to the same left coset of $G_x$, i.e., $\pi G_x = \sigma G_x$. It follows that in the multiset of elements $\pi(x)$, where $\pi$ ranges over all elements of $G$ and $x$ is fixed, every element $y$ in the orbit $Gx$ appears $\#G_x$ times, and no other elements appear. In other words,

$$\sum_{\pi \in G} \pi(x) = |G_x| \cdot v_{Gx}.$$

(Do not confuse the orbit $Gx$ with the subgroup $G_x$!) Now apply $\pi$ to $v$ and sum on all $\pi \in G$. Since $\pi(v) = v$ (because $v \in \mathbb{R}(B_n)_i^G$), we get

$$|G| \cdot v = \sum_{\pi \in G} \pi(v)$$

$$= \sum_{\pi \in G} \left( \sum_{x \in (B_n)_i} c_x \pi(x) \right)$$

$$= \sum_{x \in (B_n)_i} c_x \left( \sum_{\pi \in G} \pi(x) \right)$$

$$= \sum_{x \in (B_n)_i} c_x \cdot (\#G_x) \cdot v_{Gx}.$$

Dividing by $|G|$ expresses $v$ as a linear combination of the elements $v_{Gx}$ (or $v_o$), as desired.                                                                                                □

Now let us consider the effect of applying the order-raising operator $U_i$ to an element $v$ of $\mathbb{R}(B_n)_i^G$.

**5.7 Lemma.** If $v \in \mathbb{R}(B_n)_i^G$, then $U_i(v) \in \mathbb{R}(B_n)_{i+1}^G$.

*Proof.* Note that since $\pi \in G$ is an automorphism of $B_n$, we have $x < y$ in $B_n$ if and only if $\pi(x) < \pi(y)$ in $B_n$. It follows [why?] that if $x \in (B_n)_i$ then

$$U_i(\pi(x)) = \pi(U_i(x)).$$

Since $U_i$ and $\pi$ are linear transformations, it follows by linearity that $U_i \pi(u) = \pi U_i(u)$ for all $u \in \mathbb{R}(B_n)_i$. In other words, $U_i \pi = \pi U_i$. Then

$$\pi(U_i(v)) = U_i(\pi(v))$$

$$= U_i(v),$$

so $U_i(v) \in \mathbb{R}(B_n)_{i+1}^G$, as desired.                                                                □

We come to the main result of this chapter and indeed our main result on the Sperner property.

**5.8 Theorem.** *Let $G$ be a subgroup of $\mathfrak{S}_n$. Then the quotient poset $B_n/G$ is graded of rank $n$, rank-symmetric, rank-unimodal, and Sperner.*

*Proof.* Let $P = B_n/G$. We have already seen in Proposition 5.5 that $P$ is graded of rank $n$ and rank-symmetric. We want to define order-raising operators $\hat{U}_i : \mathbb{R}P_i \to \mathbb{R}P_{i+1}$ and order-lowering operators $\hat{D}_i : \mathbb{R}P_i \to \mathbb{R}P_{i-1}$. Let us first consider just $\hat{U}_i$. The idea is to identify the basis element $v_{\mathfrak{o}}$ of $\mathbb{R}B_n^G$ with the basis element $\mathfrak{o}$ of $\mathbb{R}P$, and to let $\hat{U}_i : \mathbb{R}P_i \to \mathbb{R}P_{i+1}$ correspond to the usual order-raising operator $U_i : \mathbb{R}(B_n)_i \to \mathbb{R}(B_n)_{i+1}$. More precisely, suppose that the order-raising operator $U_i$ for $B_n$ given by (4.2) satisfies

$$U_i(v_{\mathfrak{o}}) = \sum_{\mathfrak{o}' \in (B_n)_{i+1}/G} c_{\mathfrak{o},\mathfrak{o}'} v_{\mathfrak{o}'}, \tag{5.2}$$

where $\mathfrak{o} \in (B_n)_i/G$. (Note that by Lemma 5.7, $U_i(v_{\mathfrak{o}})$ does indeed have the form given by (5.2).) Then define the linear operator $\hat{U}_i : \mathbb{R}((B_n)_i/G) \to \mathbb{R}((B_n)_i/G)$ by

$$\hat{U}_i(\mathfrak{o}) = \sum_{\mathfrak{o}' \in (B_n)_{i+1}/G} c_{\mathfrak{o},\mathfrak{o}'} \mathfrak{o}'.$$

NOTE. We can depict the "transport of $U_i$ to $\hat{U}_i$" by a *commutative diagram*:

$$
\begin{array}{ccc}
(\mathbb{R}B_n)_i^G & \xrightarrow{\;\;U_i\;\;} & (\mathbb{R}B_n)_{i+1}^G \\
\cong \downarrow & & \downarrow \cong \\
\mathbb{R}(B_n/G)_i & \xrightarrow{\;\;\hat{U}_i\;\;} & \mathbb{R}(B_n/G)_{i+1}
\end{array}
$$

The arrows pointing down are the linear transformations induced by $v_{\mathfrak{o}} \mapsto \mathfrak{o}$. The map obtained by applying the top arrow followed by the rightmost down arrow is the same as applying the leftmost down arrow followed by the bottom arrow.

We claim that $\hat{U}_i$ is order-raising. We need to show that if $c_{\mathfrak{o},\mathfrak{o}'} \neq 0$, then $\mathfrak{o}' > \mathfrak{o}$ in $B_n/G$. Since $v_{\mathfrak{o}'} = \sum_{x' \in \mathfrak{o}'} x'$, the only way $c_{\mathfrak{o},\mathfrak{o}'} \neq 0$ in (5.2) is for some $x' \in \mathfrak{o}'$ to satisfy $x' > x$ for some $x \in \mathfrak{o}$. But this is just what it means for $\mathfrak{o}' > \mathfrak{o}$, so $\hat{U}_i$ is order-raising.

Now comes the heart of the argument. We want to show that $\hat{U}_i$ is one-to-one for $i < n/2$. Now by Theorem 4.7, $U_i$ is one-to-one for $i < n/2$. Thus the restriction of $U_i$ to the subspace $\mathbb{R}(B_n)_i^G$ is one-to-one. (The restriction of a one-to-one function is always one-to-one.) But $U_i$ and $\hat{U}_i$ are exactly the same transformation, except for the names of the basis elements on which they act. Thus $\hat{U}_i$ is also one-to-one for $i < n/2$.

**Fig. 5.1** The poset $B_X/G$ of
nonisomorphic graphs with
four vertices

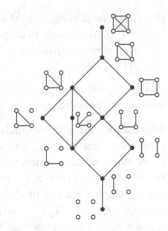

An exactly analogous argument can be applied to $D_i$ instead of $U_i$. We obtain
one-to-one order-lowering operators $\hat{D}_i \colon \mathbb{R}(B_n)_i^G \to \mathbb{R}(B_n)_{i-1}^G$ for $i > n/2$. It
follows from Proposition 4.4, Lemma 4.5, and (4.4) that $B_n/G$ is rank-unimodal
and Sperner, completing the proof.                                               □

We will consider two interesting applications of Theorem 5.8. For our first
application, we let $n = \binom{m}{2}$ for some $m \geq 1$ and let $M = \{1, \dots, m\}$. Set $X = \binom{M}{2}$,
the set of all two-element subsets of $M$. Think of the elements of $X$ as (possible)
edges of a simple graph with vertex set $M$. If $B_X$ is the Boolean algebra of all subsets
of $X$ (so $B_X$ and $B_n$ are isomorphic), then an element $x$ of $B_X$ is a collection of edges
on the vertex set $M$, in other words, just a simple graph on $M$. Define a subgroup $G$
of $\mathfrak{S}_X$ as follows. Informally, $G$ consists of all permutations of the edges $\binom{M}{2}$ that
are induced from permutations of the vertices $M$. More precisely, if $\pi \in \mathfrak{S}_m$, then
define $\hat{\pi} \in \mathfrak{S}_X$ by $\hat{\pi} \cdot \{i, j\} = \{\pi \cdot i, \pi \cdot j\}$. Thus $G$ is isomorphic to $\mathfrak{S}_m$.

When are two graphs $x, y \in B_X$ in the same orbit of the action of $G$ on $B_X$?
Since the elements of $G$ just permute vertices, we see that $x$ and $y$ are in the same
orbit if we can obtain $x$ from $y$ by permuting vertices. This is just what it means for
two simple graphs $x$ and $y$ to be *isomorphic*—they are the same graph except for the
names of the vertices (thinking of edges as pairs of vertices). Thus the elements of
$B_X/G$ are *isomorphism classes* of simple graphs on the vertex set $M$. In particular,
$\#(B_X/G)$ is the number of nonisomorphic $m$-vertex simple graphs, and $\#(B_X/G)_i$
is the number of nonisomorphic such graphs with $i$ edges. We have $x \leq y$ in $B_X/G$
if there is some way of labelling the vertices of $x$ and $y$ so that every edge of $x$
is an edge of $y$. Equivalently, some *spanning subgraph* of $y$ (i.e., a subgraph of $y$
with all the vertices of $y$) is isomorphic to $x$, as illustrated in Figure 5.1 for the case
$m = 4$. Hence by Theorem 5.8 there follows the following result, which is by no
means obvious and has no known non-algebraic proof.

**5.9 Theorem.** *(a) Fix $m \geq 1$. Let $p_i$ be the number of nonisomorphic simple
graphs with $m$ vertices and $i$ edges. Then the sequence $p_0, p_1, \dots, p_{\binom{m}{2}}$ is
symmetric and unimodal.*

(b) *Let T be a collection of simple graphs with m vertices such that no element of T is isomorphic to a spanning subgraph of another element of T. Then #T is maximized by taking T to consist of all nonisomorphic simple graphs with $\lfloor \frac{1}{2}\binom{m}{2} \rfloor$ edges.*

Our second example of the use of Theorem 5.8 is more subtle and will be the topic of the next chapter.

**Digression.** Edge reconstruction. Much work has been done on "reconstruction problems," that is, trying to reconstruct a mathematical structure such as a graph from some of its substructures. The most famous of such problems is *vertex reconstruction*: given a simple graph $G$ on $p$ vertices $v_1, \ldots, v_p$, let $G_i$ be the subgraph obtained by deleting vertex $v_i$ (and all incident edges). Given the multiset $\{G_1, \ldots, G_p\}$ of vertex-deleted subgraphs graphs, can $G$ be uniquely reconstructed? It is important to realize that the vertices are *unlabelled*, so given $G_i$ we don't know for any $j$ which vertex is $v_j$. The famous *vertex-reconstruction conjecture* (still open) states that for $p \geq 3$ any graph $G$ can be reconstructed from the multiset $\{G_1, \ldots, G_p\}$.

Here we will be concerned with *edge* reconstruction, another famous open problem. Given a simple graph $G$ with edges $e_1, \ldots, e_q$, let $H_i = G - e_i$, the graph obtained from $G$ by removing the edge $e_i$.

**Edge-Reconstruction Conjecture.** A simple graph $G$ can be uniquely reconstructed from its number of vertices and the multiset $\{H_1, \ldots, H_q\}$ of edge-deleted subgraphs.

NOTE. As in the case of vertex reconstruction, the subgraphs $H_i$ are unlabelled. The reason for including the number of vertices is that for any graph with no edges, we have $\{H_1, \ldots, H_q\} = \emptyset$, so we need to specify the number of vertices to obtain $G$.

NOTE. It can be shown that if $G$ can be vertex-reconstructed, then $G$ can be edge reconstructed. Hence the vertex-reconstruction conjecture implies the edge-reconstruction conjecture.

The techniques developed above to analyze group actions on Boolean algebra can be used to prove a special case of the edge-reconstruction conjecture. Note that a simple graph with $p$ vertices has at most $\binom{p}{2}$ edges.

**5.10 Theorem.** *Let G be a simple graph with p vertices and $q > \frac{1}{2}\binom{p}{2}$ edges. Then G is edge-reconstructible.*

*Proof.* Let $P_q$ be the set of all simple graphs with $q$ edges on the vertex set $[p] = \{1, 2, \ldots, p\}$, so $\#P_q = \binom{\binom{p}{2}}{q}$. Let $\mathbb{R}P_q$ denote the real vector space with basis $P_q$. Define a linear transformation $\psi_q : \mathbb{R}P_q \to \mathbb{R}P_{q-1}$ by

$$\psi_q(\Gamma) = \Gamma_1 + \cdots + \Gamma_q,$$

where $\Gamma_1, \ldots, \Gamma_q$ are the (labelled) graphs obtained from $\Gamma$ by deleting a single edge. By Theorem 4.7, $\psi_q$ is injective for $q > \frac{1}{2}\binom{p}{2}$. (Think of $\psi_q$ as adding edges to the *complement* of $\Gamma$, i.e., the graph with vertex set $[p]$ and edge set $\binom{[p]}{2} - E(\Gamma)$.)

The symmetric group $\mathfrak{S}_p$ acts on $P_q$ by permuting the vertices, and hence acts on $\mathbb{R}P_q$, the real vector space with basis $P_q$. A basis for the fixed space $(\mathbb{R}P_q)^{\mathfrak{S}_p}$ consists of the distinct sums $\tilde{\Gamma} = \sum_{\pi \in \mathfrak{S}_p} \pi(\Gamma)$, where $\Gamma \in P_q$. We may identify $\tilde{\Gamma}$ with the *unlabelled* graph isomorphic to $\Gamma$, since $\tilde{\Gamma} = \tilde{\Gamma}'$ if and only if $\Gamma$ and $\Gamma'$ are isomorphic. Just as in the proof of Theorem 5.8, when we restrict $\psi_q$ to $(\mathbb{R}P_q)^{\mathfrak{S}_p}$ for $q > \frac{1}{2}\binom{p}{2}$ we obtain an injection $\psi_q : (\mathbb{R}P_q)^{\mathfrak{S}_p} \to (\mathbb{R}P_{q-1})^{\mathfrak{S}_p}$. In particular, for nonisomorphic unlabelled graphs $\tilde{\Gamma}, \tilde{\Gamma}'$ with $p$ vertices, we have

$$\tilde{\Gamma}_1 + \cdots + \tilde{\Gamma}_q = \psi_q(\tilde{\Gamma}) \neq \psi_q(\tilde{\Gamma}') = \tilde{\Gamma}'_1 + \cdots + \tilde{\Gamma}'_q.$$

Hence the unlabelled graphs $\tilde{\Gamma}_1, \ldots, \tilde{\Gamma}_q$ determine $\tilde{\Gamma}$, as desired.                □

**Polynomials with Real Zeros.** There are many techniques other than the linear algebra used to prove Theorem 5.8 for showing that sequences are unimodal. Here we will discuss a technique based on simple analysis (calculus) for showing that sequences are unimodal. In fact, we will consider some stronger properties than unimodality.

A sequence $a_0, a_1, \ldots, a_n$ of real numbers is called *logarithmically concave*, or *log-concave* for short, if $a_i^2 \geq a_{i-1}a_{i+1}$ for $1 \leq i \leq n-1$. We say that $a_0, a_1, \ldots, a_n$ is *strongly log-concave* if $b_i^2 \geq b_{i-1}b_{i+1}$ for $1 \leq i \leq n-1$, where $b_i = a_i / \binom{n}{i}$. Strong log-concavity is equivalent to [why?]

$$a_i^2 \geq \left(1 + \frac{1}{i}\right)\left(1 + \frac{1}{n-i}\right)a_{i-1}a_{i+1}, \quad 1 \leq i \leq n-1,$$

from which it follows that strong log-concavity implies log-concavity.

Assume now that each $a_i \geq 0$. Does log-concavity then imply unimodality? The answer is *no*, a counterexample being $1, 0, 0, 1$. However, only this type of counterexample can occur, as we now explain. We say that the sequence $a_0, a_1, \ldots, a_n$ has *no internal zeros* if whenever we have $i < j < k$, $a_i \neq 0$, and $a_k \neq 0$, then $a_j \neq 0$.

**5.11 Proposition.** *Let $\alpha = (a_0, a_1, \ldots, a_n)$ be a sequence of nonnegative real numbers with no internal zeros. If $\alpha$ is log-concave, then $\alpha$ is unimodal.*

*Proof.* If there are at most two values of $j$ for which $a_j \neq 0$ then we always have $a_{i-1}a_{i+1} = 0$ so the conclusion is clear. Now assume that there are at least three values of $j$ for which $a_j \neq 0$ and assume that the proposition is false. Then there exists $1 \leq i \leq n-1$ for which $a_{i-1} > a_i \leq a_{i+1}$ and $a_{i+1} > 0$, so $a_i^2 < a_{i-1}a_{i+1}$, a contradiction.                □

Now we come to a fundamental method for proving log-concavity.

**5.12 Theorem** (I. Newton). *Let*

$$P(x) = \sum_{i=0}^{n} b_i x^i = \sum_{i=0}^{n} \binom{n}{i} a_i x^i$$

*be a real polynomial all of whose zeros are real numbers. Then the sequence* $b_0, b_1, \ldots, b_n$ *is strongly log-concave, or equivalently, the sequence* $a_0, a_1, \ldots, a_n$ *is log-concave. Moreover, if each* $b_i \geq 0$ *(so the zeros of* $P(x)$ *are nonpositive [why?]) then the sequence* $b_0, b_1, \ldots, b_n$ *has no internal zeros.*

*Proof.* Let $\deg P(x) = m \leq n$. By the Fundamental Theorem of Algebra, $P(x)$ has exactly $m$ real zeros, counting multiplicities. Suppose that $\alpha$ is a zero of multiplicity $r > 1$, so $P(x) = (x - \alpha)^r L(x)$ for some polynomial $L(x)$ satisfying $L(\alpha) \neq 0$. A simple computation shows that $\alpha$ is a zero of $P'(x)$ (the derivative of $P(x)$) of multiplicity $r - 1$. Moreover, if $\alpha < \beta$ are both zeros of $P(x)$, then Rolle's theorem shows that $P'(x)$ has a zero $\gamma$ satisfying $\alpha < \gamma < \beta$. It follows [why?] that $P'(x)$ has at least $m - 1$ real zeros. Since $\deg P'(x) = m - 1$ we see that $P'(x)$ has exactly $m - 1$ real zeros and no other zeros.

Let $Q(x) = \frac{d^{i-1}}{dx^{i-1}} P(x)$. Thus $Q(x)$ is a polynomial of degree at most $m - i + 1$ with only real zeros. Let $R(x) = x^{m-i+1} Q(1/x)$, a polynomial of degree at most $m - i + 1$. The zeros of $R(x)$ are just reciprocals of those zeros of $Q(x)$ not equal to 0, with possible new zeros at 0. At any rate, all zeros of $R(x)$ are real. Now let $S(x) = \frac{d^{m-i-1}}{dx^{m-i-1}} R(x)$, a polynomial of degree at most two. By Rolle's theorem (with a suitable handling of multiple zeros as above), every zero of $S(x)$ is real. An explicit computation yields

$$S(x) = \frac{m!}{2}(a_{i-1}x^2 + 2a_i x + a_{i+1}).$$

If $a_{i-1} = 0$ then trivially $a_i^2 \geq a_{i-1} a_{i+1}$. Otherwise $S(x)$ is a quadratic polynomial. Since it has real zeros, its discriminant $\Delta$ is nonnegative. But

$$\Delta = (2a_i)^2 - 4a_{i-1}a_{i+1} = 4(a_i^2 - a_{i-1}a_{i+1}) \geq 0,$$

so the sequence $a_0, a_1, \ldots, a_n$ is log-concave as claimed.

It remains to show that if each $a_i \geq 0$ then the sequence $a_0, a_1, \ldots, a_n$ has no internal zeros. Suppose to the contrary that for some $i < j < k$ we have $a_i > 0, a_j = 0, a_k > 0$. By arguing as in the previous paragraph we will obtain a polynomial of the form $c + dx^{k-i}$ with only real zeros, where $c, d > 0$. But since $k - i \geq 2$ we have that every such polynomial has a nonreal zero [why?], a contradiction which completes the proof.                                                   □

In order to give combinatorial applications of Theorem 5.12 we need to find polynomials with real zeros whose coefficients are of combinatorial interest. One such example appears in Exercise 9.10, based on the fact that the characteristic polynomial of a symmetric matrix has only real zeros.

## Notes for Chapter 5

The techniques developed in this chapter had their origins in papers of Harper [62] and Pouzet and Rosenberg [104]. The closest treatment to ours appears in a paper of Stanley [126]. This latter paper also contains the proof of Theorem 5.10 (edge reconstruction) given here. This result was first proved by Lovász [83] by an inclusion–exclusion argument. The condition $q > \frac{1}{2}\binom{p}{2}$ in Theorem 5.10 was improved to $q > p(\log_2 p - 1)$ by Müller [95] (generalizing the argument of Lovász) and by Krasikov and Roditty [78] (generalizing the argument of Stanley).

For further information on Newton's Theorem 5.12, see, e.g., Hardy et al. [61, p. 52]. For a general survey on unimodality, log-concavity, etc., see Stanley [128], with a sequel by Brenti [13].

## Exercises for Chapter 5

1. (a) Let $G = \{\iota, \pi\}$ be a group of order two (with identity element $\iota$). Let $G$ act on $\{1, 2, 3, 4\}$ by $\pi \cdot 1 = 2, \pi \cdot 2 = 1, \pi \cdot 3 = 3$, and $\pi \cdot 4 = 4$. Draw the Hasse diagram of the quotient poset $B_4/G$.
   (b) Do the same for the action $\pi \cdot 1 = 2, \pi \cdot 2 = 1, \pi \cdot 3 = 4$, and $\pi \cdot 4 = 3$.

2. Draw the Hasse diagram of the poset of nonisomorphic simple graphs with five vertices (with the subgraph ordering). What is the size of the largest antichain? How many antichains have this size?

3. Give an example of a finite graded poset $P$ with the Sperner property, together with a group $G$ acting on $P$, such that the quotient poset $P/G$ is not Sperner. (By Theorem 5.8, $P$ cannot be a Boolean algebra.)

4. Consider the poset $P$ whose Hasse diagram is given by

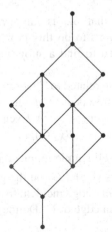

Find a subgroup $G$ of the symmetric group $\mathfrak{S}_7$ for which $P \cong B_7/G$ or else prove that such a group does not exist.

5. A $(0, 1)$-*necklace* of *length n* and *weight i* is a circular arrangement of $i$ 1's and
   $n-i$ 0's. For instance, the $(0, 1)$-necklaces of length 6 and weight 3 are (writing
   a circular arrangement linearly) 000111, 001011, 010011, and 010101. (Cyclic
   shifts of a linear word represent the same necklace, e.g., 000111 is the same
   as 110001.) Let $N_n$ denote the set of all $(0, 1)$-necklaces of length $n$. Define a
   partial order on $N_n$ by letting $u \leq v$ if we can obtain $v$ from $u$ by changing
   some 0's to 1's. It's easy to see (you may assume it) that $N_n$ is graded of rank
   $n$, with the rank of a necklace being its weight.

   (a)  (*) Show that $N_n$ is rank-symmetric, rank-unimodal, and Sperner.
   (b)  (difficult) Show that $N_n$ has a symmetric chain decomposition, as defined
        on page 40.

6. (unsolved) Show that every quotient poset $B_n/G$ has a symmetric chain
   decomposition.
7. Let $M$ be a finite multiset, say with $a_i$ $i$'s for $1 \leq i \leq k$. Let $B_M$ denote the
   poset of all submultisets of $M$, ordered by multiset inclusion. For instance, the
   figure below illustrates the case $a_1 = 2, a_2 = 3$.

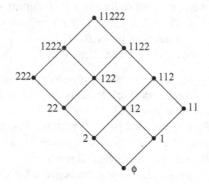

   Use Theorem 5.8 to show that $B_M$ is rank-symmetric, rank-unimodal, and
   Sperner. (There are other ways to do this problem, but you are asked to use
   Theorem 5.8. Thus you need to find a subgroup $G$ of $\mathfrak{S}_n$ for suitable $n$ for
   which $B_M \cong B_n/G$.)
8. (unsolved) Let $G$ be the group related to Theorem 5.9, so $G$ acts on $B_X$ where $X$
   consists of all two-element subsets of an $m$-set. Find an explicit order-matching
   $\mu : (B_X/G)_i \to (B_X/G)_{i+1}$ for $i < \frac{1}{2}\binom{m}{2}$. Even the weaker problem of finding
   an explicit injection $(B_X/G)_i \to (B_X/G)_{i+1}$ is open.

9. (a)  (*) Let $\mathcal{G}_p$ be the set of all simple graphs on the vertex set $[p]$, so $\#\mathcal{G}_p =$
        $2^{\binom{p}{2}}$. Given a graph $G \in \mathcal{G}_p$, let $G_i$ be the graph obtained by *switching* at
        vertex $i$, i.e., deleting every edge incident to vertex $i$ and adding every edge
        from vertex $i$ that isn't an edge of $G$. Define a linear transformation

$$\phi : \mathbb{R}\mathcal{G}_p \to \mathbb{R}\mathcal{G}_p$$

by $\phi(G) = G_1 + \cdots + G_p$. Show that $\phi$ is invertible if and only if $p \not\equiv$ 0 (mod 4).

(b) The graph $G$ is *switching-reconstructible* if it can be uniquely reconstructed from the multiset of *unlabelled* vertex switches $G_i$. That is, we are given each $G_i$ as an unlabelled graph. Show that $G$ is switching-reconstructible if $p \not\equiv 0 \pmod{4}$.

(c) (unsolved) Show that $G$ is switching-reconstructible if $p \neq 4$.

(d) Show that the number of edges of $G$ can be determined from the multiset of unlabelled $G_i$'s if $p \neq 4$. Find two graphs with four vertices and a different number of edges, but with the same unlabelled $G_i$'s.

(e) Define $G$ to be *weakly switching-reconstructible* if it can be uniquely reconstructed from the multiset of *labelled* vertex switches $G_i$. That is, we are given each $G_i$ as a labelled graph, but we aren't told the vertex $i$ that was switched. Show that $G$ is weakly switching-reconstructible if $p \neq 4$, but that $G$ need not be weakly switching-reconstructible if $p = 4$.

10. Suppose $X$ is a finite $n$-element set and $G$ a group of permutations of $X$. Thus $G$ acts on the subsets of $X$. We say that $G$ acts *transitively* on the $j$-element subsets if for every two $j$-element subsets $S$ and $T$, there is a $\pi \in G$ for which $\pi \cdot S = T$. Show that if $G$ acts transitively on $j$-element subsets for some $j \leq n/2$, then $G$ acts transitively on $i$-element subsets for all $0 \leq i \leq j$. (Although this can be done directly, there is a very easy proof using results proven in the text.)

11. In Example 5.4(b) is drawn the Hasse diagram of $B_5/G$, where $G$ is generated by the cycle $(1, 2, 3, 4, 5)$. Using the vertex labels shown in this figure, compute explicitly $\widehat{U}_2(12)$ and $\widehat{U}_2(13)$ as linear combinations of 123 and 124, where $\widehat{U}_2$ is defined as in the proof of Theorem 5.8. What is the matrix of $\widehat{U}_2$ with respect to the bases $(B_5/G)_2$ and $(B_5/G)_3$?

12. A real polynomial $F(x) = \sum_{i=0}^{n} a_i x^i$ is called *log-concave* if the sequence $a_0, a_1, \ldots, a_n$ of coefficients is log-concave. Let $F(x)$ and $G(x)$ be log-concave polynomials whose coefficients are positive. Show that the same is true for $F(x)G(x)$.

13. (*) Let $F(x)$ be a real polynomial with positive leading coefficient whose zeros all have the polar form $re^{i\theta}$, where $\frac{2\pi}{3} \leq \theta \leq \frac{4\pi}{3}$. Show that $F(x)$ has positive, log-concave coefficients.

# Chapter 6
# Young Diagrams and $q$-Binomial Coefficients

A *partition* $\lambda$ of an integer $n \geq 0$ is a sequence $\lambda = (\lambda_1, \lambda_2, \dots)$ of integers $\lambda_i \geq 0$ satisfying $\lambda_1 \geq \lambda_2 \geq \cdots$ and $\sum_{i \geq 1} \lambda_i = n$. Thus all but finitely many $\lambda_i$ are equal to 0. Each $\lambda_i > 0$ is called a *part* of $\lambda$. We sometimes suppress 0's from the notation for $\lambda$, e.g., $(5, 2, 2, 1)$, $(5, 2, 2, 1, 0, 0, 0)$, and $(5, 2, 2, 1, 0, 0, \dots)$ all represent the same partition $\lambda$ (of 10, with four parts). If $\lambda$ is a partition of $n$, then we denote this by $\lambda \vdash n$ or $|\lambda| = n$.

**6.1 Example.** There are seven partitions of 5, namely (writing, e.g., 221 as short for $(2, 2, 1)$): 5, 41, 32, 311, 221, 2111, and 11111.

The subject of partitions of integers has been extensively developed, but we will only be concerned here with a small part related to our previous discussion. Given positive integers $m$ and $n$, let $L(m, n)$ denote the set of all partitions with at most $m$ parts and with largest part at most $n$. For instance, $L(2, 3) = \{\emptyset, 1, 2, 3, 11, 21, 31, 22, 32, 33\}$. (Note that we are denoting by $\emptyset$ the unique partition $(0, 0, \dots)$ with no parts.) If $\lambda = (\lambda_1, \lambda_2, \dots)$ and $\mu = (\mu_1, \mu_2, \dots)$ are partitions, then define $\lambda \leq \mu$ if $\lambda_i \leq \mu_i$ for all $i$. This makes the set of all partitions into a very interesting poset, denoted $Y$ and called *Young's lattice* (named after the British mathematician Alfred Young, 1873–1940). (It is called "Young's lattice" rather than "Young's poset" because it turns out to have certain properties which define a *lattice*. However, these properties are irrelevant to us here, so we will not bother to define the notion of a lattice.) We will be looking at some properties of $Y$ in Chapter 8. The partial ordering on $Y$, when restricted to $L(m, n)$, makes $L(m, n)$ into a poset which also has some fascinating properties. Figure 6.1 shows $L(1, 4)$, $L(2, 2)$, and $L(2, 3)$, while Figure 6.2 shows $L(3, 3)$.

© Springer International Publishing AG, part of Springer Nature 2018
R. P. Stanley, *Algebraic Combinatorics*, Undergraduate Texts in Mathematics,
https://doi.org/10.1007/978-3-319-77173-1_6

**Fig. 6.1** The posets $L(1, 4)$,
$L(2, 2)$, and $L(2, 3)$

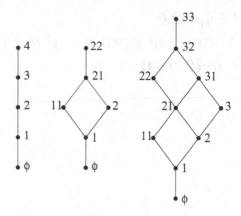

There is a nice geometric way of viewing partitions and the poset $L(m, n)$. The *Young diagram* (sometimes just called the *diagram*) of a partition $\lambda$ is a left-justified array of squares, with $\lambda_i$ squares in the $i$th row. For instance, the Young diagram of $(4, 3, 1, 1)$ looks like:

If dots are used instead of boxes, then the resulting diagram is called a *Ferrers diagram*. Thus the Ferrers diagram of $(4, 3, 1, 1)$ looks like

```
•   •   •   •

•   •   •

•

•
```

The advantage of Young diagrams over Ferrers diagrams is that we can put numbers in the boxes of a Young diagram, which we will do in Chapter 8. Observe that $L(m, n)$ is simply the set of Young diagrams $D$ fitting in an $m \times n$ rectangle (where the upper-left (northwest) corner of $D$ is the same as the northwest corner of the rectangle), ordered by inclusion. *We will always assume that when a Young diagram D is contained in a rectangle R, the northwest corners agree.* It is also clear from the Young diagram point of view that $L(m, n)$ and $L(n, m)$ are isomorphic partially ordered sets, the isomorphism being given by transposing the diagram (i.e., interchanging rows and columns). If $\lambda$ has Young diagram $D$, then the partition whose diagram is $D^t$ (the transpose of $D$) is called the *conjugate* of $\lambda$ and is denoted $\lambda'$. For instance, $(4, 3, 1, 1)' = (4, 2, 2, 1)$, with diagram

**Fig. 6.2** The poset $L(3, 3)$

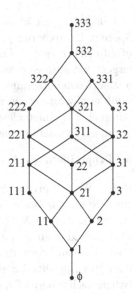

**6.2 Proposition.** *The poset $L(m, n)$ is graded of rank $mn$ and rank-symmetric. The rank of a partition $\lambda$ is just $|\lambda|$ (the sum of the parts of $\lambda$ or the number of squares in its Young diagram).*

*Proof.* As in the proof of Proposition 5.5, we leave to the reader everything except rank-symmetry. To show rank-symmetry, consider the complement $\bar{\lambda}$ of $\lambda$ in an $m \times n$ rectangle $R$, i.e., all the squares of $R$ except for $\lambda$. (Note that $\bar{\lambda}$ depends on $m$ and $n$ and not just $\lambda$.) For instance, in $L(5, 4)$, the complement of $(4, 3, 1, 1)$ looks like

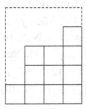

If we rotate the diagram of $\bar{\lambda}$ by $180°$ then we obtain the diagram of a partition $\tilde{\lambda} \in L(m, n)$ satisfying $|\lambda| + |\tilde{\lambda}| = mn$. This correspondence between $\lambda$ and $\tilde{\lambda}$ shows that $L(m, n)$ is rank-symmetric.                                      $\square$

Our main goal in this chapter is to show that $L(m, n)$ is rank-unimodal and Sperner. Let us write $p_i(m, n)$ as short for $p_i(L(m, n))$, the number of elements of $L(m, n)$ of rank $i$. Equivalently, $p_i(m, n)$ is the number of partitions of $i$ with largest part at most $n$ and with at most $m$ parts, or, in other words, the number of distinct Young diagrams with $i$ squares which fit inside an $m \times n$ rectangle (with the same northwest corner, as explained previously). Though not really necessary for our goal, it is nonetheless interesting to obtain some information on these numbers $p_i(m, n)$. First let us consider the total number $\#L(m, n)$ of elements in $L(m, n)$.

**6.3 Proposition.** *We have* $\#L(m, n) = \binom{m+n}{m}$.

*Proof.* We will give an elegant combinatorial proof, based on the fact that $\binom{m+n}{m}$ is equal to the number of sequences $a_1, a_2, \ldots, a_{m+n}$, where each $a_j$ is either $N$ or $E$, and there are $m$ $N$'s (and hence $n$ $E$'s) in all. We will associate a Young diagram $D$ contained in an $m \times n$ rectangle $R$ with such a sequence as follows. Begin at the lower left-hand corner of $R$ and trace out the southeast boundary of $D$, ending at the upper right-hand corner of $R$. This is done by taking a sequence of unit steps (where each square of $R$ is one unit in length), each step either north or east. Record the sequence of steps, using $N$ for a step to the north and $E$ for a step to the east.

*Example.* Let $m = 5$, $n = 6$, and $\lambda = (4, 3, 1, 1)$. Then $R$ and $D$ are given by:

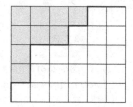

The corresponding sequence of $N$'s and $E$'s is $NENNEENENEE$.

It is easy to see (left to the reader) that the above correspondence gives a bijection between Young diagrams $D$ fitting in an $m \times n$ rectangle $R$ and sequences of $m$ $N$'s and $nE$'s. Hence the number of diagrams is equal to $\binom{m+n}{m}$, the number of sequences. $\qquad\square$

We now consider how many elements of $L(m, n)$ have rank $i$. To this end, let $q$ be an indeterminate; and given $j \geq 1$ define $(j) = 1 + q + q^2 + \cdots + q^{j-1}$. Thus $(1) = 1$, $(2) = 1 + q$, $(3) = 1 + q + q^2$, etc. Note that $(j)$ is a polynomial in $q$ whose value at $q = 1$ is just $j$ (denoted $(j)_{q=1} = j$). Next define $(j)! = (1)(2) \cdots (j)$ for $j \geq 1$ and set $(0)! = 1$. Thus $(1)! = 1$, $(2)! = 1 + q$, $(3)! = (1 + q)(1 + q + q^2) = 1 + 2q + 2q^2 + q^3$, etc., and $(j)!_{q=1} = j!$. Finally define for $k \geq j \geq 0$

$$\binom{k}{j} = \frac{(k)!}{(j)!(k - j)!}.$$

The expression $\binom{k}{j}$ is called a *q-binomial coefficient* (or *Gaussian coefficient*).

When $q$ is regarded as a prime power rather than as an indeterminate, then Exercise 4 in Chapter 4 gives a definition of $\binom{n}{k}$ in terms of the field $\mathbb{F}_q$. In this chapter we have no need of this algebraic interpretation of $\binom{n}{k}$.

Since $(r)!_{q=1} = r!$, it is clear that

$$\binom{k}{j}_{q=1} = \binom{k}{j}.$$

One sometimes says that $\binom{k}{j}$ is a "$q$-analogue" of the binomial coefficient $\binom{k}{j}$. There is no precise definition of a $q$-analogue $P(q)$ of some mathematical object $P$ (such as a formula or definition). It should have the property that there is a reasonable way to interpret $P(1)$ as being $P$. Ideally $P(q)$ should have some interpretation involving $\mathbb{F}_q$ when $q$ is regarded as a prime power. The $q$-analogue of the set $\{1\}$ is the finite field $\mathbb{F}_q$, and the $q$-analogue of the set $[n] = \{1, 2, \ldots, n\}$ is the vector space $\mathbb{F}_q^n$.

**6.4 Example.** We have $\binom{k}{j} = \binom{k}{k-j}$ [why?]. Moreover,

$$\binom{k}{0} = \binom{k}{k} = 1$$

$$\binom{k}{1} = \binom{k}{k-1} = (k) = 1 + q + q^2 + \cdots + q^{k-1}$$

$$\binom{4}{2} = \frac{(4)(3)(2)(1)}{(2)(1)(2)(1)} = 1 + q + 2q^2 + q^3 + q^4$$

$$\binom{5}{2} = \binom{5}{3} = 1 + q + 2q^2 + 2q^3 + 2q^4 + q^5 + q^6.$$

In the above example, $\binom{k}{j}$ was always a polynomial in $q$ (and with nonnegative integer coefficients). It is not obvious that this is always the case, but it will follow easily from the following lemma.

**6.5 Lemma.** *We have*

$$\binom{k}{j} = \binom{k-1}{j} + q^{k-j}\binom{k-1}{j-1}, \tag{6.1}$$

*whenever $k \geq 1$, with the initial conditions $\binom{0}{0} = 1$, $\binom{k}{j} = 0$ if $j < 0$ or $j > k$ (the same initial conditions satisfied by the binomial coefficients $\binom{k}{j}$).*

*Proof.* This is a straightforward computation. Specifically, we have

$$\binom{k-1}{j} + q^{k-j}\binom{k-1}{j-1} = \frac{(k-1)!}{(j)!(k-1-j)!} + q^{k-j}\frac{(k-1)!}{(j-1)!(k-j)!}$$

$$= \frac{(k-1)!}{(j-1)!(k-1-j)!}\left(\frac{1}{(j)} + \frac{q^{k-j}}{(k-j)}\right)$$

$$= \frac{(k-1)!}{(j-1)!(k-1-j)!} \cdot \frac{(k-j)+q^{k-j}(j)}{(j)(k-j)}$$

$$= \frac{(k-1)!}{(j-1)!(k-1-j)!} \cdot \frac{(k)}{(j)(k-j)}$$

$$= \binom{k}{j}.$$

$\square$

Note that if we put $q = 1$ in (6.1) we obtain the well-known formula

$$\binom{k}{j} = \binom{k-1}{j} + \binom{k-1}{j-1},$$

which is just the recurrence defining Pascal's triangle. Thus (6.1) may be regarded as a $q$-analogue of the Pascal triangle recurrence.

We can regard equation (6.1) as a recurrence relation for the $q$-binomial coefficients. Given the initial conditions of Lemma 6.5, we can use (6.1) inductively to compute $\binom{k}{j}$ for any $k$ and $j$. From this it is obvious by induction that the $q$-binomial coefficient $\binom{k}{j}$ is a polynomial in $q$ with nonnegative integer coefficients. The following theorem gives an even stronger result, namely, an explicit combinatorial interpretation of the coefficients.

**6.6 Theorem.** *Let $p_i(m, n)$ denote the number of elements of $L(m, n)$ of rank $i$. Then*

$$\sum_{i \geq 0} p_i(m, n) q^i = \binom{m+n}{m}. \tag{6.2}$$

NOTE. *The sum on the left-hand side is really a finite sum, since $p_i(m, n) = 0$ if $i > mn$.*

*Proof.* Let $P(m, n)$ denote the left-hand side of (6.2). We will show that

$$P(0, 0) = 1, \text{ and } P(m, n) = 0 \text{ if } m < 0 \text{ or } n < 0 \tag{6.3}$$

$$P(m, n) = P(m, n-1) + q^n P(m-1, n). \tag{6.4}$$

Note that (6.3) and (6.4) completely determine $P(m, n)$. On the other hand, substituting $k = m + n$ and $j = m$ in (6.1) shows that $\binom{m+n}{m}$ also satisfies (6.4). Moreover, the initial conditions of Lemma 6.5 show that $\binom{m+n}{m}$ also satisfies (6.3). Hence (6.3) and (6.4) imply that $P(m, n) = \binom{m+n}{m}$, so to complete the proof we need only establish (6.3) and (6.4).

Equation (6.3) is clear, since $L(0, n)$ consists of a single point (the empty partition $\emptyset$), so $\sum_{i \geq 0} p_i(0, n)q^i = 1$; while $L(m, n)$ is empty (or undefined, if you prefer) if $m < 0$ or $n < 0$.

The crux of the proof is to show (6.4). Taking the coefficient of $q^i$ of both sides of (6.4), we see [why?] that (6.4) is equivalent to

$$p_i(m, n) = p_i(m, n - 1) + p_{i-n}(m - 1, n). \tag{6.5}$$

Consider a partition $\lambda \vdash i$ whose Young diagram $D$ fits in an $m \times n$ rectangle $R$. If $D$ does not contain the upper right-hand corner of $R$, then $D$ fits in an $m \times (n - 1)$ rectangle, so there are $p_i(m, n - 1)$ such partitions $\lambda$. If on the other hand $D$ does contain the upper right-hand corner of $R$, then $D$ contains the whole first row of $R$. When we remove the first row of $R$, we have left a Young diagram of size $i - n$ which fits in an $(m - 1) \times n$ rectangle. Hence there are $p_{i-n}(m - 1, n)$ such $\lambda$, and the proof follows [why?]. $\qquad \square$

Note that if we set $q = 1$ in (6.2), then the left-hand side becomes $\#L(m, n)$ and the right-hand side $\binom{m+n}{m}$, agreeing with Proposition 6.3.

As the reader may have guessed by now, the poset $L(m, n)$ is isomorphic to a quotient poset $B_s/G$ for a suitable integer $s > 0$ and finite group $G$ acting on $B_s$. Actually, it is clear that we must have $s = mn$ since $L(m, n)$ has rank $mn$ and in general $B_s/G$ has rank $s$. What is not so clear is the right choice of $G$. To this end, let $R = R_{mn}$ denote an $m \times n$ rectangle of squares. For instance, $R_{35}$ is given by the 15 squares of the diagram

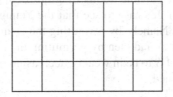

We now define the group $G = G_{mn}$ as follows. It is a subgroup of the group $\mathfrak{S}_R$ of all permutations of the squares of $R$. A permutation $\pi$ in $G$ is allowed to permute the elements in each row of $R$ in any way and then to permute the rows among themselves in any way. The elements of each row can be permuted in $n!$ ways, so since there are $m$ rows there are a total of $n!^m$ permutations preserving the rows. Then the $m$ rows can be permuted in $m!$ ways, so it follows that the order of $G_{mn}$ is given by $m!n!^m$. The group $G_{mn}$ is called the *wreath product* of $\mathfrak{S}_n$ and $\mathfrak{S}_m$, denoted $\mathfrak{S}_n \wr \mathfrak{S}_m$ or $\mathfrak{S}_n \text{ wr } \mathfrak{S}_m$. However, we will not discuss the general theory of wreath products here.

**6.7 Example.** Suppose $m = 4$ and $n = 5$, with the boxes of $R$ labelled as follows.

| 1  | 2  | 3  | 4  | 5  |
|----|----|----|----|----|
| 6  | 7  | 8  | 9  | 10 |
| 11 | 12 | 13 | 14 | 15 |
| 16 | 17 | 18 | 19 | 20 |

Then a typical permutation $\pi$ in $G_{45}$ looks like

| 16 | 20 | 17 | 19 | 18 |
|----|----|----|----|----|
| 4  | 1  | 5  | 2  | 3  |
| 12 | 13 | 15 | 14 | 11 |
| 7  | 9  | 6  | 10 | 8  |

i.e., $\pi(16) = 1$, $\pi(20) = 2$, etc.

We have just defined a group $G_{mn}$ of permutations of the set $R = R_{mn}$ of squares of an $m \times n$ rectangle. Hence $G_{mn}$ acts on the boolean algebra $B_R$ of all subsets of the set $R$. The next lemma describes the orbits of this action.

**6.8 Lemma.** *Every orbit $\mathfrak{o}$ of the action of $G_{mn}$ on $B_R$ contains exactly one Young diagram $D$, i.e., exactly one subset $D \subseteq R$ such that $D$ is left-justified, and if $\lambda_i$ is the number of elements of $D$ in row $i$ of $R$, then $\lambda_1 \geq \lambda_2 \geq \cdots \geq \lambda_m$.*

*Proof.* Let $S$ be a subset of $R$, and suppose that $S$ has $\alpha_i$ elements in row $i$. If $\pi \in G_{mn}$ and $\pi \cdot S$ has $\beta_i$ elements in row $i$, then $\beta_1, \ldots, \beta_m$ is just some permutation of $\alpha_1, \ldots, \alpha_m$ [why?]. There is a unique ordering $\lambda_1, \ldots, \lambda_m$ of $\alpha_1, \ldots, \alpha_m$ satisfying $\lambda_1 \geq \cdots \geq \lambda_m$, so the only possible Young diagram $D$ in the orbit $\pi \cdot S$ is the one of shape $\lambda = (\lambda_1, \ldots, \lambda_m)$. It's easy to see that the Young diagram $D_\lambda$ of shape $\lambda$ is indeed in the orbit $\pi \cdot S$. Namely, by permuting the elements in the rows of $R$ we can left-justify the rows of $S$, and then by permuting the rows of $R$ themselves we can arrange the row sizes of $S$ to be in weakly decreasing order. Thus we obtain the Young diagram $D_\lambda$ as claimed. $\qquad\square$

We are now ready for the main result of this chapter.

**6.9 Theorem.** *Set $R = R_{mn}$. Then the quotient poset $B_R/G_{mn}$ is isomorphic to $L(m, n)$.*

*Proof.* Each element of $B_R/G_{mn}$ contains a unique Young diagram $D_\lambda$ by Lemma 6.8. Moreover, two different orbits cannot contain the same Young diagram $D$ since orbits are disjoint. Thus the map $\varphi \colon B_R/G_{mn} \to L(m, n)$ defined by $\varphi(D_\lambda) = \lambda$ is a bijection (one-to-one and onto). We claim that in fact $\varphi$ is an isomorphism of partially ordered sets. We need to show the following: let $\mathfrak{o}$ and $\mathfrak{o}^*$ be orbits of $G_{mn}$ (i.e., elements of $B_R/G_{mn}$). Let $D_\lambda$ and $D_{\lambda^*}$ be the unique Young diagrams in $\mathfrak{o}$ and $\mathfrak{o}^*$, respectively. Then there exist $D \in \mathfrak{o}$ and $D^* \in \mathfrak{o}^*$ satisfying $D \subseteq D^*$ if and only if $\lambda \leq \lambda^*$ in $L(m, n)$.

The "if" part of the previous sentence is clear, for if $\lambda \leq \lambda^*$ then $D_\lambda \subseteq D_{\lambda^*}$. So assume there exist $D \in \mathfrak{o}$ and $D^* \in \mathfrak{o}^*$ satisfying $D \subseteq D^*$. The lengths of the rows of $D$, written in decreasing order, are $\lambda_1, \ldots, \lambda_m$, and similarly for $D^*$. Since each row of $D$ is contained in a row of $D^*$, it follows that for each $1 \leq j \leq m$, $D^*$ has at least $j$ rows of size at least $\lambda_j$. Thus the length $\lambda_j^*$ of the $j$th largest row of $D^*$ is at least as large as $\lambda_j$. In other words, $\lambda_j \leq \lambda_j^*$, as was to be proved.                   $\square$

Combining the previous theorem with Theorem 5.8 yields the following result.

**6.10 Corollary.** *The posets $L(m, n)$ are rank-symmetric, rank-unimodal, and Sperner.*

Note that the rank-symmetry and rank-unimodality of $L(m, n)$ can be rephrased as follows: the $q$-binomial coefficient $\binom{m+n}{m}$ has symmetric and unimodal coefficients. While rank-symmetry is easy to prove (see Proposition 6.2), the unimodality of the coefficients of $\binom{m+n}{m}$ is by no means apparent. It was first proved by J. Sylvester in 1878 by a proof similar to the one above, though stated in the language of the invariant theory of binary forms. For a long time it was an open problem to find a combinatorial proof that the coefficients of $\binom{m+n}{m}$ are unimodal. Such a proof would give an explicit injection (one-to-one function) $\mu \colon L(m, n)_i \to L(m, n)_{i+1}$ for $i < \frac{1}{2}mn$. (One difficulty in finding such maps $\mu$ is to make use of the hypothesis that $i < \frac{1}{2}mn$.) Finally around 1989 such a proof was found by K. M. O'Hara. However, O'Hara's proof has the defect that the maps $\mu$ are not order-matchings. Thus her proof does not prove that $L(m, n)$ is Sperner, but only that it's rank-unimodal. It is an outstanding open problem in algebraic combinatorics to find an explicit order-matching $\mu \colon L(m, n)_i \to L(m, n)_{i+1}$ for $i < \frac{1}{2}mn$.

Note that the Sperner property of $L(m, n)$ (together with the fact that the largest level is in the middle) can be stated in the following simple terms: the largest possible collection $\mathcal{C}$ of Young diagrams fitting in an $m \times n$ rectangle such that no diagram in $\mathcal{C}$ is contained in another diagram in $\mathcal{C}$ is obtained by taking all the diagrams of size $\lfloor \frac{1}{2}mn \rfloor$. Although the statement of this fact requires almost no mathematics to understand, there is no known proof that doesn't use algebraic machinery. The several known algebraic proofs are all closely related, and the one we have given is the simplest. Corollary 6.10 is a good example of the efficacy of algebraic combinatorics.

**An Application to Number Theory.** There is an interesting application of Corollary 6.10 to a number-theoretic problem. Fix a positive integer $k$. For a finite subset $S$ of $\mathbb{R}^+ = \{\beta \in \mathbb{R} \colon \beta > 0\}$, and for a real number $\alpha > 0$, define

$$f_k(S, \alpha) = \#\left\{ T \in \binom{S}{k} \colon \sum_{t \in T} t = \alpha \right\}.$$

In other words, $f_k(S, \alpha)$ is the number of $k$-element subsets of $S$ whose elements sum to $\alpha$. For instance, $f_3(\{1, 3, 4, 6, 7\}, 11) = 2$, since $1+3+7 = 1+4+6 = 11$.

Given positive integers $k < n$, our object is to maximize $f_k(S, \alpha)$ subject to the condition that $\#S = n$. We are free to choose both $S$ and $\alpha$, but $k$ and $n$ are fixed. Call this maximum value $h_k(n)$. Thus

$$h_k(n) = \max_{\substack{\alpha \in \mathbb{R}^+ \\ S \subset \mathbb{R}^+ \\ \#S = n}} f_k(S, \alpha).$$

What sort of behavior can we expect of the maximizing set $S$? If the elements of $S$ are "spread out," say $S = \{1, 2, 4, 8, \ldots, 2^{n-1}\}$, then all the subset sums of $S$ are distinct. Hence for any $\alpha \in \mathbb{R}^+$ we have $f_k(S, \alpha) = 0$ or $1$. Similarly, if the elements of $S$ are "unrelated" (e.g., linearly independent over the rationals, such as $S = \{1, \sqrt{2}, \sqrt{3}, \pi, \pi^2\}$), then again all subset sums are distinct and $f_k(S, \alpha) = 0$ or $1$. These considerations make it plausible that we should take $S = [n] = \{1, 2, \ldots, n\}$ and then choose $\alpha$ appropriately. In other words, we are led to the conjecture that for any $S \in \binom{\mathbb{R}^+}{n}$ and $\alpha \in \mathbb{R}^+$, we have

$$f_k(S, \alpha) \le f_k([n], \beta), \tag{6.6}$$

for some $\beta \in \mathbb{R}^+$ to be determined.

First let us evaluate $f_k([n], \alpha)$ for any $\alpha$. This will enable us to determine the value of $\beta$ in (6.6). Let $S = \{i_1, \ldots, i_k\} \subseteq [n]$ with

$$1 \le i_1 < i_2 < \cdots < i_k \le n, \quad i_1 + \cdots + i_k = \alpha. \tag{6.7}$$

Let $j_r = i_r - r$. Then (since $1 + 2 + \cdots + k = \binom{k+1}{2}$)

$$n - k \ge j_k \ge j_{k-1} \ge \cdots \ge j_1 \ge 0, \quad j_1 + \cdots + j_k = \alpha - \binom{k+1}{2}. \tag{6.8}$$

Conversely, given $j_1, \ldots, j_k$ satisfying (6.8) we can recover $i_1, \ldots, i_k$ satisfying (6.7). Hence $f_k([n], \alpha)$ is equal to the number of sequences $j_1, \ldots, j_k$ satisfying (6.8). Now let

$$\lambda(S) = (j_k, j_{k-1}, \ldots, j_1).$$

Note that $\lambda(S)$ is a partition of the integer $\alpha - \binom{k+1}{2}$ with at most $k$ parts and with largest part at most $n - k$. Thus

$$f_k([n], \alpha) = p_{\alpha - \binom{k+1}{2}}(k, n - k), \tag{6.9}$$

or equivalently,

$$\sum_{\alpha \ge \binom{k+1}{2}} f_k([n], \alpha) q^{\alpha - \binom{k+1}{2}} = \binom{n}{k}.$$

By the rank-unimodality (and rank-symmetry) of $L(n - k, k)$ (Corollary 6.10), the largest coefficient of $\binom{n}{k}$ is the middle one, that is, the coefficient of $\lfloor k(n - k)/2 \rfloor$.

It follows that for fixed $k$ and $n$, $f_k([n], \alpha)$ is maximized for $\alpha = \lfloor k(n-k)/2 \rfloor + \binom{k+1}{2} = \lfloor k(n+1)/2 \rfloor$. Hence the following result is plausible.

**6.11 Theorem.** *Let $S \in \binom{\mathbb{R}^+}{n}$, $\alpha \in \mathbb{R}^+$, and $k \in \mathbb{P}$. Then*

$$f_k(S, \alpha) \le f_k([n], \lfloor k(n+1)/2 \rfloor).$$

*Proof.* Let $S = \{a_1, \ldots, a_n\}$ with $0 < a_1 < \cdots < a_n$. Let $T$ and $U$ be distinct $k$-element subsets of $S$ with the same element sums, say $T = \{a_{i_1}, \ldots, a_{i_k}\}$ and $U = \{a_{j_1}, \ldots, a_{j_k}\}$ with $i_1 < i_2 < \cdots < i_k$ and $j_1 < j_2 < \cdots < j_k$. Define $T^* = \{i_1, \ldots, i_k\}$ and $U^* = \{j_1, \ldots, j_k\}$, so $T^*, U^* \in \binom{[n]}{k}$. The crucial observation is the following:

**Claim.** The elements $\lambda(T^*)$ and $\lambda(U^*)$ are incomparable in $L(k, n-k)$, i.e., neither $\lambda(T^*) \le \lambda(U^*)$ nor $\lambda(U^*) \le \lambda(T^*)$.

**Proof of claim.** Suppose not, say $\lambda(T^*) \le \lambda(U^*)$ to be definite. Thus by definition of $L(k, n-k)$ we have $i_r - r \le j_r - r$ for $1 \le r \le k$. Hence $i_r \le j_r$ for $1 \le r \le k$, so also $a_{i_r} \le a_{j_r}$ (since $a_1 < \cdots < a_n$). But $a_{i_1} + \cdots + a_{i_k} = a_{j_1} + \cdots + a_{j_k}$ by assumption, so $a_{i_r} = a_{j_r}$ for all $r$. This contradicts the assumption that $T$ and $U$ are distinct and proves the claim.

It is now easy to complete the proof of Theorem 6.11. Suppose that $S_1, \ldots, S_r$ are distinct $k$-element subsets of $S$ with the same element sums. By the claim, $\{\lambda(S_1^*), \ldots, \lambda(S_r^*)\}$ is an antichain in $L(k, n-k)$. Hence $r$ cannot exceed the size of the largest antichain in $L(k, n-k)$. By Theorem 6.6 and Corollary 6.10, the size of the largest antichain in $L(k, n-k)$ is given by $p_{\lfloor k(n-k)/2 \rfloor}(k, n-k)$. By (6.9) this number is equal to $f_k([n], \lfloor k(n+1)/2 \rfloor)$. In other words,

$$r \le f_k([n], \lfloor k(n+1)/2 \rfloor),$$

which is what we wanted to prove.    □

Note that an equivalent statement of Theorem 6.11 is that $h_k(n)$ is equal to the coefficient of $q^{\lfloor k(n-k)/2 \rfloor}$ in $\binom{n}{k}$ [why?].

**Variation on a theme.** Suppose that in Theorem 6.11 we do not want to specify the cardinality of the subsets of $S$. In other words, for any $\alpha \in \mathbb{R}$ and any finite subset $S \subset \mathbb{R}^+$, define

$$f(S, \alpha) = \#\{T \subseteq S : \sum_{t \in T} t = \alpha\}.$$

How large can $f(S, \alpha)$ be if we require $\#S = n$? Call this maximum value $h(n)$. Thus

$$h(n) = \max_{\substack{\alpha \in \mathbb{R}^+ \\ S \subset \mathbb{R}^+ \\ \#S = n}} f(S, \alpha). \tag{6.10}$$

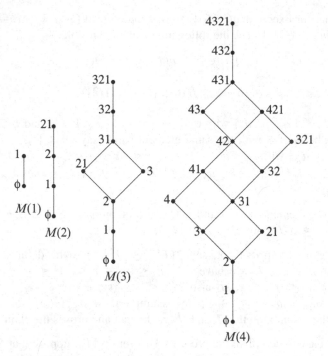

**Fig. 6.3** The posets $M(1)$, $M(2)$, $M(3)$, and $M(4)$

For instance, if $S = \{1, 2, 3\}$ then $f(S, 3) = 2$ (coming from the subsets $\{1, 2\}$ and $\{3\}$). This is easily seen to be best possible, i.e., $h(3) = 2$.

We will find $h(n)$ in a manner analogous to the proof of Theorem 6.11. The big difference is that the relevant poset $M(n)$ is *not* of the form $B_n/G$, so we will have to prove the injectivity of the order-raising operator $U_i$ from scratch. Our proofs will be somewhat sketchy; it shouldn't be difficult for the reader who has come this far to fill in the details.

Let $M(n)$ be the set of all subsets of $[n]$, with the ordering $A \leq B$ if the elements of $A$ are $a_1 > a_2 > \cdots > a_j$ and the elements of $B$ are $b_1 > b_2 > \cdots > b_k$, where $j \leq k$ and $a_i \leq b_i$ for $1 \leq i \leq j$. (The empty set $\emptyset$ is the bottom element of $M(n)$.) Figure 6.3 shows $M(1)$, $M(2)$, $M(3)$, and $M(4)$.

It is easy to see that $M(n)$ is graded of rank $\binom{n+1}{2}$. The rank of the subset $T = \{a_1, \ldots, a_k\}$ is

$$\text{rank}(T) = a_1 + \cdots + a_k. \tag{6.11}$$

It follows [why?] that the rank-generating function of $M(n)$ is given by

$$F(M(n), q) = \sum_{i=0}^{\binom{n+1}{2}} (\#M(n)_i)q^i = (1+q)(1+q^2)\cdots(1+q^n).$$

Define linear transformations

$$U_i : \mathbb{R}M(n)_i \to \mathbb{R}M(n)_{i+1}, \quad D_i : \mathbb{R}M(n)_i \to \mathbb{R}M(n)_{i-1}$$

by

$$U_i(x) = \sum_{\substack{y \in M(n)_{i+1} \\ x < y}} y, \quad x \in M(n)_i$$

$$D_i(x) = \sum_{\substack{v \in M(n)_{i-1} \\ v < x}} c(v, x)v, \quad x \in M(n)_i,$$

where the coefficient $c(v, x)$ is defined as follows. Let the elements of $v$ be $a_1 > \cdots > a_j > 0$ and the elements of $x$ be $b_1 > \cdots > b_k > 0$. Since $x$ covers $v$, there is a unique $r$ for which $a_r = b_r - 1$ (and $a_k = b_k$ for all other $k$). In the case $b_r = 1$ we set $a_r = 0$. (e.g., if $x$ is given by $5 > 4 > 1$ and $v$ by $5 > 4$, then $r = 3$ and $a_3 = 0$.) Set

$$c(v, x) = \begin{cases} \binom{n+1}{2}, & \text{if } a_r = 0 \\ (n - a_r)(n + a_r + 1), & \text{if } a_r > 0. \end{cases}$$

It is a straightforward computation (proof omitted) to obtain the commutation relation

$$D_{i+1}U_i - U_{i-1}D_i = \left( \binom{n+1}{2} - 2i \right) I_i, \tag{6.12}$$

where $I_i$ denotes the identity linear transformation on $\mathbb{R}M(n)_i$. Clearly by definition $U_i$ is order-raising. We want to show that $U_i$ is injective (one-to-one) for $i < \frac{1}{2}\binom{n+1}{2}$. We can't argue as in the proof of Lemma 4.6 that $U_{i-1}D_i$ is semidefinite since the matrices of $U_{i-1}$ and $D_i$ are no longer transposes of one another. Instead we use the following result from linear algebra.

**6.12 Lemma.** *Let $V$ and $W$ be finite-dimensional vector spaces over a field. Let $A: V \to W$ and $B: W \to V$ be linear transformations. Then*

$$x^{\dim V} \det(AB - xI) = x^{\dim W} \det(BA - xI).$$

*In other words, $AB$ and $BA$ have the same nonzero eigenvalues.*

We can now prove the key linear algebraic result.

**6.13 Lemma.** *The linear transformation $U_i$ is injective for $i < \frac{1}{2}\binom{n+1}{2}$ and surjective (onto) for $i \geq \frac{1}{2}\binom{n+1}{2}$.*

*Proof.* We prove by induction on $i$ that $D_{i+1}U_i$ has positive real eigenvalues for $i < \frac{1}{2}\binom{n+1}{2}$. For $i = 0$ this is easy to check since $\dim \mathbb{R}M(n)_0 = 1$. Assume the induction hypothesis for some $i < \frac{1}{2}\binom{n+1}{2} - 1$, i.e., assume that $D_i U_{i-1}$ has positive eigenvalues. By Lemma 6.12, $U_{i-1} D_i$ has nonnegative eigenvalues. By (6.12), we have

$$D_{i+1} U_i = U_{i-1} D_i + \left( \binom{n+1}{2} - 2i \right) I_i.$$

Thus the eigenvalues of $D_{i+1} U_i$ are $\binom{n+1}{2} - 2i$ more than those of $U_{i-1} D_i$. Since $\binom{n+1}{2} - 2i > 0$, it follows that $D_{i+1} U_i$ has positive eigenvalues. Hence it is invertible, so $U_i$ is injective. Similarly (or by "symmetry") $U_i$ is surjective for $i \geq \frac{1}{2}\binom{n+1}{2}$. □

The main result on the posets $M(n)$ now follows by a familiar argument.

**6.14 Theorem.** *The poset $M(n)$ is graded of rank $\binom{n+1}{2}$, rank-symmetric, rank-unimodal, and Sperner.*

*Proof.* We have already seen that $M(n)$ is graded of rank $\binom{n+1}{2}$ and rank-symmetric. By the previous lemma, $U_i$ is injective for $i < \frac{1}{2}\binom{n+1}{2}$ and surjective for $i \geq \frac{1}{2}\binom{n+1}{2}$. The proof follows from Proposition 4.4 and Lemma 4.5. □

NOTE. As a consequence of Theorem 6.14, the polynomial $F(M(n), q) = (1 + q)(1 + q^2) \cdots (1 + q^n)$ has unimodal coefficients. No combinatorial proof of this fact is known, unlike the situation for $L(m, n)$ (where we mentioned the proof of O'Hara above).

We can now determine $h(n)$ (as defined by (6.10)) by an argument analogous to the proof of Theorem 6.11.

**6.15 Theorem.** *Let $S \in \binom{\mathbb{R}^+}{n}$ and $\alpha \in \mathbb{R}^+$. Then*

$$f(S, \alpha) \leq f\left( [n], \left\lfloor \frac{1}{2} \binom{n+1}{2} \right\rfloor \right) = h(n).$$

*Proof.* Let $S = \{a_1, \ldots, a_n\}$ with $0 < a_1 < \cdots < a_n$. Let $T$ and $U$ be distinct subsets of $S$ with the same element sums, say $T = \{a_{r_1}, \ldots, a_{r_j}\}$ and $U = \{a_{s_1}, \ldots, a_{s_k}\}$ with $r_1 < r_2 < \cdots < r_j$ and $s_1 < s_2 < \cdots < s_k$. Define $T^* = \{r_1, \ldots, r_j\}$ and $U^* = \{s_1, \ldots, s_k\}$, so $T^*, U^* \in M(n)$. The following fact is proved exactly in the same way as the analogous fact for $L(m, n)$ (the claim in the proof of Theorem 6.11) and will be omitted here.

**Fact.** The elements $T^*$ and $U^*$ are incomparable in $M(n)$, i.e., neither $T^* \leq U^*$ nor $U^* \leq T^*$.

It is now easy to complete the proof of Theorem 6.15. Suppose that $S_1, \ldots, S_t$ are distinct subsets of $S$ with the same element sums. By the above fact, $\{S_1^*, \ldots, S_t^*\}$ is an antichain in $M(n)$. Hence $t$ cannot exceed the size of the largest antichain in $M(n)$. By Theorem 6.14, the size of the largest antichain in $M(n)$ is the size

$p_{\left\lfloor \frac{1}{2}\binom{n+1}{2}\right\rfloor}$ of the middle rank. By (6.11) this number is equal to $f([n], \lfloor\frac{1}{2}\binom{n+1}{2}\rfloor)$.
In other words,

$$t \leq f\left([n], \left\lfloor \frac{1}{2}\binom{n+1}{2}\right\rfloor\right),$$

which is what we wanted to prove.                                                                    □

NOTE. Theorem 6.15 is known as the *weak Erdős–Moser conjecture*. The
original (strong) Erdős–Moser conjecture deals with the case $S \subset \mathbb{R}$ rather than
$S \subset \mathbb{R}^+$. There is a difference between these two cases; for instance, $h(3) = 2$
(corresponding to $S = \{1, 2, 3\}$ and $\alpha = 3$), while the set $\{-1, 0, 1\}$ has *four* subsets
whose elements sum to 0 (including the empty set). (Can you see where the proof
of Theorem 6.15 breaks down if we allow $S \subset \mathbb{R}$?) The original Erdős–Moser
conjecture asserts that if $\#S = 2m + 1$, then

$$f(S, \alpha) \leq f(\{-m, -m + 1, \ldots, m\}, 0). \tag{6.13}$$

This result can be proved by a somewhat tricky modification of the proof given
above for the weak case; see Exercise 6.5. No proof of the Erdős–Moser conjecture
(weak or strong) is known other than the one indicated here (sometimes given in a
more sophisticated context, as explained in the next Note).

NOTE. The key to the proof of Theorem 6.15 is the definition of $U_i$ and $D_i$ which
gives the commutation relation (6.12). The reader may be wondering how anyone
managed to discover these definitions (especially that of $D_i$). In fact, the original
proof of Theorem 6.15 was based on the representation theory of the orthogonal Lie
algebra $\mathfrak{o}(2n + 1, \mathbb{C})$. In this context, the definitions of $U_i$ and $D_i$ are built into the
theory of the "principal subalgebras" of $\mathfrak{o}(2n + 1, \mathbb{C})$. R. A. Proctor was the first
to remove the representation theory from the proof and present it solely in terms of
linear algebra.

# Notes for Chapter 6

For an undergraduate level introduction to the theory of partitions, see Andrews and
Eriksson [3]. A more extensive treatment is given by Andrews [2], while a brief
introduction appears in [130, Section 1.8].

As already mentioned in the text, the rank-unimodality of $L(m, n)$, that is, of the
coefficients of the $q$-binomial coefficient $\binom{m+n}{m}$, is due to J. J. Sylvester [135], with
a combinatorial proof later given by K. M. O'Hara [98]. An explication of O'Hara's
work was given by D. Zeilberger [147].

The unimodality of the coefficients of the polynomial $(1+q)(1+q^2)\cdots(1+q^n)$
is implicit in the work of E.B. Dynkin [37], [38, p. 332]. J.W.B. Hughes was the first

to observe explicitly that this polynomial arises as a special case of Dynkin's work. The Spernicity of $L(m, n)$ and $M(n)$, and a proof of the Erdős–Moser conjecture, were first given by Stanley [123]. It was mentioned in the text above that R.A. Proctor [105] was the first to remove the representation theory from the proof and present it solely in terms of linear algebra.

For two proofs of Lemma 6.12, see W.V. Parker [99] and J. Schmid [116].

## Exercises for Chapter 6

1. (a) Let $A(m, n)$ denote the adjacency matrix (over $\mathbb{R}$) of the Hasse diagram of $L(m, n)$. Show that if $A(m, n)$ is nonsingular, then $\binom{m+n}{m}$ is even.
   (b) (unsolved) For which $m$ and $n$ is $A(m, n)$ nonsingular? The pairs $(m, n)$ with this property for $m \le n$ and $m+n \le 13$ are $(1, 1)$, $(1, 3)$, $(1, 5)$, $(3, 3)$, $(1, 7)$, $(1, 9)$, $(3, 7)$, $(5, 5)$, $(1, 11)$, $(3, 9)$, and $(5, 7)$.
   (c) (very difficult) Show that every irreducible (over $\mathbb{Q}$) factor of the characteristic polynomial of the matrix $A(m, n)$ has degree at most $\frac{1}{2}\phi(2(m + n + 1))$, where $\phi$ is the Euler phi-function (defined on page 85).

2. (a) (moderately difficult) Show that the number $c(m, n)$ of cover relations in $L(m, n)$, i.e., the number of pairs $(\lambda, \mu)$ of partitions in $L(m, n)$ for which $\mu$ covers $\lambda$, is given by

$$c(m, n) = \frac{(m + n - 1)!}{(m - 1)! \, (n - 1)!}.$$

   (b) (considerably more difficult) (*) Show that the number $d(m, n)$ of pairs $(\lambda, \mu)$ of elements in $L(m, n)$ for which $\lambda \le \mu$ is given by

$$d(m, n) = \frac{(m + n)! \, (m + n + 1)!}{m! \, (m + 1)! \, n! \, (n + 1)!}.$$

3. (difficult) (*) Note that $L(m, n)$ is the set of all partitions $\lambda$ in Young's lattice $Y$ satisfying $\lambda \le \langle n^m \rangle$, the partition with $m$ parts equal to $n$. Let $Y_\mu$ denote the set of all partitions $\lambda \le \mu$. Is $Y_\mu$ always rank-unimodal?

4. (a) Find an explicit order matching $\mu : L(2, n)_i \to L(2, n)_{i+1}$ for $i < n$.
   (b) (more difficult) Do the same for $L(3, n)_i \to L(3, n)_{i+1}$ for $i < 3n/2$.
   (c) (even more difficult) Do the same for $L(4, n)_i \to L(4, n)_{i+1}$ for $i < 2n$.
   (d) (unsolved) Do the same for $L(5, n)_i \to L(5, n)_{i+1}$ for $i < 5n/2$.

5. Assume that $M(j) \times M(k)^*$ is rank-symmetric, rank-unimodal, and Sperner. Here $M(k)^*$ denotes the dual of $M(k)$, i.e., $x \le y$ in $M(k)^*$ if and only if $y \le x$ in $M(k)$. (Actually $M(k) \cong M(k)^*$, but this is not needed here.) Deduce the original Erdős–Moser conjecture given by (6.13), namely, if $S \subset \mathbb{R}$ and $\#S = 2m + 1$, then

$$f(S, \alpha) \leq f(\{-m, -m+1, \ldots, m\}, 0).$$

NOTE. If $P$ and $Q$ are posets, then the *direct product* $P \times Q$ is the poset on the set $\{(x, y) : x \in P, \; y \in Q\}$ satisfying $(x, y) \leq (x', y')$ if and only if $x \leq x'$ in $P$ and $y \leq y'$ in $Q$.

6. (unsolved) Show that $L(m, n)$ has a symmetric chain decomposition. This is known to be true for $m \leq 4$.

# Chapter 7
# Enumeration Under Group Action

In Chapters 5 and 6 we considered the quotient poset $B_n/G$, where $G$ is a subgroup of the symmetric group $\mathfrak{S}_n$. If $p_i$ is the number of elements of rank $i$ of this poset, then the sequence $p_0, p_1, \ldots, p_n$ is rank-symmetric and rank-unimodal. Thus it is natural to ask whether there is some nice formula for the numbers $p_i$. For instance, in Theorem 5.9 $p_i$ is the number of nonisomorphic graphs with $m$ vertices (where $n = \binom{m}{2}$) and $i$ edges; is there some nice formula for this number? For the group $G_{mn} = \mathfrak{S}_n \wr \mathfrak{S}_m$ of Theorem 6.6 we obtained a simple generating function for $p_i$ (i.e., a formula for the rank-generating function $F(B_{mn}/G_{mn}, q) = \sum_i p_i q^i$), but this was a very special situation. In this chapter we will present a general theory for enumerating inequivalent objects subject to a group of symmetries, which will include a formula for the rank-generating functions $F(B_n/G, q)$. The chief architect of this theory is G. Pólya (though much of it was anticipated by J. H. Redfield) and hence is often called *Pólya's theory of enumeration* or just *Pólya theory*. See the references at the end of this chapter for further historical information.

Pólya theory is most easily understood in terms of "colorings" of some geometric or combinatorial object. For instance, consider a row of five squares:

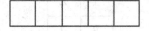

In how many ways can we color the squares using $n$ colors? Each square can be colored any of the $n$ colors, so there are $n^5$ ways in all. These colorings can by indicated as

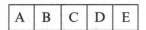

where $A, B, C, D$, and $E$ are the five colors. Now assume that we are allowed to rotate the row of five squares $180°$ and that two colorings are considered the same if one can be obtained from the other by such a rotation. (We may think that we have cut the row of five squares out of paper and colored them on one side.) We say that two colorings are *equivalent* if they are the same or can be transformed into

© Springer International Publishing AG, part of Springer Nature 2018                    75
R. P. Stanley, *Algebraic Combinatorics*, Undergraduate Texts in Mathematics,
https://doi.org/10.1007/978-3-319-77173-1_7

one another by a 180° rotation. The first naive assumption is that every coloring is equivalent to exactly one other (besides itself), so the number of inequivalent colorings is $n^5/2$. Clearly this reasoning cannot be correct since $n^5/2$ is not always an integer! The problem, of course, is that some colorings stay the same when we rotate 180°. In fact, these are exactly the colorings

| A | B | C | B | A |
|---|---|---|---|---|

where $A$, $B$, and $C$ are any three colors. There are $n^3$ such colorings, so the total number of inequivalent colorings is given by

$$\frac{1}{2}(\text{number of colorings which don't equal their 180° rotation})$$

$$+(\text{number of colorings which equal their 180° rotation})$$

$$= \frac{1}{2}(n^5 - n^3) + n^3$$

$$= \frac{1}{2}(n^5 + n^3).$$

Pólya theory gives a systematic method for obtaining formulas of this sort for any underlying symmetry group.

The general setup is the following. Let $X$ be a finite set, and $G$ a subgroup of the symmetric group $\mathfrak{S}_X$. Think of $G$ as a group of symmetries of $X$. Let $C$ be another set (which may be infinite), which we think of as a set of "colors." A *coloring* of $X$ is a function $f : X \to C$. For instance, $X$ could be the set of four squares of a $2 \times 2$ chessboard, labelled as follows:

| 1 | 2 |
|---|---|
| 3 | 4 |

Let $C = \{r, b, y\}$ (the colors red, blue, and yellow). A typical coloring of $X$ would then look like

| r | b |
|---|---|
| y | r |

The above diagram thus indicates the function $f : X \to C$ given by $f(1) = r$, $f(2) = b$, $f(3) = y$, $f(4) = r$.

NOTE. We could work in the slightly greater generality of a group $G$ acting on the set $X$, i.e., we are given a homomorphism $\varphi: G \to \mathfrak{S}_X$ that need not be injective. However, we then have a well-defined induced injective homomorphism $\psi: H \to \mathfrak{S}_X$, where $H = G/(\ker \varphi)$. The results obtained below for $H$ are identical to those we get for $G$, so nothing is lost by assuming that $\varphi$ is injective. In this case we can identify $G$ with its image $\varphi(G)$.

We define two colorings $f$ and $g$ to be *equivalent* (or *G-equivalent*, when it is necessary to specify the group), denoted $f \sim g$ or $f \overset{G}{\sim} g$, if there exists an element $\pi \in G$ such that

$$g(\pi(x)) = f(x) \text{ for all } x \in X.$$

We may write this condition more succinctly as $g\pi = f$, where $g\pi$ denotes the composition of functions (from right to left). It is easy to check, using the fact that $G$ is a group, that $\sim$ is an equivalence relation. One should think that equivalent functions are the same "up to symmetry."

**7.1 Example.** Let $X$ be the $2 \times 2$ chessboard and $C = \{r, b, y\}$ as above. There are many possible choices of a symmetry group $G$, and this will affect when two colorings are equivalent. For instance, consider the following groups:

- $G_1$ consists of only the identity permutation $(1)(2)(3)(4)$.
- $G_2$ is the group generated by a vertical reflection. It consists of the two elements $(1)(2)(3)(4)$ (the identity element) and $(1, 2)(3, 4)$ (the vertical reflection).
- $G_3$ is the group generated by a reflection in the main diagonal. It consists of the two elements $(1)(2)(3)(4)$ (the identity element) and $(1)(4)(2, 3)$ (the diagonal reflection).
- $G_4$ is the group of all rotations of $X$. It is a cyclic group of order four with elements $(1)(2)(3)(4)$, $(1, 2, 4, 3)$, $(1, 4)(2, 3)$, and $(1, 3, 4, 2)$.
- $G_5$ is the dihedral group of all rotations and reflections of $X$. It has eight elements, namely, the four elements of $G_4$ and the four reflections $(1, 2)(3, 4)$, $(1, 3)(2, 4)$, $(1)(4)(2, 3)$, and $(2)(3)(1, 4)$.
- $G_6$ is the symmetric group of *all* 24 permutations of $X$. Although this is a perfectly valid group of symmetries, it no longer has any connection with the geometric representation of $X$ as the squares of a $2 \times 2$ chessboard.

Consider the inequivalent colorings of $X$ with two red squares, one blue square, and one yellow square, in each of the six cases above.

($G_1$)  There are 12 colorings in all with two red squares, one blue square, and one yellow square, and all are inequivalent under the trivial group (the group with one element). In general, whenever $G$ is the trivial group then two colorings are equivalent if and only if they are the same [why?].

($G_2$) There are now six inequivalent colorings, represented by

| r | r |
|---|---|
| b | y |

| r | b |
|---|---|
| r | y |

| r | y |
|---|---|
| r | b |

| b | y |
|---|---|
| r | r |

| r | b |
|---|---|
| y | r |

| r | y |
|---|---|
| b | r |

Each equivalence class contains two elements.

($G_3$) Now there are seven classes, represented by

| r | r |
|---|---|
| b | y |

| r | r |
|---|---|
| y | b |

| b | y |
|---|---|
| r | r |

| y | b |
|---|---|
| r | r |

| r | b |
|---|---|
| y | r |

| b | r |
|---|---|
| r | y |

| y | r |
|---|---|
| r | b |

The first five classes contain two elements each and the last two classes only one element. Although $G_2$ and $G_3$ are isomorphic as abstract groups, as permutation groups they have a different structure. Specifically, the generator $(1, 2)(3, 4)$ of $G_2$ has two cycles of length two, while the generator $(1)(4)(2, 3)$ has two cycles of length one and one of length two. As we will see below, it is the lengths of the cycles of the elements of $G$ that determine the sizes of the equivalence classes. This explains why the number of classes for $G_2$ and $G_3$ is different.

($G_4$) There are three classes, each with four elements. The size of each class is equal to the order of the group because of the colorings have any symmetry with respect to the group, i.e., for any coloring $f$ using the colors $r, r, y, b$, the only group element $\pi$ that fixes $f$ (so $f\pi = f$) is the identity ($\pi = (1)(2)(3)(4)$).

| r | r |
|---|---|
| y | b |

| r | r |
|---|---|
| b | y |

| r | b |
|---|---|
| y | r |

($G_5$) Under the full dihedral group there are now two classes.

| r | r |
|---|---|
| b | y |

| r | b |
|---|---|
| y | r |

The first class has eight elements and the second four elements. In general, the size of a class is the index in $G$ of the subgroup fixing some fixed coloring in that class [why?]. For instance, the subgroup fixing the second coloring above is $\{(1)(2)(3)(4), (1, 4)(2)(3)\}$, which has index four in the dihedral group of order eight.

($G_6$) Under the group $\mathfrak{S}_4$ of all permutations of the squares there is clearly only one class, with all 12 colorings. In general, for any set $X$ if the group is the symmetric group $\mathfrak{S}_X$ then two colorings are equivalent if and only if each color appears the same number of times [why?].

Our object in general is to count the number of equivalence classes of colorings which use each color a specified number of times. We will put the information into a *generating function*—a polynomial whose coefficients are the numbers we seek. Consider for example the set $X$, the group $G = G_5$ (the dihedral group), and the set $C = \{r, b, y\}$ of colors in Example 7.1 above. Let $\kappa(i, j, k)$ be the number of inequivalent colorings using red $i$ times, blue $j$ times, and yellow $k$ times. Think of the colors $r, b, y$ as *variables*, and form the polynomial

$$F_G(r, b, y) = \sum_{i+j+k=4} \kappa(i, j, k) r^i b^j y^k.$$

Note that we sum only over $i, j, k$ satisfying $i + j + k = 4$ since a total of four colors will be used to color the four-element set $X$. The reader should check that

$$F_G(r, b, y) = (r^4 + b^4 + y^4) + (r^3 b + r b^3 + r^3 y + r y^3 + b^3 y + b y^3)$$
$$+ 2(r^2 b^2 + r^2 y^2 + b^2 y^2) + 2(r^2 b y + r b^2 y + r b y^2).$$

For instance, the coefficient of $r^2 b y$ is two because, as we have seen above, there are two inequivalent colorings using the colors $r, r, b, y$. Note that $F_G(r, b, y)$ is a *symmetric function* of the variables $r, b, y$ (i.e., it stays the same if we permute the variables in any way), because insofar as counting inequivalent colorings goes, it makes no difference what *names* we give the colors. As a special case we may ask for the *total* number of inequivalent colorings with four colors. This is obtained by setting $r = b = y = 1$ in $F_G(r, b, y)$ [why?], yielding $F_G(1, 1, 1) = 3 + 6 + 2 \cdot 3 + 2 \cdot 3 = 21$.

What happens to the generating function $F_G$ in the above example when we use the $n$ colors $r_1, r_2, \ldots, r_n$ (which can be thought of as different shades of red)? Clearly all that matters are the *multiplicities* of the colors, without regard for their order. In other words, there are five cases: (a) all four colors the same, (b) one color used three times and another used once, (c) two colors used twice each, (d) one color used twice and two others once each, and (e) four colors used once each. These five cases correspond to the five partitions of 4, i.e., the five ways of writing 4 as a sum of positive integers without regard to order: $4, 3 + 1, 2 + 2, 2 + 1 + 1, 1 + 1 + 1 + 1$. Our generating function becomes

$$F_G(r_1, r_2, \ldots, r_n) = \sum_i r_i^4 + \sum_{i \neq j} r_i^3 r_j$$

$$+ 2 \sum_{i < j} r_i^2 r_j^2 + 2 \sum_{\substack{i \neq j \\ i \neq k \\ j < k}} r_i^2 r_j r_k + 3 \sum_{i < j < k < l} r_i r_j r_k r_l,$$

where the indices in each sum lie between 1 and $n$. If we set all variables equal to one (obtaining the total number of colorings with $n$ colors), then simple combinatorial reasoning yields

$$F_G(1, 1, \ldots, 1) = n + n(n-1) + 2\binom{n}{2} + 2n\binom{n-1}{2} + 3\binom{n}{4}$$

$$= \frac{1}{8}(n^4 + 2n^3 + 3n^2 + 2n). \tag{7.1}$$

Note that the polynomial (7.1) has the following description: the denominator 8 is the order of the group $G_5$, and the coefficient of $n^i$ in the numerator is just the number of permutations in $G_5$ with $i$ cycles! For instance, the coefficient of $n^2$ is 3, and $G_5$ has the three elements $(1, 2)(3, 4)$, $(1, 3)(2, 4)$, and $(1, 4)(2, 3)$ with two cycles. We want to prove a general result of this nature.

The basic tool which we will use is a simple result from the theory of permutation groups known as *Burnside's lemma*. It was actually first proved by Cauchy when $G$ is transitive (i.e., $|Y/G| = 1$ in Lemma 7.2 below) and by Frobenius in the general case, and is sometimes called the *Cauchy–Frobenius lemma*.

**7.2 Lemma** (Burnside's lemma). *Let $Y$ be a finite set and $G$ a subgroup of $\mathfrak{S}_Y$. For each $\pi \in G$, let*

$$\text{Fix}(\pi) = \{y \in Y : \pi(y) = y\},$$

*so $\#\text{Fix}(\pi)$ is the number of cycles of length one in the permutation $\pi$. Let $Y/G$ be the set of orbits of $G$. Then*

$$|Y/G| = \frac{1}{\#G} \sum_{\pi \in G} \#\text{Fix}(\pi).$$

An equivalent form of Burnside's lemma is the statement that the average number of elements of $Y$ fixed by an element of $G$ is equal to the number of orbits. Before proceeding to the proof, let us consider an example.

**7.3 Example.** Let $Y = \{a, b, c, d\}$,

$$G = \{(a)(b)(c)(d), (a, b)(c, d), (a, c)(b, d), (a, d)(b, c)\},$$

and

$$G' = \{(a)(b)(c)(d), (a, b)(c)(d), (a)(b)(c, d), (a, b)(c, d)\}.$$

Both groups are isomorphic to $\mathbb{Z}_2 \times \mathbb{Z}_2$ (compare Example 5.1(c) and (d)). By Burnside's lemma the number of orbits of $G$ is $\frac{1}{4}(4 + 0 + 0 + 0) = 1$. Indeed, given any two elements $i, j \in Y$, it is clear by inspection that there is a $\pi \in G$ (which happens to be unique) such that $\pi(i) = j$. On the other hand, the number of orbits of $G'$ is $\frac{1}{4}(4 + 2 + 2 + 0) = 2$. Indeed, the two orbits are $\{a, b\}$ and $\{c, d\}$.

**Proof of Burnside's lemma.** For $y \in Y$ let $G_y = \{\pi \in G : \pi \cdot y = y\}$ (the set of permutations fixing $y$). Then

$$\frac{1}{\#G} \sum_{\pi \in G} \#\mathrm{Fix}(\pi) = \frac{1}{\#G} \sum_{\pi \in G} \sum_{\substack{y \in Y \\ \pi \cdot y = y}} 1$$

$$= \frac{1}{\#G} \sum_{y \in Y} \sum_{\substack{\pi \in G \\ \pi \cdot y = y}} 1$$

$$= \frac{1}{\#G} \sum_{y \in Y} \#G_y.$$

Now (as in the proof of Lemma 5.6) the multiset of elements $\pi \cdot y$, $\pi \in G$, contains every element in the orbit $Gy$ the same number of times, namely $\#G/\#Gy$ times. Thus $y$ occurs $\#G/\#Gy$ times among the $\pi \cdot y$, so

$$\frac{\#G}{\#Gy} = \#G_y.$$

Therefore

$$\frac{1}{\#G} \sum_{\pi \in G} \#\mathrm{Fix}(\pi) = \frac{1}{\#G} \sum_{y \in Y} \frac{\#G}{\#Gy}$$

$$= \sum_{y \in Y} \frac{1}{\#Gy}.$$

How many times does a term $1/\#\mathcal{O}$ appear in the above sum, where $\mathcal{O}$ is a fixed orbit? We are asking for the number of $y$ such that $Gy = \mathcal{O}$. But $Gy = \mathcal{O}$ if and only if $y \in \mathcal{O}$, so $1/\#\mathcal{O}$ appears $\#\mathcal{O}$ times. Thus each orbit gets counted exactly once, so the above sum is equal to the number of orbits. $\square$

**7.4 Example.** How many inequivalent colorings of the vertices of a regular hexagon $H$ are there using $n$ colors, under cyclic symmetry? Let $C_n$ be the set of all $n$-colorings of $H$. Let $G$ be the group of all permutations of $C_n$ which permute the colors cyclically, so $G \cong \mathbb{Z}_6$. We are asking for the number of orbits of $G$ [why?]. We want to apply Burnside's lemma, so for each of the six elements $\sigma$ of $G$ we need to compute the number of colorings fixed by that element. Let $\pi$ be a generator of $G$.

- $\sigma = 1$ (the identity): All $n^6$ colorings are fixed by $\sigma$.
- $\sigma = \pi, \pi^{-1}$: Only the $n$ colorings with all colors equal are fixed.
- $\sigma = \pi^2, \pi^4$: Any coloring of the form $ababab$ is fixed (writing the colors linearly in the order they appear around the hexagon, starting at any fixed vertex). There are $n$ choices for $a$ and $n$ for $b$, so $n^2$ colorings in all.
- $\sigma = \pi^3$: The fixed colorings are of the form $abcabc$, so $n^3$ in all.

Hence by Burnside's lemma, we have

$$\text{number of orbits} = \frac{1}{6}(n^6 + n^3 + 2n^2 + 2n).$$

The reader who has followed the preceding example will have no trouble understanding the following result.

**7.5 Theorem.** *Let G be a group of permutations of a finite set X. Then the number $N_G(n)$ of inequivalent (with respect to G) n-colorings of X is given by*

$$N_G(n) = \frac{1}{\#G} \sum_{\pi \in G} n^{c(\pi)}, \tag{7.2}$$

*where $c(\pi)$ denotes the number of cycles of $\pi$.*

*Proof.* Let $\pi_n$ denote the action of $\pi \in G$ on the set $\mathcal{C}_n$ of $n$-colorings of $X$. We want to determine the set $\text{Fix}(\pi_n)$, so that we can apply Burnside's lemma. Let $C$ be the set of $n$ colors. If $f: X \to C$ is a coloring fixed by $\pi$, then for all $x \in X$ we have

$$f(x) = \pi_n \cdot f(x) = f(\pi(x)).$$

Thus $f \in \text{Fix}(\pi_n)$ if and only if $f(x) = f(\pi(x))$. Hence $f(x) = f(\pi^k(x))$ for any $k \geq 1$ [why?]. The elements $y$ of $X$ of the form $\pi^k(x)$ for $k \geq 1$ are just the elements of the cycle of $\pi$ containing $x$. Thus to obtain $f \in \text{Fix}(\pi_n)$, we should take the cycles $\sigma_1, \ldots, \sigma_{c(\pi)}$ of $\pi$ and color each element of $\sigma_i$ the same color. There are $n$ choices for each $\sigma_i$, so $n^{c(\pi)}$ colorings in all fixed by $\pi$. In other words, $\#\text{Fix}(\pi_n) = n^{c(\pi)}$, and the proof follows by Burnside's lemma.    $\square$

We would now like not just to count the *total* number of inequivalent colorings with $n$ colors but more strongly to specify the number of occurrences of each color. We will need to use not just the number $c(\pi)$ of cycles of each $\pi \in G$, but rather the lengths of each of the cycles of $\pi$. Thus given a permutation $\pi$ of an $n$-element set $X$, define the *type* of $\pi$ to be

$$\text{type}(\pi) = (c_1, c_2, \ldots, c_n),$$

where $\pi$ has $c_i$ $i$-cycles. For instance, if $\pi = 4, 7, 3, 8, 2, 10, 11, 1, 6, 9, 5$, then

$$\text{type}(\pi) = \text{type } (1, 4, 8)(2, 7, 11, 5)(3)(6, 10, 9)$$

$$= (1, 0, 2, 1, 0, 0, 0, 0, 0, 0, 0).$$

Note that we always have $\sum_i i c_i = n$ [why?]. Define the *cycle indicator* of $\pi$ to be the monomial

$$Z_\pi = z_1^{c_1} z_2^{c_2} \cdots z_n^{c_n}.$$

(Many other notations are used for the cycle indicator. The use of $Z_\pi$ comes from the German word *Zyklus* for cycle. The original paper of Pólya was written in German.) Thus for the example above, we have $Z_\pi = z_1 z_3^2 z_4$.

Now given a subgroup $G$ of $\mathfrak{S}_X$, the *cycle indicator* (or *cycle index polynomial* or *cycle enumerator*) of $G$ is defined by

$$Z_G = Z_G(z_1, \ldots, z_n) = \frac{1}{\#G} \sum_{\pi \in G} Z_\pi.$$

Thus $Z_G$ (also denoted $P_G$, $\mathrm{Cyc}(G)$, etc.) is a polynomial in the variables $z_1, \ldots, z_n$.

**7.6 Example.** If $X$ consists of the vertices of a square and $G$ is the group of rotations of $X$ (a cyclic group of order 4), then

$$Z_G = \frac{1}{4}(z_1^4 + z_2^2 + 2z_4).$$

If reflections are also allowed (so $G$ is the dihedral group of order 8), then

$$Z_G = \frac{1}{8}(z_1^4 + 3z_2^2 + 2z_1^2 z_2 + 2z_4).$$

We are now ready to state the main result of this chapter.

**7.7 Theorem** (Pólya's Theorem, 1937). *Let $G$ be a group of permutations of the $n$-element set $X$. Let $C = \{r_1, r_2, \ldots\}$ be a set of colors. Let $\kappa(i_1, i_2, \ldots)$ be the number of inequivalent (under the action of $G$) colorings $f : X \to C$ such that color $r_j$ is used $i_j$ times. Define*

$$F_G(r_1, r_2, \ldots) = \sum_{i_1, i_2, \ldots} \kappa(i_1, i_2, \ldots) r_1^{i_1} r_2^{i_2} \cdots.$$

*(Thus $F_G$ is a polynomial or a power series in the variables $r_1, r_2, \ldots$, depending on whether or not $C$ is finite or infinite.) Then*

$$F_G(r_1, r_2, \ldots) =$$

$$Z_G(r_1 + r_2 + r_3 + \cdots, r_1^2 + r_2^2 + r_3^2 + \cdots, \ldots, r_1^j + r_2^j + r_3^j + \cdots, \ldots).$$

*(In other words, substitute $\sum_i r_i^j$ for $z_j$ in $Z_G$.)*

Before giving the proof let us consider an example.

**7.8 Example.** Suppose that in Example 7.6 our set of colors is $C = \{a, b, c, d\}$, and that we take $G$ to be the group of cyclic symmetries. Then

$$F_G(a, b, c, d) = \frac{1}{4}\left((a+b+c+d)^4 + (a^2+b^2+c^2+d^2)^2 + 2(a^4+b^4+c^4+d^4)\right)$$

$$= (a^4 + \cdots) + (a^3 b + \cdots) + 2(a^2 b^2 + \cdots) + 3(a^2 bc + \cdots) + 6abcd.$$

An expression such as $(a^2b^2 + \cdots)$ stands for the sum of all monomials in the variables $a, b, c, d$ with exponents $2, 2, 0, 0$ (in some order). The coefficient of all such monomials is 2, indicating two inequivalent colorings using one color twice and another color twice. If instead $G$ were the full dihedral group, we would get

$$F_G(a, b, c, d) = \frac{1}{8}\left((a + b + c + d)^4 + 3(a^2 + b^2 + c^2 + d^2)^2\right.$$

$$+ 2(a + b + c + d)^2(a^2 + b^2 + c^2 + d^2) + 2(a^4 + b^4 + c^4 + d^4)\big)$$

$$= (a^4 + \cdots) + (a^3b + \cdots) + 2(a^2b^2 + \cdots) + 2(a^2bc + \cdots) + 3abcd.$$

**Proof of Pólya's theorem.** Let $\#X = t$ and $i_1 + i_2 + \cdots = t$, where each $i_j \geq 0$. Let $\boldsymbol{i} = (i_1, i_2, \ldots)$, and let $\mathcal{C}_{\boldsymbol{i}}$ denote the set of all colorings of $X$ with color $r_j$ used $i_j$ times. The group $G$ acts on $\mathcal{C}_{\boldsymbol{i}}$, since if $f \in \mathcal{C}_{\boldsymbol{i}}$ and $\pi \in G$, then $\pi \cdot f \in \mathcal{C}_{\boldsymbol{i}}$. ("Rotating" a colored object does not change how many times each color appears.) Let $\pi_{\boldsymbol{i}}$ denote the action of $\pi$ on $\mathcal{C}_{\boldsymbol{i}}$. We want to apply Burnside's lemma to compute the number of orbits, so we need to find $\#\mathrm{Fix}(\pi_{\boldsymbol{i}})$.

In order for $f \in \mathrm{Fix}(\pi_{\boldsymbol{i}})$, we must color $X$ so that (a) in any cycle of $\pi$, all the elements get the same color, and (b) the color $r_j$ appears $i_j$ times. Consider the product

$$H_\pi = \prod_j (r_1^j + r_2^j + \cdots)^{c_j(\pi)},$$

where $c_j(\pi)$ is the number of $j$-cycles (cycles of length $j$) of $\pi$. When we expand this product as a sum of monomials $r_1^{j_1} r_2^{j_2} \cdots$, we get one of these monomials by choosing a term $r_k^j$ from each factor of $H_\pi$ and multiplying these terms together. Choosing $r_k^j$ corresponds to coloring all the elements of some $j$-cycle with $r_k$. Since a factor $r_1^j + r_2^j + \cdots$ occurs precisely $c_j(\pi)$ times in $H_\pi$, choosing a term $r_k^j$ from every factor corresponds to coloring $X$ so that every cycle is monochromatic (i.e., all the elements of that cycle get the same color). The product of these terms $r_k^j$ will be the monomial $r_1^{j_1} r_2^{j_2} \cdots$, where we have used color $r_k$ a total of $j_k$ times. It follows that the coefficient of $r_i^{i_1} r_2^{i_2} \cdots$ in $H_\pi$ is equal to $\#\mathrm{Fix}(\pi_{\boldsymbol{i}})$. Thus

$$H_\pi = \sum_{\boldsymbol{i}} \#\mathrm{Fix}(\pi_{\boldsymbol{i}}) r_1^{i_1} r_2^{i_2} \cdots. \tag{7.3}$$

Now sum both sides of (7.3) over all $\pi \in G$ and divide by $\#G$. The left-hand side becomes

$$\frac{1}{\#G} \sum_{\pi \in G} \prod_j (r_1^j + r_2^j + \cdots)^{c_j(\pi)} = Z_G(r_1 + r_2 + \cdots, r_1^2 + r_2^2 + \cdots, \ldots).$$

On the other hand, the right-hand side becomes

$$\sum_i \left[ \frac{1}{\#G} \sum_{\pi \in G} \#\mathrm{Fix}(\pi_i) \right] r_1^{i_1} r_2^{i_2} \cdots .$$

By Burnside's lemma, the expression in brackets is just the number of orbits of $\pi_i$ acting on $C_i$, i.e., the number of inequivalent colorings using color $r_j$ a total of $i_j$ times, as was to be proved.                                                       $\square$

**7.9 Example** (necklaces). A *necklace* of length $\ell$ is a circular arrangement of $\ell$ (colored) beads. Two necklaces are considered the same if they are cyclic rotations of one another. Let $X$ be a set of $\ell$ (uncolored) beads, say $X = \{1, 2, \ldots, \ell\}$. Regarding the beads as being placed equidistantly on a circle in the order $1, 2, \ldots, \ell$, let $G$ be the cyclic group of rotations of $X$. Thus if $\pi$ is the cycle $(1, 2, \ldots, \ell)$, then $G = \{1, \pi, \pi^2, \ldots, \pi^{\ell-1}\}$. For example, if $\ell = 6$ then the elements of $G$ are

$$\pi^0 = (1)(2)(3)(4)(5)(6)$$

$$\pi = (1, 2, 3, 4, 5, 6)$$

$$\pi^2 = (1, 3, 5)(2, 4, 6)$$

$$\pi^3 = (1, 4)(2, 5)(3, 6)$$

$$\pi^4 = (1, 5, 3)(2, 6, 4)$$

$$\pi^5 = (1, 6, 5, 4, 3, 2).$$

In general, if $d$ is the greatest common divisor of $m$ and $\ell$ (denoted $d = \gcd(m, \ell)$), then $\pi^m$ has $d$ cycles of length $\ell/d$. An integer $m$ satisfies $1 \le m \le \ell$ and $\gcd(m, \ell) = d$ if and only if $1 \le m/d \le \ell/d$ and $\gcd(m/d, \ell/d) = 1$. Hence the number of such integers $m$ is given by the Euler phi-function (or totient function) $\phi(\ell/d)$, which by definition is equal to the number of integers $1 \le i \le \ell/d$ such that $\gcd(i, \ell/d) = 1$. As an aside, recall that $\phi(k)$ can be computed by the formula

$$\phi(k) = k \prod_{\substack{p|k \\ p \text{ prime}}} \left(1 - \frac{1}{p}\right). \tag{7.4}$$

For instance, $\phi(1000) = 1000(1 - \frac{1}{2})(1 - \frac{1}{5}) = 400$. Putting all this together gives the following formula for the cycle enumerator $Z_G(z_1, \ldots, z_\ell)$:

$$Z_G(z_1, \ldots, z_\ell) = \frac{1}{\ell} \sum_{d|\ell} \phi(\ell/d) z_{\ell/d}^d,$$

or (substituting $\ell/d$ for $d$),

$$Z_G(z_1, \ldots, z_\ell) = \frac{1}{\ell} \sum_{d \mid \ell} \phi(d) z_d^{\ell/d}.$$

There follows from Pólya's theorem the following result (originally proved by MacMahon (1854–1929) before Pólya discovered his general result).

**7.10 Theorem.** *(a) The number $N_\ell(n)$ of n-colored necklaces of length $\ell$ is given by*

$$N_\ell(n) = \frac{1}{\ell} \sum_{d \mid \ell} \phi(\ell/d) n^d. \tag{7.5}$$

*(b) We have*

$$F_G(r_1, r_2, \ldots) = \frac{1}{\ell} \sum_{d \mid \ell} \phi(d)(r_1^d + r_2^d + \cdots)^{\ell/d}.$$

NOTE.  (b) reduces to (a) if $r_1 = r_2 = \cdots = 1$. Moreover, since clearly $N_\ell(1) = 1$, putting $n = 1$ in (7.5) yields the well-known identity

$$\sum_{d \mid \ell} \phi(\ell/d) = \ell.$$

What if we are allowed to flip necklaces over, not just rotate them? Now the group becomes the dihedral group of order $2\ell$, and the corresponding inequivalent colorings are called *dihedral necklaces*. We leave to the reader to work out the cycle enumerators

$$\frac{1}{2\ell} \left( \sum_{d \mid \ell} \phi(d) z_d^{\ell/d} + m z_1^2 z_2^{m-1} + m z_2^m \right), \text{ if } \ell = 2m$$

$$\frac{1}{2\ell} \left( \sum_{d \mid \ell} \phi(d) z_d^{\ell/d} + \ell z_1 z_2^m \right), \text{ if } \ell = 2m + 1. \tag{7.6}$$

**7.11 Example.** Let $G = \mathfrak{S}_\ell$, the group of all permutations of $\{1, 2, \ldots, \ell\} = X$. Thus for instance

$$Z_{\mathfrak{S}_3}(z_1, z_2, z_3) = \frac{1}{6}(z_1^3 + 3z_1 z_2 + 2z_3)$$

$$Z_{\mathfrak{S}_4}(z_1, z_2, z_3, z_4) = \frac{1}{24}(z_1^4 + 6z_1^2 z_2 + 3z_2^2 + 8z_1 z_3 + 6z_4).$$

It is easy to count the number of inequivalent colorings in $C_i$. If two colorings of $X$ use each color the same number of times, then clearly there is *some* permutation of $X$ which sends one of the colorings to the other. Hence $C_i$ consists of a single orbit. Thus

$$F_{\mathfrak{S}_\ell}(r_1, r_2, \dots) = \sum_{i_1 + i_2 + \cdots = \ell} r_1^{i_1} r_2^{i_2} \cdots,$$

the sum of all monomials of degree $\ell$.

To count the total number of inequivalent $n$-colorings, note that

$$\sum_{\ell \geq 0} F_{\mathfrak{S}_\ell}(r_1, r_2, \dots) x^\ell = \frac{1}{(1 - r_1 x)(1 - r_2 x) \cdots}, \qquad (7.7)$$

since if we expand each factor on the right-hand side into the series $\sum_{j \geq 0} r_i^j x^j$ and multiply, the coefficient of $x^\ell$ will just be the sum of all monomials of degree $\ell$. For fixed $n$, let $f_n(\ell)$ denote the number of inequivalent $n$-colorings of $X$. Since $f_n(\ell) = F_{\mathfrak{S}_\ell}(1, 1, \dots, 1)$ ($n$ 1's in all), there follows from (7.7) that

$$\sum_{\ell \geq 0} f_n(\ell) x^\ell = \frac{1}{(1 - x)^n}. \qquad (7.8)$$

The right-hand side can be expanded (e.g., by Taylor's theorem or by the binomial theorem for the exponent $-n$) as

$$\frac{1}{(1 - x)^n} = \sum_{\ell \geq 0} \binom{n + \ell - 1}{\ell} x^\ell.$$

Hence

$$f_n(\ell) = \binom{n + \ell - 1}{\ell}.$$

It is natural to ask whether there might be a more direct proof of such a simple result. This is actually a standard result in elementary enumerative combinatorics. For fixed $\ell$ and $n$ we want the number of solutions to $i_1 + i_2 + \cdots + i_n = \ell$ in nonnegative integers. Suppose that we arrange $n - 1$ vertical bars and $\ell$ dots in a line. There are $\binom{n + \ell - 1}{\ell}$ such arrangements since there are a total of $n + \ell - 1$ positions, and we choose $\ell$ of them in which to place a dot. An example of such an arrangement for $\ell = 8$ and $n = 7$ is

$$| \; | \; \bullet \bullet \; | \; \bullet \; | \; | \; \bullet \bullet \bullet \; | \; \bullet \bullet$$

The number of dots in each "compartment," read from left to right, gives the numbers $i_1, \ldots, i_n$. For the example above, we get $(i_1, \ldots, i_7) = (0, 0, 2, 1, 0, 3, 2)$. Since this correspondence between solutions to $i_1 + i_2 + \cdots + i_n = \ell$ and arrangements of bars and dots is clearly a bijection, we get $\binom{n+\ell-1}{\ell}$ solutions as claimed.

Recall (Theorem 7.5) that the number of inequivalent $n$-colorings of $X$ (with respect to any group $G$ of permutations of $X$) is given by

$$\frac{1}{\#G} \sum_{\pi \in G} n^{c(\pi)},$$

where $c(\pi)$ denotes the number of cycles of $\pi$. Hence for $G = \mathfrak{S}_\ell$ we get the identity

$$\frac{1}{\ell!} \sum_{\pi \in \mathfrak{S}_\ell} n^{c(\pi)} = \binom{n+\ell-1}{\ell}$$

$$= \frac{1}{\ell!} n(n+1)(n+2) \cdots (n+\ell-1).$$

Multiplying by $\ell!$ yields

$$\sum_{\pi \in \mathfrak{S}_\ell} n^{c(\pi)} = n(n+1)(n+2) \cdots (n+\ell-1). \tag{7.9}$$

Equivalently [why?], if we define $c(\ell, k)$ to be the number of permutations in $\mathfrak{S}_\ell$ with $k$ cycles (called a *signless Stirling number of the first kind*), then

$$\sum_{k=1}^{\ell} c(\ell, k) x^k = x(x+1)(x+2) \cdots (x+\ell-1).$$

For instance, $x(x+1)(x+2)(x+3) = x^4 + 6x^3 + 11x^2 + 6x$, so (taking the coefficient of $x^2$) 11 permutations in $\mathfrak{S}_4$ have two cycles, namely, $(123)(4), (132)(4), (124)(3)$, $(142)(3), (134)(2), (143)(2), (234)(1), (243)(1), (12)(34), (13)(24), (14)(23)$.

Although it was easy to compute the generating function $F_{\mathfrak{S}_\ell}(r_1, r_2, \ldots)$ directly without the necessity of computing the cycle indicator $Z_{\mathfrak{S}_\ell}(z_1, \ldots, z_\ell)$, we can still ask whether there is a formula of some kind for this polynomial. First we determine explicitly its coefficients.

**7.12 Theorem.** *Let $\sum i c_i = \ell$. The number of permutations $\pi \in \mathfrak{S}_\ell$ with $c_i$ cycles of length $i$ (or equivalently, the coefficient of $z_1^{c_1} z_2^{c_2} \cdots$ in $\ell! Z_{\mathfrak{S}_\ell}(z_1, \ldots, z_\ell)$) is equal to $\ell!/1^{c_1} c_1! 2^{c_2} c_2! \cdots$.*

*Example* The number of permutations in $\mathfrak{S}_{15}$ with three 1-cycles, two 2-cycles, and two 4-cycles is $15!/1^3 \cdot 3! \cdot 2^2 \cdot 2! \cdot 4^2 \cdot 2! = 851{,}350{,}500$.

**Proof of Theorem 7.12.** Fix $c = (c_1, c_2, \ldots)$ and let $X_c$ be the set of all permutations $\pi \in \mathfrak{S}_\ell$ with $c_i$ cycles of length $i$. Given a permutation $\sigma = a_1 a_2 \cdots a_\ell$ in $\mathfrak{S}_\ell$, construct a permutation $f(\sigma) \in X_c$ as follows. Let the 1-cycles of $f(\sigma)$ be $(a_1), (a_2), \ldots, (a_{c_1})$. Then let the 2-cycles of $f(\sigma)$ be $(a_{c_1+1}, a_{c_1+2}), (a_{c_1+3}, a_{c_1+4}), \ldots, (a_{c_1+2c_2-1}, a_{c_1+2c_2})$. Then let the 3-cycles of $f(\sigma)$ be $(a_{c_1+2c_2+1}, a_{c_1+2c_2+2}, a_{c_1+2c_2+3}), (a_{c_1+2c_2+4}, a_{c_1+2c_2+5}, a_{c_1+2c_2+6}), \ldots, (a_{c_1+2c_2+3c_3-2}, a_{c_1+2c_2+3c_3-1}, a_{c_1+2c_2+3c_3})$, etc., continuing until we reach $a_\ell$ and have produced a permutation in $X_c$. For instance, if $\ell = 11, c_1 = 3, c_2 = 2, c_4 = 1$, and $\sigma = 4, 9, 6, 11, 7, 1, 3, 8, 10, 2, 5$, then

$$f(\sigma) = (4)(9)(6)(11, 7)(1, 3)(8, 10, 2, 5).$$

We have defined a function $f: \mathfrak{S}_\ell \to X_c$. Given $\pi \in X_c$, what is $\#f^{-1}(\pi)$, the number of permutations sent to $\pi$ by $f$? A cycle of length $i$ can be written in $i$ ways, namely,

$$(b_1, b_2, \ldots, b_i) = (b_2, b_3, \ldots, b_i, b_1) = \cdots = (b_i, b_1, b_2, \ldots, b_{i-1}).$$

Moreover, there are $c_i!$ ways to order the $c_i$ cycles of length $i$. Hence

$$\#f^{-1}(\pi) = c_1! \, c_2! \, c_3! \cdots 1^{c_1} 2^{c_2} 3^{c_3} \cdots,$$

the same number for any $\pi \in X_c$. It follows that

$$\#X_c = \frac{\#\mathfrak{S}_\ell}{c_1! c_2! \cdots 1^{c_1} 2^{c_2} \cdots}$$

$$= \frac{\ell!}{c_1! c_2! \cdots 1^{c_1} 2^{c_2} \cdots},$$

as was to be proved.  □

As for the polynomial $Z_{\mathfrak{S}_\ell}$ itself, we have the following result. Write $\exp y = e^y$.

**7.13 Theorem.** *We have*

$$\sum_{\ell \geq 0} Z_{\mathfrak{S}_\ell}(z_1, z_2, \ldots) x^\ell = \exp\left(z_1 x + z_2 \frac{x^2}{2} + z_3 \frac{x^3}{3} + \cdots\right).$$

*Proof.* There are some sophisticated ways to prove this theorem which "explain" why the exponential function appears, but we will be content here with a "naive" computational proof. Write

$$\exp\left(z_1 x + z_2 \frac{x^2}{2} + z_3 \frac{x^3}{3} + \cdots\right)$$

$$= e^{z_1 x} \cdot e^{z_2 \frac{x^2}{2}} \cdot e^{z_3 \frac{x^3}{3}} \cdots$$

$$= \left( \sum_{n \geq 0} \frac{z_1^n x^n}{n!} \right) \left( \sum_{n \geq 0} \frac{z_2^n x^{2n}}{2^n n!} \right) \left( \sum_{n \geq 0} \frac{z_3^n x^{3n}}{3^n n!} \right) \cdots .$$

When we multiply this product out, the coefficient of $z_1^{c_1} z_2^{c_2} \cdots x^{\ell}$, where $\ell = c_1 + 2c_2 + \cdots$, is given by

$$\frac{1}{1^{c_1} c_1! \, 2^{c_2} c_2! \cdots} = \frac{1}{\ell!} \left( \frac{\ell!}{1^{c_1} c_1! \, 2^{c_2} c_2! \cdots} \right).$$

By Theorem 7.12 this is just the coefficient of $z_1^{c_1} z_2^{c_2} \cdots$ in $Z_{\mathfrak{S}_\ell}(z_1, z_2, \dots)$, as was to be proved. $\qquad \square$

As a check of Theorem 7.13, set each $z_i = n$ to obtain

$$\sum_{\ell \geq 0} Z_{\mathfrak{S}_\ell}(n, n, \dots) x^\ell = \exp \left( nx + n \frac{x^2}{2} + n \frac{x^3}{3} + \cdots \right)$$

$$= \exp \left( n \left( x + \frac{x^2}{2} + \frac{x^3}{3} + \cdots \right) \right)$$

$$= \exp \left( n \log(1 - x)^{-1} \right)$$

$$= \frac{1}{(1 - x)^n},$$

agreeing with Theorem (7.5) and (7.8).

Theorem 7.13 has many enumerative applications. We give one such result here as an example.

**7.14 Proposition.** *Let $f(n)$ be the number of permutations $\pi \in \mathfrak{S}_n$ of odd order. Equivalently, $\pi^k = \iota$ (the identity permutation) for some odd $k$. Then*

$$f(n) = \begin{cases} 1^2 \cdot 3^2 \cdot 5^2 \cdots (n - 1)^2, & n \text{ even} \\ 1^2 \cdot 3^2 \cdot 5^2 \cdots (n - 2)^2 \cdot n, & n \text{ odd}. \end{cases}$$

*Proof.* A permutation has odd order if and only if all its cycle lengths are odd. Hence [why?]

$$f(n) = n! \, Z_{\mathfrak{S}_n}(z_i = 1, \ i \text{ odd}; \ z_i = 0, \ i \text{ even}).$$

Making this substitution in Theorem 7.13 gives

$$\sum_{n \geq 0} f(n) \frac{x^n}{n!} = \exp \left( x + \frac{x^3}{3} + \frac{x^5}{5} + \cdots \right).$$

Since $-\log(1-x) = x + \frac{x^2}{2} + \frac{x^3}{3} + \cdots$, we get [why?]

$$\sum_{n \geq 0} f(n)\frac{x^n}{n!} = \exp\left(\frac{1}{2}\left(-\log(1-x) + \log(1+x)\right)\right)$$

$$= \exp\frac{1}{2}\log\left(\frac{1+x}{1-x}\right)$$

$$= \sqrt{\frac{1+x}{1-x}}.$$

We therefore need to find the coefficients in the power series expansion of $\sqrt{(1+x)/(1-x)}$ at $x = 0$. There is a simple trick for doing so:

$$\sqrt{\frac{1+x}{1-x}} = (1+x)(1-x^2)^{-1/2}$$

$$= (1+x) \sum_{m \geq 0} \binom{-1/2}{m}(-x^2)^m$$

$$= \sum_{m \geq 0}(-1)^m \binom{-1/2}{m}(x^{2m} + x^{2m+1}),$$

where by definition

$$\binom{-1/2}{m} = \frac{1}{m!}\left(-\frac{1}{2}\right)\left(-\frac{3}{2}\right)\cdots\left(-\frac{2m-1}{2}\right).$$

It is now a routine computation to check that the coefficient of $x^n/n!$ in $\sqrt{(1+x)/(1-x)}$ agrees with the claimed value of $f(n)$.                    $\square$

**Quotients of Boolean Algebras.** We will show how to apply Pólya theory to the problem of counting the number of elements of given rank in a quotient poset $B_X/G$. Here $X$ is a finite set, $B_X$ is the Boolean algebra of all subsets of $X$, and $G$ is a group of permutations of $X$ (with an induced action on $B_X$). What do colorings of $X$ have to do with subsets? The answer is very simple: a 2-coloring $f: X \rightarrow \{0, 1\}$ corresponds to a subset $S_f$ of $X$ by the usual rule

$$s \in S_f \Longleftrightarrow f(s) = 1.$$

Note that two 2-colorings $f$ and $g$ are $G$-equivalent if and only if $S_f$ and $S_g$ are in the same orbit of $G$ (acting on $B_X$). Thus the number of inequivalent 2-colorings $f$ of $X$ with $i$ values equal to 1 is just $\#(B_X/G)_i$, the number of elements of $B_X/G$ of rank $i$. As an immediate application of Pólya's theorem (Theorem 7.7) we obtain the following result.

**7.15 Corollary.** *We have*

$$\sum_i \#(B_X/G)_i \, q^i = Z_G(1+q, 1+q^2, 1+q^3, \dots).$$

*Proof.* If $\kappa(i, j)$ denotes the number of inequivalent 2-colorings of $X$ with the colors 0 and 1 such that 0 is used $j$ times and 1 is used $i$ times (so $i + j = \#X$), then by Pólya's theorem we have

$$\sum_{i,j} \kappa(i, j) x^i y^j = Z_G(x + y, x^2 + y^2, x^3 + y^3, \dots).$$

Setting $x = q$ and $y = 1$ yields the desired result [why?].                        □

Combining Corollary 7.15 with the rank-unimodality of $B_X/G$ (Theorem 5.8) yields the following corollary.

**7.16 Corollary.** *For any finite group $G$ of permutations of a finite set $X$, the polynomial $Z_G(1 + q, 1 + q^2, 1 + q^3, \dots)$ has symmetric, unimodal, integer coefficients.*

**7.17 Example.** (a) For the poset $P$ of Example 5.4(a) we have $G = \{(1)(2)(3), (1, 2)(3)\}$, so $Z_G(z_1, z_2, z_3) = \frac{1}{2}(z_1^3 + z_1 z_2)$. Hence

$$\sum_{i=0}^{3} (\#P_i) q^i = \frac{1}{2}\left((1+q)^3 + (1+q)(1+q^2)\right)$$

$$= 1 + 2q + 2q^2 + q^3.$$

(b) For the poset $P$ of Example 5.4(b) we have $G = \{(1)(2)(3)(4)(5), (1, 2, 3, 4, 5), (1, 3, 5, 2, 4), (1, 4, 2, 5, 3), (1, 5, 4, 3, 2)\}$, so $Z_G(z_1, z_2, z_3, z_4, z_5) = \frac{1}{5}(z_1^5 + 4z_5)$. Hence

$$\sum_{i=0}^{5} (\#P_i) q^i = \frac{1}{5}\left((1+q)^5 + 4(1+q^5)\right)$$

$$= 1 + q + 2q^2 + 2q^3 + q^4 + q^5.$$

Note that we are equivalently counting two-colored necklaces (as defined in Example 7.9), say with colors red and blue, of length five according to the number of blue beads.

(c) Let $X$ be the squares of a $2 \times 2$ chessboard, labelled as follows:

| 1 | 2 |
|---|---|
| 3 | 4 |

Let $G$ be the wreath product $\mathfrak{S}_2 \wr \mathfrak{S}_2$, as defined in Chapter 6. Then

$$G = \{(1)(2)(3)(4), (1,2)(3)(4), (1)(2)(3,4), (1,2)(3,4),$$

$$(1,3)(2,4), (1,4)(2,3), (1,3,2,4), (1,4,2,3)\},$$

so

$$Z_G(z_1, z_2, z_3, z_4) = \frac{1}{8}(z_1^4 + 2z_1^2 z_2 + 3z_2^2 + 2z_4).$$

Hence

$$\sum_{i=0}^{4}(\#P_i)q^i = \frac{1}{8}\left((1+q)^4 + 2(1+q)^2(1+q^2) + 3(1+q^2)^2 + 2(1+q^4)\right)$$

$$= 1 + q + 2q^2 + q^3 + q^4$$

$$= \binom{4}{2},$$

agreeing with Theorem 6.6.

Using more sophisticated methods (such as the representation theory of the symmetric group), the following generalization of Corollary 7.16 can be proved: let $P(q)$ be any polynomial with symmetric, unimodal, nonnegative, integer coefficients, such as $1 + q + 3q^2 + 3q^3 + 8q^4 + 3q^5 + 3q^6 + q^7 + q^8$ or $q^5 + q^6$ ($= 0 + 0q + \cdots + 0q^4 + q^5 + q^6 + 0q^7 + \cdots + 0q^{11}$). Then the polynomial $Z_G(P(q), P(q^2), P(q^3), \ldots)$ has symmetric, unimodal, nonnegative, integer coefficients.

**Graphs.** A standard application of Pólya theory is to the enumeration of nonisomorphic graphs. We saw at the end of Chapter 5 that if $M$ is an $m$-element vertex set, $X = \binom{M}{2}$, and $\mathfrak{S}_m^{(2)}$ is the group of permutations of $X$ induced by permutations of $M$, then an orbit of $i$-element subsets of $X$ may be regarded as an isomorphism class of graphs on the vertex set $M$ with $i$ edges. Thus $\#(B_X/\mathfrak{S}_m^{(2)})_i$ is the number of nonisomorphic graphs (without loops or multiple edges) on the vertex set $M$ with $i$ edges. It follows from Corollary 7.15 that if $g_i(m)$ denotes the number of nonisomorphic graphs with $m$ vertices and $i$ edges, then

$$\sum_{i=0}^{\binom{m}{2}} g_i(m)q^i = Z_{\mathfrak{S}_m^{(2)}}(1+q, 1+q^2, 1+q^3, \ldots).$$

Thus we would like to compute the cycle enumerator $Z_{\mathfrak{S}_m^{(2)}}(z_1, z_2, \ldots)$. If two permutations $\pi$ and $\sigma$ of $M$ have the same cycle type (number of cycles of each length), then their actions on $X$ also have the same cycle type [why?]. Thus for each possible cycle type of a permutation of $M$ (i.e., for each partition of $m$) we need to compute the induced cycle type on $X$. We also know from Theorem 7.12 the number of permutations of $M$ of each type. For small values of $m$ we can pick some permutation $\pi$ of each type and compute directly its action on $X$ in order to determine the induced cycle type. For $m = 4$ we have:

| Cycle lengths of $\pi$ | Number | $\pi$ | Induced permutation $\pi'$ | Cycle lengths of $\pi'$ |
|---|---|---|---|---|
| 1, 1, 1, 1 | 1 | (1)(2)(3)(4) | (12)(13)(14)(23)(24)(34) | 1, 1, 1, 1, 1, 1 |
| 2, 1, 1 | 6 | (1, 2)(3)(4) | (12)(13, 23)(14, 24)(34) | 2, 2, 1, 1 |
| 3, 1 | 8 | (1, 2, 3)(4) | (12, 23, 13)(14, 24, 34) | 3, 3 |
| 2, 2 | 3 | (1, 2)(3, 4) | (12)(13, 24)(14, 23)(34) | 2, 2, 1, 1 |
| 4 | 6 | (1, 2, 3, 4) | (12, 23, 34, 14)(13, 24) | 4, 2 |

It follows that

$$Z_{\mathfrak{S}_4^{(2)}}(z_1, z_2, z_3, z_4, z_5, z_6) = \frac{1}{24}(z_1^6 + 9z_1^2 z_2^2 + 8z_3^2 + 6z_2 z_4).$$

If we set $z_i = 1 + q^i$ and simplify, we obtain the polynomial

$$\sum_{i=0}^{6} g_i(4)q^i = 1 + q + 2q^2 + 3q^3 + 2q^4 + q^5 + q^6.$$

Indeed, this polynomial agrees with the rank-generating function of the poset of Figure 5.1.

Suppose that we instead want to count the number $h_i(4)$ of nonisomorphic graphs with four vertices and $i$ edges, where now we allow at most *two* edges between any two vertices. We can take $M$, $X$, and $G = \mathfrak{S}_4^{(2)}$ as before, but now we have three colors: red for no edges, blue for one edge, and yellow for two edges. A monomial $r^i b^j y^k$ corresponds to a coloring with $i$ pairs of vertices having no edges between them, $j$ pairs having one edge, and $k$ pairs having two edges. The total number $e$ of edges is $j + 2k$. Hence if we let $r = 1, b = q, y = q^2$, then the monomial $r^i b^j y^k$ becomes $q^{j+2k} = q^e$. It follows that

$$\sum_{i=0}^{i(i-1)} h_i(4)q^i = Z_{\mathfrak{S}_4^{(2)}}(1 + q + q^2, 1 + q^2 + q^4, 1 + q^3 + q^6, \dots)$$

$$= \frac{1}{24}\left((1 + q + q^2)^6 + 9(1 + q + q^2)^2(1 + q^2 + q^4)^2\right.$$

$$\left. + 8(1 + q^3 + q^6)^2 + 6(1 + q^2 + q^4)(1 + q^4 + q^8)\right)$$

$$= 1 + q + 3q^2 + 5q^3 + 8q^4 + 9q^5 + 12q^6 + 9q^7 + 8q^8 + 5q^9$$
$$+ 3q^{10} + q^{11} + q^{12}.$$

The total number of nonisomorphic graphs on four vertices with edge multiplicities at most two is $\sum_i h_i(4) = 66$.

It should now be clear that if we restrict the edge multiplicity to be $r$, then the corresponding generating function is $Z_{\mathfrak{S}_4^{(2)}}(1 + q + q^2 + \cdots + q^{r-1}, 1 + q^2 + q^4 + \cdots + q^{2r-2}, \ldots)$. In particular, to obtain the *total* number $N(r, 4)$ of nonisomorphic graphs on four vertices with edge multiplicity at most $r$, we simply set each $z_i = r$, obtaining

$$N(r, 4) = Z_{\mathfrak{S}_4^{(2)}}(r, r, r, r, r, r)$$

$$= \frac{1}{24}(r^6 + 9r^4 + 14r^2).$$

This is the same as number of inequivalent $r$-colorings of the set $X = \binom{M}{2}$ (where $\#M = 4$) [why?].

Of course the same sort of reasoning can be applied to any number of vertices. For five vertices our table becomes the following (using such notation as $1^5$ to denote a sequence of five 1's).

| Cycle lengths of $\pi$ | Number | $\pi$ | Induced permutation $\pi'$ | Cycle lengths of $\pi'$ |
|---|---|---|---|---|
| $1^5$ | 1 | $(1)(2)(3)(4)(5)$ | $(12)(13)\cdots(45)$ | $1^{10}$ |
| $2, 1^3$ | 10 | $(1, 2)(3)(4)(5)$ | $(12)(13, 23)(14, 25)(15, 25)(34)(35)(45)$ | $2^3, 1^4$ |
| $3, 1^2$ | 20 | $(1, 2, 3)(4)(5)$ | $(12, 23, 13)(14, 24, 34)(15, 25, 35)(45)$ | $3^3, 1$ |
| $2^2, 1$ | 15 | $(1, 2)(3, 4)(5)$ | $(12)(13, 24)(14, 23)(15, 25)(34)(35, 45)$ | $2^4, 1^2$ |
| $4, 1$ | 30 | $(1, 2, 3, 4)(5)$ | $(12, 23, 34, 14)(13, 24)(15, 25, 35, 45)$ | $4^2, 2$ |
| $3, 2$ | 20 | $(1, 2, 3)(4, 5)$ | $(12, 23, 13)(14, 25, 34, 15, 24, 35)(45)$ | $6, 3, 1$ |
| $5$ | 24 | $(1, 2, 3, 4, 5)$ | $(12, 23, 34, 45, 15)(13, 24, 35, 14, 25)$ | $5^2$ |

Thus

$$Z_{\mathfrak{S}_5^{(2)}}(z_1, \ldots, z_{10}) = \frac{1}{120}(z_1^{10} + 10z_1^4 z_2^3 + 20z_1 z_3^3 + 15z_1^2 z_2^4 + 30z_2 z_4^2 + 20z_1 z_3 z_6 + 24z_5^2),$$

from which we compute

$$\sum_{i=0}^{10} g_i(5)q^i = Z_{\mathfrak{S}_5^{(2)}}(1 + q, 1 + q^2, \ldots, 1 + q^{10})$$

$$= 1 + q + 2q^2 + 4q^3 + 6q^4 + 6q^5 + 6q^6 + 4q^7 + 2q^8 + q^9 + q^{10}.$$

For an arbitrary number $m = \#M$ of vertices there exist explicit formulas for the cycle indicator of the induced action of $\pi \in \mathfrak{S}_M$ on $\binom{M}{2}$, thereby obviating the need to compute $\pi'$ explicitly as we did in the above tables, but the overall expression for $Z_{\mathfrak{S}_m^{(2)}}$ cannot be simplified significantly or put into a simple generating function as we did in Theorem 7.13. For reference we record

$$Z_{\mathfrak{S}_6^{(2)}} = \frac{1}{6!}(z_1^{15} + 15z_1^7z_2^4 + 40z_1^3z_3^4 + 45z_1^3z_2^6 + 90z_1z_2z_4^3 + 120z_1z_2z_3^2z_6$$

$$+144z_5^3 + 15z_1^3z_2^6 + 90z_1z_2z_4^3 + 40z_3^5 + 120z_3z_6^2)$$

$$(g_0(6), g_1(6), \ldots, g_{15}(6)) = (1, 1, 2, 5, 9, 15, 21, 24, 24, 21, 15, 9, 5, 2, 1, 1).$$

Moreover if $u(n)$ denotes the number of nonisomorphic simple graphs with $n$ vertices, then

$$(u(0), u(1), \ldots, u(11))$$

$$= (1, 1, 2, 4, 11, 34, 156, 1044, 12346, 274668, 12005168, 1018997864).$$

A table of $u(n)$ for $n \le 75$ is given at

$$\texttt{http://oeis.org/A000088/b000088.txt}$$

In particular,

$$u(75) = 91965776790545918117055311393231179873443957239$$
$$0555232344598910500368551136102062542965342147$$
$$8723210428876893185920222186100317580740213865$$
$$7140377683043095632048495393006440764501648363$$
$$4760490012493552274952950606265577383468983364$$
$$6883724923654397496226869104105041619919159586$$
$$8518775275216748149124234654756641508154401414$$
$$8480274454866344981385848105320672784068407907$$
$$1134767688676890584660201791139593590722767979$$
$$8617445756819562952590259920801220117529208077$$
$$0705444809177422214784902579514964768094933848$$
$$3173060596932480677345855848701061537676603425$$
$$1254842843718829212212327337499413913712750831$$

0550986833980707875560051306072520155744624852

0263616216031346723897074759199703968653839368

7763608064327592656680387259609907<sub></sub>2,

a number of 726 digits! Compare

$$\frac{2^{\binom{75}{2}}}{75!} = .9196577679054591809 \cdots \times 10^{726},$$

which agrees with $u(75)$ to 17 significant digits [why?].

# Notes for Chapter 7

Burnside's lemma (Lemma 7.2) was actually first stated and proved by Frobenius [47, end of §4]. Frobenius in turn credits Cauchy [22, p. 286] for proving the lemma in the transitive case. Burnside, in the first edition of his book [16, §118–119], attributes the lemma to Frobenius, but in the second edition [17] this citation is absent. For more on the history of Burnside's lemma, see [96] and [145]. Many authors now call this result the Cauchy–Frobenius lemma. The cycle indicator $Z_G(z_1, z_2, \dots)$ (where $G$ is a subgroup of $\mathfrak{S}_n$) was first considered by Redfield [108], who called it the *group reduction function*, denoted $\mathrm{Grf}(G)$. Pólya [101] independently defined the cycle indicator, proved the fundamental Theorem 7.7, and gave numerous applications. For an English translation of Pólya's paper, see [102]. Much of Pólya's work was anticipated by Redfield. For interesting historical information about the work of Redfield and its relation to Pólya theory, see [58, 60, 82, 109] (all in the same issue of *Journal of Graph Theory*). The Wikipedia article "John Howard Redfield" also gives information and references on the interesting story of the rediscovery and significance of Redfield's work.

The application of Pólya's theorem to the enumeration of nonisomorphic graphs appears in Pólya's original paper [101]. For much additional work on graphical enumeration, see the text of Harary and Palmer [59].

Subsequent to Pólya's work there have been a huge number of expositions, applications, and generalizations of Pólya theory. An example of such a generalization appears in Exercise 7.14. We mention here only the nice survey [31] by de Bruijn.

Theorem 7.13 (the generating function for the cycle indicator $Z_{\mathfrak{S}_\ell}$ of the symmetric group $\mathfrak{S}_\ell$) goes back to Frobenius (see [48, bottom of p. 152 of GA]) and Hurwitz [70, §4]. It is clear that they were aware of Theorem 7.13, even if they did not state it explicitly. For a more conceptual approach and further aspects see Stanley [131, §§5.1–5.2].

## Exercises for Chapter 7

1.  Verify (7.6), i.e., the formula for the cycle enumerator of the dihedral group with its defining action.
2.  For a simple graph $\Gamma$ with vertex set $V$, we can define an *automorphism* of $\Gamma$ to be a bijection $\varphi\colon V \to V$ such that $u$ and $v$ are adjacent if and only if $\varphi(u)$ and $\varphi(v)$ are adjacent. The automorphisms form a group under composition, called the *automorphism group* Aut($\Gamma$) of $\Gamma$. Let $\Gamma$ be the graph shown below.

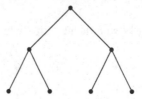

    Let $G$ be the automorphism group of $\Gamma$, so $G$ has order eight.

    (a)  What is the cycle index polynomial of $G$, acting on the vertices of $\Gamma$?
    (b)  In how many ways can one color the vertices of $\Gamma$ in $n$ colors, up to the symmetry of $\Gamma$?

3.  (*) What is the total number of inequivalent ways to color the vertices of a regular dodecahedron with the colors dark red and burgundy, up to the 120 rotations and reflections of the dodecahedron? Do this computation in your head.
4.  A regular tetrahedron $T$ has four vertices, six edges, and four triangles. The rotational symmetries of $T$ (no reflections allowed) form a group $G$ of order 12.

    (a)  What is the cycle index polynomial of $G$ acting on the *vertices* of $T$?
    (b)  In how many ways can the vertices of $T$ be colored in $n$ colors, up to rotational symmetry?
    (c)  What about coloring the six *edges* of $T$, up to rotational symmetry?

5.  Let $G$ be the graph on the vertex set $\{1, 2, \ldots, 4r\}$ which is a cycle of length $4r, r \geq 2$, together with an edge from some vertex to its antipode. (Thus $G$ has $4r + 1$ edges.) How many ways are there to color the vertices of $G$ from a set of $n$ colors, up to isomorphism? (Colors may be repeated. Not all $n$ colors need to be used. Adjacent vertices are allowed to have the same color.)
6.  How many necklaces (up to cyclic symmetry) have $n$ red beads and $n$ blue beads? (Express your answer as a sum over all divisors $d$ of $n$.)
7.  (not directly related to the text) A *primitive* necklace is a necklace with no symmetries, i.e., no nonidentity rotation of the necklace preserves the necklace. Let $M_\ell(n)$ denote the number of primitive $n$-colored necklaces with $\ell$ beads. Show that

$$M_\ell(n) = \frac{1}{\ell} \sum_{d|\ell} \mu(\ell/d)n^d,$$

where $\mu$ denotes the Möbius function from number theory. (Compare (7.5).)

8. Ten balls are stacked in a triangular array with 1 atop 2 atop 3 atop 4. (Think of billiards.) The triangular array is free to rotate in two dimensions.

   (a) Find the generating function for the number of inequivalent colorings using the ten colors $r_1, r_2, \ldots, r_{10}$. (You don't need to simplify your answer.)
   (b) How many inequivalent colorings have four red balls, three green balls, and three chartreuse balls? How many have four red balls, four turquoise balls, and two aquamarine balls?

9. The dihedral group $D_4$ of order 8 acts on the set $X$ of 64 squares of an $8 \times 8$ chessboard $B$. Find the number of ways to choose two subsets $S \subseteq T$ of $X$, up to the action of $D_4$. For instance, all eight ways to choose $S$ to be a single corner square $s$ and $T$ to be $\{s, t\}$, where $t$ is adjacent to $s$ (i.e., has an edge in common with $s$), belong to the same orbit of $D_4$. Write your answer as a (short) finite sum.

10. For any finite group $G$ of permutations of an $\ell$-element set $X$, let $f(n)$ be the number of inequivalent (under the action of $G$) colorings of $X$ with $n$ colors. Find $\lim_{n \to \infty} f(n)/n^\ell$. Interpret your answer as saying that "most" colorings of $X$ are asymmetric (have no symmetries).

11. Let $X$ be a finite set, and let $G$ be a subgroup of the symmetric group $\mathfrak{S}_X$. Suppose that the number of orbits of $G$ acting on $n$-colorings of $X$ is given by the polynomial

$$f(n) = \frac{1}{a}(n^p + bn^{p-2} + \cdots + (p-1)n),$$

where $p$ is prime.

   (a) What is the order (number of elements) of $G$?
   (b) What is the size $\#X$ of $X$?
   (c) How many transpositions are in $G$? A *transposition* is a permutation that transposes (interchanges) two elements of $X$ and leaves the remaining elements fixed.
   (d) How many orbits does $G$ have acting on $X$?
   (e) Show that $G$ is either a cyclic group or is not a simple group. A group $G$ with more than one element is *simple* if its only normal subgroups are $G$ and $\{1\}$.

12. It is known that there exists a *nonabelian* group $G$ of order 27 such that $x^3 = 1$ for all $x \in G$. Use this fact to give an example of two nonisomorphic finite subgroups $G$ and $H$ of $\mathfrak{S}_X$ for some finite set $X$ such that $Z_G = Z_H$.

13. (somewhat difficult) Let $N_G(n)$ be the polynomial of Theorem 7.5, and let $\#X = d$. Show that $(-1)^d N_G(-n)$ is equal to the number of inequivalent $n$-

colorings $f: X \to [n]$ of $X$ such that the subgroup $H$ of $G$ fixing $f$ (i.e., $\pi \cdot f = f$ for all $\pi \in H$) is contained in the alternating group $\mathfrak{A}_X$. This result could be called a *reciprocity theorem* for the polynomial $N_G(n)$.

14. (difficult) Suppose that a finite group $G$ acts on a finite set $X$ and another finite group $H$ acts on a set $C$ of colors. Call two colorings $f, g: X \to C$ *equivalent* if there are permutations $\pi \in G$ and $\sigma \in H$ such that

$$f(x) = \sigma \cdot g(\pi \cdot x), \quad \text{for all } x \in X.$$

Thus we are allowed not only to permute the elements of $X$ by some element of $G$ but also to permute the colors by some element of $H$. Show that the total number of inequivalent colorings is given by

$$Z_G\left(\frac{\partial}{\partial z_1}, \frac{\partial}{\partial z_2}, \frac{\partial}{\partial z_3}, \cdots\right) Z_H(e^{z_1+z_2+z_3+\cdots}, e^{2(z_2+z_4+z_6+\cdots)}, e^{3(z_3+z_6+z_9+\cdots)}, \ldots),$$

evaluated at $z_1 = z_2 = z_3 = \cdots = 0$.

*Example.* Let $n$ be the number of two-colored necklaces of four beads, where we may also interchange the two colors to get an equivalent coloring. Thus $Z_G = \frac{1}{4}(z_1^4 + z_2^2 + 2z_4)$ and $Z_H = \frac{1}{2}(z_1^2 + z_2)$. Hence

$$n = \frac{1}{4} \cdot \frac{1}{2}\left(\frac{\partial^4}{\partial z_1^4} + \frac{\partial^2}{\partial z_2^2} + 2\frac{\partial}{\partial z_4}\right)(e^{2(z_1+z_2+z_3+\cdots)} + e^{2(z_2+z_4+z_6+\cdots)})|_{z_i=0}$$

$$= \frac{1}{8}\left(\frac{\partial^4}{\partial z_1^4} + \frac{\partial^2}{\partial z_2^2} + 2\frac{\partial}{\partial z_4}\right)\left(\frac{(2z_1)^4}{4!} + \frac{(2z_2)^2}{2!} + \frac{2z_4}{1!} + \frac{(2z_2)^4}{4!} + \frac{2z_4}{1!}\right)\Bigg|_{z_i=0}$$

$$= \frac{1}{8}(16 + 4 + 4 + 4 + 4)$$

$$= 4.$$

The four different necklaces are 0000, 1000, 1100, and 1010.

15. (a) Let $e_6(n)$ denote the number of permutations $\pi \in \mathfrak{S}_n$ satisfying $\pi^6 = \iota$ (the identity permutation). Find a simple formula for the generating function

$$E_6(x) = \sum_{n \geq 0} e_6(n)\frac{x^n}{n!}.$$

(b) Generalize to $e_k(n) = \#\{\pi \in \mathfrak{S}_n : \pi^k = \iota\}$ for any $k \geq 1$.

16. Let $f(n)$ be the number of permutations in the symmetric group $\mathfrak{S}_n$ all of whose cycles have even length. For instance, $f(4) = 9$ and $f(11) = 0$.

(a) Let

$$F(x) = \sum_{n=0}^{\infty} f(n) \frac{x^n}{n!}.$$

Find a simple expression for $F(x)$. Your answer should not involve any summation symbols (or their equivalent), logarithms, or the function $e^x$.

(b) Use (a) to find a simple formula for $f(n)$.

(c) Give a combinatorial proof of (b).

17. (difficult) Give a combinatorial proof of Proposition 7.14. Despite the similarity between Proposition 7.14 and Exercise 7.16, the latter is much easier to prove combinatorially than the former.

18. Let $c(w)$ denote the number of cycles of a permutation $w \in \mathfrak{S}_n$. Let $f(n)$ denote the average value of $c(w)(c(w) - 1)$ for $w \in \mathfrak{S}_n$, i.e.,

$$f(n) = \frac{1}{n!} \sum_{w \in \mathfrak{S}_n} c(w)(c(w) - 1).$$

(Set $f(0) = 1$.) Find a simple formula for the generating function $\sum_{n \geq 0} f(n) t^n$.

19. (a) (*) Let $n \geq 1$, and let $G$ be a subgroup of $\mathfrak{S}_n$ of odd order. Show that the quotient poset $B_n/G$ has the same number of elements of even rank as of odd rank.

(b) Generalize (a) as follows: give a necessary and sufficient condition on a subgroup $G$ of $\mathfrak{S}_n$, in terms of the cycle lengths of elements of $G$, for $B_n/G$ to have the same number of elements of even rank as of odd rank.

20. Let $c(\ell, k)$ denote the number of permutations in $\mathfrak{S}_\ell$ with $k$ cycles. Show that the sequence

$$c(\ell, 1), c(\ell, 2), \ldots, c(\ell, \ell)$$

is strongly log-concave.

# Chapter 8
# A Glimpse of Young Tableaux

We defined in Chapter 6 Young's lattice $Y$, the poset of all partitions of all nonnegative integers, ordered by containment of their Young diagrams.

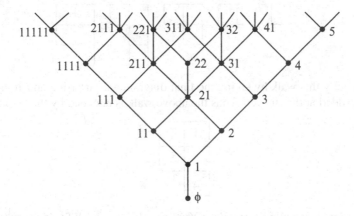

Here we will be concerned with the counting of certain walks in the Hasse diagram (considered as a graph) of $Y$. Note that since $Y$ is infinite, we cannot talk about its eigenvalues and eigenvectors. We need different techniques for counting walks. It will be convenient to denote the length of a walk by $n$, rather than by $\ell$ as in previous chapters.

Note that $Y$ is a graded poset (of infinite rank), with $Y_i$ consisting of all partitions of $i$. In other words, we have $Y = Y_0 \cup Y_1 \cup \cdots$ (disjoint union), where every maximal chain intersects each level $Y_i$ exactly once. We call $Y_i$ the $i$th *level* of $Y$, just as we did for finite graded posets.

Since the Hasse diagram of $Y$ is a simple graph (no loops or multiple edges), a walk of length $n$ is specified by a sequence $\lambda^0, \lambda^1, \ldots, \lambda^n$ of vertices of $Y$. We will call a walk in the Hasse diagram of a poset a *Hasse walk*. Each $\lambda^i$ is a partition of some integer, and we have either (a) $\lambda^i < \lambda^{i+1}$ and $|\lambda^i| = |\lambda^{i+1}| - 1$, or

© Springer International Publishing AG, part of Springer Nature 2018

R. P. Stanley, *Algebraic Combinatorics*, Undergraduate Texts in Mathematics,

https://doi.org/10.1007/978-3-319-77173-1_8

(b) $\lambda^i > \lambda^{i+1}$ and $|\lambda^i| = |\lambda^{i+1}| + 1$. (Recall that for a partition $\lambda$, we write $|\lambda|$ for the sum of the parts of $\lambda$.) A step of type (a) is denoted by $U$ (for "up," since we move up in the Hasse diagram), while a step of type (b) is denoted by $D$ (for "down"). If the walk $W$ has steps of types $A_1, A_2, \ldots, A_n$, respectively, where each $A_i$ is either $U$ or $D$, then we say that $W$ is of *type* $A_n A_{n-1} \cdots A_2 A_1$. Note that the type of a walk is written in the *opposite* order to that of the walk. This is because we will soon regard $U$ and $D$ as linear transformations, and we multiply linear transformations *right-to-left* (opposite to the usual left-to-right reading order). For instance (abbreviating a partition $(\lambda_1, \ldots, \lambda_m)$ as $\lambda_1 \cdots \lambda_m$), the walk $\emptyset, 1, 2, 1, 11, 111, 211, 221, 22, 21, 31, 41$ is of type $UUDDUUUUDUU = U^2 D^2 U^4 D U^2$.

There is a nice combinatorial interpretation of walks of type $U^n$ which begin at $\emptyset$. Such walks are of course just saturated chains $\emptyset = \lambda^0 \lessdot \lambda^1 \lessdot \cdots \lessdot \lambda^n$. In other words, they may be regarded as sequences of Young diagrams, beginning with the empty diagram and adding one new square at each step. An example of a walk of type $U^5$ is given by

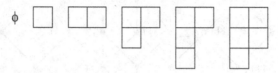

We can specify this walk by taking the final diagram and inserting an $i$ into square $s$ if $s$ was added at the $i$th step. Thus the above walk is encoded by the "tableau"

| 1 | 2 |
|---|---|
| 3 | 5 |
| 4 |   |

Such an object $\tau$ is called a *standard Young tableaux* (or SYT). It consists of the Young diagram $D$ of some partition $\lambda$ of an integer $n$, together with the numbers $1, 2, \ldots, n$ inserted into the squares of $D$, so that each number appears exactly once, and every row and column is *increasing*. We call $\lambda$ the *shape* of the SYT $\tau$, denoted $\lambda = \text{sh}(\tau)$. For instance, there are five SYT of shape $(2, 2, 1)$, given by

| 1 | 2 |   | 1 | 2 |   | 1 | 3 |   | 1 | 3 |   | 1 | 4 |
|---|---|---|---|---|---|---|---|---|---|---|---|---|---|
| 3 | 4 |   | 3 | 5 |   | 2 | 4 |   | 2 | 5 |   | 2 | 5 |
| 5 |   |   | 4 |   |   | 5 |   |   | 4 |   |   | 3 |   |

Let $f^\lambda$ denote the number of SYT of shape $\lambda$, so for instance $f^{(2,2,1)} = 5$. The numbers $f^\lambda$ have many interesting properties; for instance, there is a famous explicit formula for them known as the Frame–Robinson–Thrall hook length formula. For the sake of completeness we state this formula without proof, though it is not needed in what follows.

Let $u$ be a square of the Young diagram of the partition $\lambda$. Define the *hook* $H(u)$ of $u$ (or at $u$) to be the set of all squares directly to the right of $u$ or directly below $u$, including $u$ itself. The size (number of squares) of $H(u)$ is called the *hook length* of $u$ (or at $u$), denoted $h(u)$. In the diagram of the partition $(4, 2, 2)$ below, we have inserted the hook length $h(u)$ inside each square $u$.

| 6 | 5 | 2 | 1 |
|---|---|---|---|
| 3 | 2 |   |   |
| 2 | 1 |   |   |

**8.1 Theorem** (Hook Length Formula). *Let $\lambda \vdash n$. Then*

$$f^\lambda = \frac{n!}{\prod_{u \in \lambda} h(u)}.$$

*Here the notation $u \in \lambda$ means that $u$ ranges over all squares of the Young diagram of $\lambda$.*

For instance, the diagram of the hook lengths of $\lambda = (4, 2, 2)$ above gives

$$f^{(4,2,2)} = \frac{8!}{6 \cdot 5 \cdot 2 \cdot 1 \cdot 3 \cdot 2 \cdot 2 \cdot 1} = 56.$$

In this chapter we will be concerned with the connection between SYT and counting walks in Young's lattice. If $w = A_n A_{n-1} \cdots A_1$ is some word in $U$ and $D$ and $\lambda \vdash n$, then let us write $\alpha(w, \lambda)$ for the number of Hasse walks in $Y$ of type $w$ which start at the empty partition $\emptyset$ and end at $\lambda$. For instance, $\alpha(UDUU, 11) = 2$, the corresponding walks being $\emptyset, 1, 2, 1, 11$ and $\emptyset, 1, 11, 1, 11$. Thus in particular $\alpha(U^n, \lambda) = f^\lambda$ [why?]. In a similar fashion, since the number of Hasse walks of type $D^n U^n$ which begin at $\emptyset$, go up to a partition $\lambda \vdash n$, and then back down to $\emptyset$ is given by $(f^\lambda)^2$, we have

$$\alpha(D^n U^n, \emptyset) = \sum_{\lambda \vdash n} (f^\lambda)^2. \tag{8.1}$$

Our object is to find an explicit formula for $\alpha(w, \lambda)$ of the form $f^\lambda c_w$, where $c_w$ does not depend on $\lambda$. (It is by no means *a priori* obvious that such a formula should exist.) In particular, since $f^\emptyset = 1$, we will obtain by setting $\lambda = \emptyset$ a simple formula for the number of (closed) Hasse walks of type $w$ from $\emptyset$ to $\emptyset$ (thus including a simple formula for (8.1)).

There is an easy condition for the existence of *any* Hasse walk of type $w$ from $\emptyset$ to $\lambda$, given by the next lemma.

**8.2 Lemma.** *Suppose* $w = D^{s_k} U^{r_k} \cdots D^{s_2} U^{r_2} D^{s_1} U^{r_1}$, *where* $r_i \geq 0$ *and* $s_i \geq 0$. *Let* $\lambda \vdash n$. *Then there exists a Hasse walk of type w from $\emptyset$ to $\lambda$ if and only if:*

$$\sum_{i=1}^{k} (r_i - s_i) = n$$

$$\sum_{i=1}^{j} (r_i - s_i) \geq 0 \text{ for } 1 \leq j \leq k.$$

*Proof.* Since each $U$ moves up one level and each $D$ moves down one level, we see that $\sum_{i=1}^{k}(r_i - s_i)$ is the level at which a walk of type $w$ beginning at $\emptyset$ ends. Hence $\sum_{i=1}^{k}(r_i - s_i) = |\lambda| = n$.

After $\sum_{i=1}^{j}(r_i + s_i)$ steps we will be at level $\sum_{i=1}^{j}(r_i - s_i)$. Since the lowest level is level 0, we must have $\sum_{i=1}^{j}(r_i - s_i) \geq 0$ for $1 \leq j \leq k$.

The easy proof that the two conditions of the lemma are *sufficient* for the existence of a Hasse walk of type $w$ from $\emptyset$ to $\lambda$ is left to the reader.          $\square$

If $w$ is a word in $U$ and $D$ satisfying the conditions of Lemma 8.2, then we say that $w$ is a *valid $\lambda$-word*. Note that the condition of being a valid $\lambda$-word depends only on $|\lambda|$.

The proof of our formula for $\alpha(w, \lambda)$ will be based on linear transformations analogous to those defined by (4.2) and (4.3). As in Chapter 4 let $\mathbb{R}Y_j$ be the real vector space with basis $Y_j$. Define two linear transformations $U_i : \mathbb{R}Y_i \rightarrow \mathbb{R}Y_{i+1}$ and $D_i : \mathbb{R}Y_i \rightarrow \mathbb{R}Y_{i-1}$ by

$$U_i(\lambda) = \sum_{\substack{\mu \vdash i+1 \\ \lambda < \mu}} \mu$$

$$D_i(\lambda) = \sum_{\substack{\nu \vdash i-1 \\ \nu < \lambda}} \nu,$$

for all $\lambda \vdash i$. For instance (using abbreviated notation for partitions)

$$U_{21}(54422211) = 64422211 + 55422211 + 54432211 + 54422221 + 544222111$$

$$D_{21}(54422211) = 44422211 + 54322211 + 54422111 + 5442221.$$

It is clear [why?] that if $r$ is the number of *distinct* (i.e., unequal) parts of $\lambda$, then $U_i(\lambda)$ is a sum of $r + 1$ terms and $D_i(\lambda)$ is a sum of $r$ terms. The next lemma is an analogue for $Y$ of the corresponding result for $B_n$ (Lemma 4.6).

**8.3 Lemma.** *For any $i \geq 0$ we have*

$$D_{i+1}U_i - U_{i-1}D_i = I_i, \tag{8.2}$$

*the identity linear transformation on $\mathbb{R}Y_i$.*

*Proof.* Apply the left-hand side of (8.2) to a partition $\lambda$ of $i$, expand in terms of the basis $Y_i$, and consider the coefficient of a partition $\mu$. If $\mu \neq \lambda$ and $\mu$ can be obtained from $\lambda$ by adding one square $s$ to (the Young diagram of) $\lambda$ and then removing a (necessarily different) square $t$, then there is exactly one choice of $s$ and $t$. Hence the coefficient of $\mu$ in $D_{i+1}U_i(\lambda)$ is equal to 1. But then there is exactly one way to remove a square from $\lambda$ and then add a square to get $\mu$, namely, remove $t$ and add $s$. Hence the coefficient of $\mu$ in $U_{i-1}D_i(\lambda)$ is also 1, so the coefficient of $\mu$ when the left-hand side of (8.2) is applied to $\lambda$ is 0.

If now $\mu \neq \lambda$ and we cannot obtain $\mu$ by adding a square and then deleting a square from $\lambda$ (i.e., $\mu$ and $\lambda$ differ in more than two rows), then clearly when we apply the left-hand side of (8.2) to $\lambda$, the coefficient of $\mu$ will be 0.

Finally consider the case $\lambda = \mu$. Let $r$ be the number of distinct (unequal) parts of $\lambda$. Then the coefficient of $\lambda$ in $D_{i+1}U_i(\lambda)$ is $r + 1$, while the coefficient of $\lambda$ in $U_{i-1}D_i(\lambda)$ is $r$, since there are $r + 1$ ways to add a square to $\lambda$ and then remove it, while there are $r$ ways to remove a square and then add it back in. Hence when we apply the left-hand side of (8.2) to $\lambda$, the coefficient of $\lambda$ is equal to 1.

Combining the conclusions of the three cases just considered shows that the left-hand side of (8.2) is just $I_i$, as was to be proved. $\qquad \square$

We come to one of the main results of this chapter.

**8.4 Theorem.** *Let $\lambda$ be a partition and $w = A_n A_{n-1} \cdots A_1$ a valid $\lambda$-word. Let $S_w = \{i : A_i = D\}$. For each $i \in S_w$, let $a_i$ be the number of $D$'s in $w$ to the right of $A_i$, and let $b_i$ be the number of $U$'s in $w$ to the right of $A_i$. Thus $a_i - b_i$ is the level we occupy in $Y$ before taking the step $A_i = D$. Then*

$$\alpha(w, \lambda) = f^\lambda \prod_{i \in S_w} (b_i - a_i).$$

Before proving Theorem 8.4, let us give an example. Suppose $w = U^3 D^2 U^2 D U^3 = UUUDDUUDUUU$ and $\lambda = (2, 2, 1)$. Then $S_w = \{4, 7, 8\}$ and $a_4 = 0$, $b_4 = 3$, $a_7 = 1$, $b_7 = 5$, $a_8 = 2$, $b_8 = 5$. We have also seen earlier that $f^{221} = 5$. Thus

$$\alpha(w, \lambda) = 5(3 - 0)(5 - 1)(5 - 2) = 180.$$

*Proof.* Proof of Theorem 8.4. For notational simplicity we will omit the subscripts from the linear transformations $U_i$ and $D_i$. This should cause no confusion since the subscripts will be uniquely determined by the elements on which $U$ and $D$ act. For instance, the expression $UDUU(\lambda)$ where $\lambda \vdash i$ must mean $U_{i+1}D_{i+2}U_{i+1}U_i(\lambda)$;

otherwise it would be undefined since $U_j$ and $D_j$ can only act on elements of $\mathbb{R}Y_j$, and moreover $U_j$ raises the level by one while $D_j$ lowers it by one.

By (8.2) we can replace $DU$ in any word $y$ in the letters $U$ and $D$ by $UD+I$. This replaces $y$ by a sum of two words, one with one fewer $D$ and the other with one $D$ moved one space to the right. For instance, replacing the first $DU$ in $UUDUDDU$ by $UD + I$ yields $UUUDDDU + UUDDU$. If we begin with the word $w$ and iterate this procedure, replacing a $DU$ in any word with $UD + I$, eventually there will be no $U$'s to the right of any $D$'s and the procedure will come to an end. At this point we will have expressed $w$ as a linear combination (with integer coefficients) of words of the form $U^i D^j$. Since the operation of replacing $DU$ with $UD + I$ preserves the difference between the number of $U$'s and $D$'s in each word, all the words $U^i D^j$ which appear will have $i - j$ equal to some constant $n$ (namely, the number of $U$'s minus the number of $D$'s in $w$). Specifically, say we have

$$w = \sum_{i-j=n} r_{ij}(w)U^i D^j, \tag{8.3}$$

where each $r_{ij}(w) \in \mathbb{Z}$. (We also define $r_{ij}(w) = 0$ if $i < 0$ or $j < 0$.) We claim that the $r_{ij}(w)$'s are uniquely determined by $w$. Equivalently [why?], if we have

$$\sum_{i-j=n} d_{ij}U^i D^j = 0 \tag{8.4}$$

(as an identity of linear transformations acting on the space $\mathbb{R}Y_k$ for *any* $k$), where each $d_{ij} \in \mathbb{Z}$ (or $d_{ij} \in \mathbb{R}$, if you prefer), then each $d_{ij} = 0$. Let $j'$ be the least integer for which $d_{j'+n,j'} \neq 0$. Let $\mu \vdash j'$, and apply both sides of (8.4) to $\mu$. The left-hand side has exactly one nonzero term, namely, the term with $j = j'$ [why?]. The right-hand side, on the other hand,[1] is 0, a contradiction. Thus the $r_{ij}(w)$'s are unique.

Now apply $U$ on the left to (8.3). We get

$$Uw = \sum_{i,j} r_{ij}(w)U^{i+1} D^j.$$

Hence (using uniqueness of the $r_{ij}$'s) there follows [why?]

$$r_{ij}(Uw) = r_{i-1,j}(w). \tag{8.5}$$

We next want to apply $D$ on the left to (8.3). It is easily proved by induction on $i$ (left as an exercise) that

$$DU^i = U^i D + iU^{i-1}. \tag{8.6}$$

---

[1] The phrase "the right-hand side, on the other hand" does not mean the left-hand side!

(We interpret $U^{-1}$ as being 0 and $U^0 = I$, so that (8.6) is true for $i = 0, 1$.) Hence

$$Dw = \sum_{i,j} r_{ij}(w) D U^i D^j$$

$$= \sum_{i,j} r_{ij}(w)(U^i D + i U^{i-1}) D^j,$$

from which it follows [why?] that

$$r_{ij}(Dw) = r_{i,j-1}(w) + (i+1)r_{i+1,j}(w). \qquad (8.7)$$

Setting $j = 0$ in (8.5) and (8.7) yields

$$r_{i0}(Uw) = r_{i-1,0}(w) \qquad (8.8)$$

$$r_{i0}(Dw) = (i+1)r_{i+1,0}(w). \qquad (8.9)$$

Now let (8.3) operate on $\emptyset$. Since $D^j(\emptyset) = 0$ for all $j > 0$, we get $w(\emptyset) = r_{n0}(w)U^n(\emptyset)$. Thus the coefficient of $\lambda$ in $w(\emptyset)$ is given by

$$\alpha(w, \lambda) = r_{n0}(w)\alpha(U^n, \lambda) = r_{n0}f^\lambda,$$

where as usual $\lambda \vdash n$. It is clear from (8.8) and (8.9) that

$$r_{n0}(w) = \prod_{j \in S_w} (b_j - a_j),$$

and the proof follows. $\qquad \square$

NOTE. It is possible to give a simpler proof of Theorem 8.4, but the proof we have given is useful for generalizations not appearing here.

An interesting special case of the previous theorem allows us to evaluate (8.1).

**8.5 Corollary.** *We have*

$$\alpha(D^n U^n, \emptyset) = \sum_{\lambda \vdash n} (f^\lambda)^2 = n!.$$

*Proof.* When $w = D^n U^n$ in Theorem 8.4 we have $S_w = \{n+1, n+2, \ldots, 2n\}$, $a_i = i - n - 1$, and $b_i = n$, from which the proof is immediate. $\qquad \square$

NOTE (for those familiar with the representation theory of finite groups). It can be shown that the numbers $f^\lambda$, for $\lambda \vdash n$, are the degrees of the irreducible representations of the symmetric group $\mathfrak{S}_n$. Given this, Corollary 8.5 is a special case of the result that the sum of the squares of the degrees of the irreducible representations of a finite group $G$ is equal to the order $\#G$ of $G$. There are many other intimate connections between the representation theory of $\mathfrak{S}_n$, on the one hand, and the combinatorics of Young's lattice and Young tableaux, on the other. There is also an elegant combinatorial proof of Corollary 8.5, based on the *RSK algorithm* (after Gilbert de Beauregard Robinson, Craige Schensted, and Donald Knuth) or *Robinson–Schensted correspondence*, with many fascinating properties and with deep connections to representation theory. In the first Appendix at the end of this chapter we give a description of the RSK algorithm and the combinatorial proof of Corollary 8.5.

We now consider a variation of Theorem 8.4 in which we are not concerned with the type $w$ of a Hasse walk from $\emptyset$ to $\lambda$, but only with the number of steps. For instance, there are three Hasse walks of length three from $\emptyset$ to the partition 1, given by $\emptyset, 1, \emptyset, 1$; $\emptyset, 1, 2, 1$; and $\emptyset, 1, 11, 1$. Let $\beta(\ell, \lambda)$ denote the number of Hasse walks of length $\ell$ from $\emptyset$ to $\lambda$. Note the two following easy facts:

(F1)    $\beta(\ell, \lambda) = 0$ unless $\ell \equiv |\lambda| \pmod 2$.

(F2)    $\beta(\ell, \lambda)$ is the coefficient of $\lambda$ in the expansion of $(D + U)^\ell(\emptyset)$ as a linear combination of partitions.

Because of (F2) it is important to write $(D+U)^\ell$ as a linear combination of terms $U^i D^j$, just as in the proof of Theorem 8.4 we wrote a word $w$ in $U$ and $D$ in this form. Thus define integers $b_{ij}(\ell)$ by

$$(D + U)^\ell = \sum_{i,j} b_{ij}(\ell) U^i D^j. \tag{8.10}$$

Just as in the proof of Theorem 8.4, the numbers $b_{ij}(\ell)$ exist and are well defined.

**8.6 Lemma.**  *We have $b_{ij}(\ell) = 0$ if $\ell - i - j$ is odd. If $\ell - i - j = 2m$ then*

$$b_{ij}(\ell) = \frac{\ell!}{2^m\, i!\, j!\, m!}. \tag{8.11}$$

*Proof.* The assertion for $\ell - i - j$ odd is equivalent to (F1) above, so assume $\ell - i - j$ is even. The proof is by induction on $\ell$. It's easy to check that (8.11) holds for $\ell = 1$. Now assume true for some fixed $\ell \geq 1$. Using (8.10) we obtain

$$\sum_{i,j} b_{ij}(\ell + 1) U^i D^j = (D + U)^{\ell+1}$$

$$= (D + U) \sum_{i,j} b_{ij}(\ell) U^i D^j$$

$$= \sum_{i,j} b_{ij}(\ell)(DU^i D^j + U^{i+1} D^j).$$

In the proof of Theorem 8.4 we saw that $DU^i = U^i D + iU^{i-1}$ (see (8.6)). Hence we get

$$\sum_{i,j} b_{ij}(\ell + 1)U^i D^j = \sum_{i,j} b_{ij}(\ell)(U^i D^{j+1} + iU^{i-1}D^j + U^{i+1}D^j). \qquad (8.12)$$

As mentioned after (8.10), the expansion of $(D+U)^{\ell+1}$ in terms of $U^i D^j$ is unique. Hence equating coefficients of $U^i D^j$ on both sides of (8.12) yields the recurrence

$$b_{ij}(\ell + 1) = b_{i,j-1}(\ell) + (i + 1)b_{i+1,j}(\ell) + b_{i-1,j}(\ell). \qquad (8.13)$$

It is a routine matter to check that the function $\ell!/2^m i! j! m!$ satisfies the same recurrence (8.13) as $b_{ij}(\ell)$, with the same initial condition $b_{00}(0) = 1$. From this the proof follows by induction. $\qquad \Box$

From Lemma 8.6 it is easy to prove the following result.

**8.7 Theorem.** *Let $\ell \geq n$ and $\lambda \vdash n$, with $\ell - n$ even. Then*

$$\beta(\ell, \lambda) = \binom{\ell}{n}(1 \cdot 3 \cdot 5 \cdots (\ell - n - 1))f^\lambda.$$

*Proof.* Apply both sides of (8.10) to $\emptyset$. Since $U^i D^j(\emptyset) = 0$ unless $j = 0$, we get

$$(D + U)^\ell(\emptyset) = \sum_i b_{i0}(\ell)U^i(\emptyset)$$

$$= \sum_i b_{i0}(\ell) \sum_{\lambda \vdash i} f^\lambda \lambda.$$

Since by Lemma 8.6 we have $b_{i0}(\ell) = \binom{\ell}{i}(1 \cdot 3 \cdot 5 \cdots (\ell - i - 1))$ when $\ell - i$ is even, the proof follows from (F2). $\qquad \Box$

NOTE. The proof of Theorem 8.7 only required knowing the value of $b_{i0}(\ell)$. However, in Lemma 8.6 we computed $b_{ij}(\ell)$ for all $j$. We could have carried out the proof so as only to compute $b_{i0}(\ell)$, but the general value of $b_{ij}(\ell)$ is so simple that we have included it too.

**8.8 Corollary.** *The total number of Hasse walks in $Y$ of length $2m$ from $\emptyset$ to $\emptyset$ is given by*

$$\beta(2m, \emptyset) = 1 \cdot 3 \cdot 5 \cdots (2m - 1).$$

*Proof.* Simply substitute $\lambda = \emptyset$ (so $n = 0$) and $\ell = 2m$ in Theorem 8.7. $\qquad \Box$

The fact that we can count various kinds of Hasse walks in $Y$ suggests that there may be some finite graphs related to $Y$ whose eigenvalues we can also compute. This is indeed the case, and we will discuss the simplest case here. (See Exercise 8.21 for a generalization.) Let $Y_{j-1,j}$ denote the restriction of Young's lattice $Y$ to ranks $j-1$ and $j$. Identify $Y_{j-1,j}$ with its Hasse diagram, regarded as a (bipartite) graph. Let $p(i) = \#Y_i$, the number of partitions of $i$.

**8.9 Theorem.** *The eigenvalues of $Y_{j-1,j}$ are given as follows: 0 is an eigenvalue of multiplicity $p(j) - p(j-1)$; and for $1 \le s \le j$, the numbers $\pm\sqrt{s}$ are eigenvalues of multiplicity $p(j-s) - p(j-s-1)$.*

*Proof.* Let $A$ denote the adjacency matrix of $Y_{j-1,j}$. Since $\mathbb{R}Y_{j-1,j} = \mathbb{R}Y_{j-1} \oplus \mathbb{R}Y_j$ (vector space direct sum), any vector $v \in \mathbb{R}Y_{j-1,j}$ can be written uniquely as $v = v_{j-1} + v_j$, where $v_i \in \mathbb{R}Y_i$. The matrix $A$ acts on the vector space $\mathbb{R}Y_{j-1,j}$ as follows [why?]:

$$A(v) = D(v_j) + U(v_{j-1}). \tag{8.14}$$

Just as Theorem 4.7 followed from Lemma 4.6, we deduce from Lemma 8.3 that for any $i$ we have that $U_i : \mathbb{R}Y_i \to \mathbb{R}Y_{i+1}$ is one-to-one and $D_i : \mathbb{R}Y_i \to \mathbb{R}Y_{i-1}$ is onto. It follows in particular that

$$\dim \ker(D_i) = \dim \mathbb{R}Y_i - \dim \mathbb{R}Y_{i-1}$$
$$= p(i) - p(i-1),$$

where ker denotes kernel.

*Case 1.* Let $v \in \ker(D_j)$, so $v = v_j$. Then $Av = Dv = 0$. Thus $\ker(D_j)$ is an eigenspace of $A$ for the eigenvalue 0, so 0 is an eigenvalue of multiplicity at least $p(j) - p(j-1)$.

*Case 2.* Let $v \in \ker(D_s)$ for some $0 \le s \le j-1$. Let

$$v^* = \pm\sqrt{j-s}\, U^{j-1-s}(v) + U^{j-s}(v).$$

Note that $v^* \in \mathbb{R}Y_{j-1,j}$, with $v^*_{j-1} = \pm\sqrt{j-s}\, U^{j-1-s}(v)$ and $v^*_j = U^{j-s}(v)$. Using (8.6), we compute

$$A(v^*) = U(v^*_{j-1}) + D(v^*_j)$$
$$= \pm\sqrt{j-s}\, U^{j-s}(v) + DU^{j-s}(v)$$
$$= \pm\sqrt{j-s}\, U^{j-s}(v) + U^{j-s}D(v) + (j-s)U^{j-s-1}(v)$$
$$= \pm\sqrt{j-s}\, U^{j-s}(v) + (j-s)U^{j-s-1}(v)$$
$$= \pm\sqrt{j-s}\, v^*. \tag{8.15}$$

It's easy to verify (using the fact that $U$ is one-to-one) that if $v(1), \ldots, v(t)$ is a basis for $\ker(D_s)$, then $v(1)^*, \ldots, v(t)^*$ are linearly independent. Hence by (8.15) we have that $\pm\sqrt{j-s}$ is an eigenvalue of $A$ of multiplicity at least $t = \dim \ker(D_s) = p(s) - p(s-1)$.

We have found a total of

$$p(j) - p(j-1) + 2 \sum_{s=0}^{j-1} (p(s) - p(s-1)) = p(j-1) + p(j)$$

eigenvalues of $A$. (The factor 2 above arises from the fact that both $+\sqrt{j-s}$ and $-\sqrt{j-s}$ are eigenvalues.) Since the graph $Y_{j-1,j}$ has $p(j-1) + p(j)$ vertices, we have found all its eigenvalues.                                                               □

An elegant combinatorial consequence of Theorem 8.9 is the following.

**8.10 Corollary.** *Fix $j \geq 1$. The number of ways to choose a partition $\lambda$ of $j$, then delete a square from $\lambda$ (keeping it a partition), then insert a square, then delete a square, etc., for a total of $m$ insertions and $m$ deletions, ending back at $\lambda$, is given by*

$$\sum_{s=1}^{j} [p(j-s) - p(j-s-1)]s^m, \quad m > 0. \tag{8.16}$$

*Proof.* Exactly half the closed walks in $Y_{j-1,j}$ of length $2m$ begin at an element of $Y_j$ [why?]. Hence if $Y_{j-1,j}$ has eigenvalues $\theta_1, \ldots, \theta_r$, then by Corollary 1.3 the desired number of walks is given by $\frac{1}{2}(\theta_1^{2m} + \cdots + \theta_r^{2m})$. Using the values of $\theta_1, \ldots, \theta_r$ given by Theorem 8.9 yields (8.16).                                   □

For instance, when $j = 7$, (8.16) becomes $4 + 2 \cdot 2^m + 2 \cdot 3^m + 4^m + 5^m + 7^m$. When $m = 1$ we get 30, the number of edges of the graph $Y_{6,7}$ [why?].

## Appendix 1: The RSK Algorithm

We will describe a bijection between permutations $\pi \in \mathfrak{S}_n$ and pairs $(P, Q)$ of SYT of the same shape $\lambda \vdash n$. Define a *near Young tableau* (NYT) to be the same as an SYT, except that the entries can be any distinct integers, not necessarily the integers $1, 2, \ldots, n$. Let $P_{ij}$ denote the entry in row $i$ and column $j$ of $P$. The basic operation of the RSK algorithm consists of the *row insertion* $P \leftarrow k$ of a positive integer $k$ into an NYT $P = (P_{ij})$. The operation $P \leftarrow k$ is defined as follows: let $r$ be the least integer such that $P_{1r} > k$. If no such $r$ exists (i.e., all elements of the first row of $P$ are less than $k$), then simply place $k$ at the end of the first row. The insertion process stops, and the resulting NYT is $P \leftarrow k$. If, on the other hand, $r$ does exist then replace $P_{1r}$ by $k$. The element $k$ then "bumps" $P_{1r} := k'$ into the second row, i.e., insert $k'$ into the second row of $P$ by the insertion rule just described. Either $k'$

is inserted at the end of the second row, or else it bumps an element $k''$ to the third row. Continue until an element is inserted at the end of a row (possibly as the first element of a new row). The resulting array is $P \leftarrow k$.

**8.11 Example.** Let

$$
P = \begin{array}{lll}
3 & 7 & 9 & 14 \\
6 & 11 & 12 \\
10 & 16 \\
13 \\
15
\end{array}
$$

Then $P \leftarrow 8$ is shown below, with the elements inserted into each row (either by bumping or by the final insertion in the fourth row) in boldface. Thus the 8 bumps the 9, the 9 bumps the 11, the 11 bumps the 16, and the 16 is inserted at the end of a row. Hence

$$
(P \leftarrow 8) = \begin{array}{lll}
3 & 7 & \mathbf{8} & 14 \\
6 & \mathbf{9} & 12 \\
10 & \mathbf{11} \\
13 & \mathbf{16} \\
15
\end{array} \quad .
$$

We omit the proof, which is fairly straightforward, that if $P$ is an NYT, then so is $P \leftarrow k$. We can now describe the RSK algorithm. Let $\pi = a_1 a_2 \cdots a_n \in \mathfrak{S}_n$. We will inductively construct a sequence $(P_0, Q_0), (P_1, Q_1), \ldots, (P_n, Q_n)$ of pairs $(P_i, Q_i)$ of NYT of the same shape, where $P_i$ and $Q_i$ each have $i$ squares. First, define $(P_0, Q_0) = (\emptyset, \emptyset)$. If $(P_{i-1}, Q_{i-1})$ have been defined, then set $P_i = (P_{i-1} \leftarrow a_i)$. In other words, $P_i$ is obtained from $P_{i-1}$ by row inserting $a_i$. Now define $Q_i$ to be the NYT obtained from $Q_{i-1}$ by inserting $i$ so that $Q_i$ and $P_i$ have the same shape. (The entries of $Q_{i-1}$ don't change; we are simply placing $i$ into a certain new square and not row-inserting it into $Q_{i-1}$.) Finally let $(P, Q) = (P_n, Q_n)$. We write $\pi \xrightarrow{\text{RSK}} (P, Q)$.

**8.12 Example.** Let $\pi = 4273615 \in \mathfrak{S}_7$. The pairs $(P_1, Q_1), \ldots, (P_7, Q_7) = (P, Q)$ are as follows:

$$
\begin{array}{cc}
\underline{P_i} & \underline{Q_i} \\
\\
4 & 1 \\
\\
\\
\begin{array}{c} 2 \\ 4 \end{array} & \begin{array}{c} 1 \\ 2 \end{array}
\end{array}
$$

$$
\begin{array}{cc}
2\,7 & 1\,3 \\
4 & 2
\end{array}
$$

$$
\begin{array}{cc}
2\,3 & 1\,3 \\
4\,7 & 2\,4
\end{array}
$$

$$
\begin{array}{cc}
2\,3\,6 & 1\,3\,5 \\
4\,7 & 2\,4
\end{array}
$$

$$
\begin{array}{cc}
1\,3\,6 & 1\,3\,5 \\
2\,7 & 2\,4 \\
4 & 6
\end{array}
$$

$$
\begin{array}{cc}
1\,3\,5 & 1\,3\,5 \\
2\,6 & 2\,4 \\
4\,7 & 6\,7
\end{array}
$$

**8.13 Theorem.** *The RSK algorithm defines a bijection between the symmetric group* $\mathfrak{S}_n$ *and the set of all pairs* $(P, Q)$ *of SYT of the same shape, where the shape* $\lambda$ *is a partition of* $n$.

*Sketch.* The key step is to define the inverse of RSK. In other words, if $\pi \mapsto (P, Q)$, then how can we recover $\pi$ uniquely from $(P, Q)$? Moreover, we need to find $\pi$ for *any* $(P, Q)$. Observe that the position occupied by $n$ in $Q$ is the last position to be occupied in the insertion process. Suppose that $k$ occupies this position in $P$. It was bumped into this position by some element $j$ in the row above $k$ that is currently the largest element of its row less than $k$. Hence we can "inverse bump" $k$ into the position occupied by $j$, and now inverse bump $j$ into the row above it by the same procedure. Eventually an element will be placed in the first row, inverse bumping another element $t$ out of the tableau altogether. Thus $t$ was the last element of $\pi$ to be inserted, i.e., if $\pi = a_1 a_2 \cdots a_n$ then $a_n = t$. Now locate the position occupied by $n - 1$ in $Q$ and repeat the procedure, obtaining $a_{n-1}$. Continuing in this way, we uniquely construct $\pi$ one element at a time from right-to-left, such that $\pi \mapsto (P, Q)$. $\qquad\square$

The RSK-algorithm provides a bijective proof of Corollary 8.5, that is,

$$
\sum_{\lambda \vdash n} (f^\lambda)^2 = n!.
$$

## Appendix 2: Plane Partitions

In this appendix we show how a generalization of the RSK algorithm leads to an elegant generating function for a two-dimensional generalization of integer partitions. A *plane partition* of an integer $n \geq 0$ is a two-dimensional array

$\pi = (\pi_{ij})_{i,j \geq 1}$ of integers $\pi_{ij} \geq 0$ that is weakly decreasing in rows and columns, i.e.,

$$\pi_{ij} \geq \pi_{i+1,j}, \quad \pi_{ij} \geq \pi_{i,j+1},$$

such that $\sum_{i,j} \pi_{ij} = n$. It follows that all but finitely many $\pi_{ij}$ are 0, and these 0's are omitted in writing a particular plane partition $\pi$. Given a plane partition $\pi$, we write $|\pi| = n$ to denote that $\pi$ is a plane partition of $n$. More generally, if $L$ is any array of nonnegative integers we write $|L|$ for the sum of the parts (entries) of $L$.

There is one plane partition of 0, namely, all $\pi_{ij} = 0$, denoted $\emptyset$. The plane partitions of the integers $0 \leq n \leq 3$ are given by

$$\emptyset \quad 1 \quad 2 \quad 11 \quad \begin{matrix} 1 \\ 1 \end{matrix} \quad 3 \quad 21 \quad 111 \quad \begin{matrix} 11 \\ 1 \end{matrix} \quad \begin{matrix} 2 \\ 1 \end{matrix} \quad \begin{matrix} 1 \\ 1 \\ 1 \end{matrix}.$$

If $pp(n)$ denotes the number of plane partitions of $n$, then $pp(0) = 1$, $pp(1) = 1$, $pp(2) = 3$, and $pp(3) = 6$.

Our object is to give a formula for the generating function

$$F(x) = \sum_{n \geq 0} pp(n)x^n = 1 + x + 3x^2 + 6x^3 + 13x^4 + 24x^5 + \cdots.$$

More generally, we will consider plane partitions with at most $r$ rows and at most $s$ columns, i.e., $\pi_{ij} = 0$ for $i > r$ or $j > s$. As a simple warmup, let us first consider the case of ordinary partitions $\lambda = (\lambda_1, \lambda_2, \ldots)$ of $n$.

**8.14 Proposition.** *Let $p_s(n)$ denote the number of partitions of $n$ with at most $s$ parts. Equivalently, $p_s(n)$ is the number of plane partitions of $n$ with at most one row and at most $s$ columns [why?].Then*

$$\sum_{n \geq 0} p_s(n)x^n = \prod_{k=1}^{s}(1 - x^k)^{-1}.$$

*Proof.* First note that the partition $\lambda$ has at most $s$ parts if and only if the conjugate partition $\lambda'$ defined in Chapter 6 has largest part at most $s$. Thus it suffices to find the generating function $\sum_{n \geq 0} p'_s(n)x^n$, where $p'_s(n)$ denotes the number of partitions of $n$ whose largest part is at most $s$. Now expanding each factor $(1 - x^k)^{-1}$ as a geometric series gives

$$\prod_{k=1}^{s} \frac{1}{1 - x^k} = \prod_{k=1}^{s}\left(\sum_{m_k \geq 0} x^{m_k k}\right).$$

How do we get a coefficient of $x^n$? We must choose a term $x^{m_k k}$ from each factor of the product, $1 \le k \le s$, so that

$$n = \sum_{k=1}^{s} m_k k.$$

But such a choice is the same as choosing the partition $\lambda$ of $n$ such that the part $k$ occurs $m_k$ times. For instance, if $s = 4$ and we choose $m_1 = 5, m_2 = 0, m_3 = 1, m_4 = 2$, then we have chosen the partition $\lambda = (4, 4, 3, 1, 1, 1, 1, 1)$ of 16. Hence the coefficient of $x^n$ is the number of partitions $\lambda$ of $n$ whose largest part is at most $s$, as was to be proved.                                                                                    □

Note that Proposition 8.14 is "trivial" in the sense that it can be seen by inspection. There is an obvious correspondence between (a) the choice of terms contributing to the coefficient of $x^n$ and (b) partitions of $n$ with largest part at most $s$. Although the generating function we will obtain for plane partitions is equally simple, it will be far less obvious why it is correct.

Plane partitions have a certain similarity with standard Young tableaux, so perhaps it is not surprising that a variant of RSK will be applicable. Instead of NYT we will be dealing with *column-strict plane partitions* (CSPP). These are plane partitions for which the nonzero elements *strictly* decrease in each column. An example of a CSPP is given by

$$
\begin{array}{l}
7\,7\,4\,3\,3\,3\,1 \\
4\,3\,3\,1 \\
3\,2 \\
2\,1 \\
1
\end{array}
\tag{8.17}
$$

We say that this CSPP has *shape* $\lambda = (7, 4, 2, 2, 1)$, the shape of the Young diagram which the numbers occupy, and that it has five rows, seven columns, and 16 parts (so $\lambda \vdash 16$).

If $P = (P_{ij})$ is a CSPP and $k \ge 1$, then we define the *row insertion* $P \leftarrow k$ as follows: let $r$ be the least integer such that $P_{1,r} < k$. If no such $r$ exists (i.e., all elements of the first row of $P$ are greater than or equal to $k$), then simply place $k$ at the end of the first row. The insertion process stops, and the resulting CSPP is $P \leftarrow k$. If, on the other hand, $r$ does exist, then replace $P_{1r}$ by $k$. The element $k$ then "bumps" $P_{1r} := k'$ into the second row, i.e., insert $k'$ into the second row of $P$ by the insertion rule just described, possibly bumping a new element $k''$ into the third row. Continue until an element is inserted at the end of a row (possibly as the first element of a new row). The resulting array is $P \leftarrow k$. Note that this rule is completely analogous to row insertion for NYT: for NYT an element bumps the leftmost element greater than it, while for CSPP an element bumps the leftmost element smaller than it.

**8.15 Example.** Let $P$ be the CSPP of (8.17). Let us row insert 6 into $P$. The set of elements which get bumped are shown in bold:

$$
\begin{array}{l}
7\ 7\ \mathbf{4}\ 3\ 3\ 3\ 1 \\
4\ \mathbf{3}\ 3\ 1 \\
3\ \mathbf{2} \\
2\ \mathbf{1} \\
1
\end{array}
$$

The final 1 that was bumped is inserted at the end of the fifth row. Thus we obtain

$$
(P \leftarrow 6) = 
\begin{array}{l}
7\ 7\ 6\ 3\ 3\ 3\ 1 \\
4\ 4\ 3\ 1 \\
3\ 3 \\
2\ 2 \\
1\ 1
\end{array}
$$

We are now ready to describe the analogue of RSK needed to count plane partitions. Instead of beginning with a permutation $\pi \in \mathfrak{S}_n$, we begin with an $r \times s$ matrix $A = (a_{ij})$ of nonnegative integers, called for short an $r \times s$ $\mathbb{N}$-*matrix*. We convert $A$ into a *two-line array*

$$
w_A = \begin{pmatrix} u_1\ u_2\ \cdots\ u_N \\ v_1\ v_2\ \cdots\ v_N \end{pmatrix},
$$

where

- $u_1 \geq u_2 \geq \cdots \geq u_N$
- If $i < j$ and $u_i = u_j$, then $v_i \geq v_j$.
- The number of columns of $w_A$ equal to $\genfrac{}{}{0pt}{}{i}{j}$ is $a_{ij}$. (It follows that $N = \sum a_{ij}$.)

It is easy to see that $w_A$ is uniquely determined by $A$, and conversely. As an example, suppose that

$$
A = \begin{bmatrix} 0\ 1\ 0\ 2 \\ 1\ 1\ 1\ 0 \\ 2\ 1\ 0\ 0 \end{bmatrix}. \tag{8.18}
$$

Then

$$
w_A = \begin{pmatrix} 3\ 3\ 3\ 2\ 2\ 2\ 1\ 1\ 1 \\ 2\ 1\ 1\ 3\ 2\ 1\ 4\ 4\ 2 \end{pmatrix}.
$$

We now insert the numbers $v_1, v_2, \ldots, v_N$ successively into a CSPP. That is, we start with $P_0 = \emptyset$ and define inductively $P_i = P_{i-1} \leftarrow v_i$. We also start with

$Q_0 = \emptyset$, and at the $i$th step insert $u_i$ into $Q_{i-1}$ (without any bumping or other altering of the elements of $Q_{i-1}$) so that $P_i$ and $Q_i$ have the same shape. Finally let $(P, Q) = (P_N, Q_N)$ and write $A \xrightarrow{\text{RSK}'} (P, Q)$.

**8.16 Example.** Let $A$ be given by (8.18). The pairs $(P_1, Q_1), \ldots, (P_9, Q_9) = (P, Q)$ are as follows:

| $P_i$ | $Q_i$ |
|---|---|
| 2 | 3 |
| 2 1 | 3 3 |
| 2 1 1 | 3 3 3 |
| 3 1 1<br>2 | 3 3 3<br>2 |
| 3 2 1<br>2 1 | 3 3 3<br>2 2 |
| 3 2 1 1<br>2 1 | 3 3 3 2<br>2 2 |
| 4 2 1 1<br>3 1<br>2 | 3 3 3 2<br>2 2<br>1 |
| 4 4 1 1<br>3 2<br>2 1 | 3 3 3 2<br>2 2<br>1 1 |
| 4 4 2 1<br>3 2 1<br>2 1 | 3 3 3 2<br>2 2 1<br>1 1 |

It is straightforward to show that if $A \xrightarrow{\text{RSK}'} (P, Q)$, then $P$ and $Q$ are CSPP of the same shape. We omit the proof of the following key lemma, which is analogous to the proof of Theorem 8.13. Let us just note a crucial property (which is easy to prove) of the correspondence $A \xrightarrow{\text{RSK}'} (P, Q)$ which allows us to recover $A$ from $(P, Q)$, namely, equal entries of $Q$ are inserted from left to right. Thus the last number placed into $Q$ is the rightmost occurrence of the least entry. Hence we can inverse bump the number in this position in $P$ to back up one step in the algorithm, just as for the usual RSK correspondence $\pi \xrightarrow{\text{RSK}} (P, Q)$.

**8.17 Lemma.** *The correspondence* $A \xrightarrow{\text{RSK}'} (P, Q)$ *is a bijection from the set of* $r \times s$ *matrices of nonnegative integers to the set of pairs* $(P, Q)$ *of CSPP of the same shape, such that the largest part of P is at most s and the largest part of Q is at most r.*

The next step is to convert the pair $(P, Q)$ of CSPP of the same shape into a single plane partition $\pi$. We do this by "merging" the $i$th column of $P$ with the $i$th column of $Q$, producing the $i$th column of $\pi$. Thus we first describe how to merge two partitions $\lambda$ and $\mu$ with distinct parts and with the same number of parts into a single partition $\rho = \rho(\lambda, \mu)$. Draw the Ferrers diagram of $\lambda$ but with each row indented one space to the right of the beginning of the previous row. Such a diagram is called the *shifted* Ferrers diagram of $\lambda$. For instance, if $\lambda = (5, 3, 2)$ then we get the shifted diagram

$$
\begin{array}{ccccc}
\bullet & \bullet & \bullet & \bullet & \bullet \\
 & \bullet & \bullet & \bullet & \\
 & & \bullet & \bullet &
\end{array}
$$

Do the same for $\mu$, and then transpose the diagram. For instance, if $\mu = (6, 3, 1)$ then we get the transposed shifted diagram

$$
\begin{array}{ccc}
 & \bullet & \\
\bullet & \bullet & \\
\bullet & \bullet & \bullet \\
\bullet & \bullet & \\
\bullet & & \\
\bullet & &
\end{array}
$$

Now merge the two diagrams into a single diagram by identifying their main diagonals. For $\lambda$ and $\mu$ as above, we get the diagram (with the main diagonal drawn for clarity):

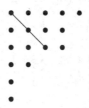

Define $\rho(\lambda, \mu)$ to be the partition for which this merged diagram is the Ferrers diagram. The above example shows that

$$
\rho(532, 631) = 544211.
$$

The map $(\lambda, \mu) \mapsto \rho(\lambda, \mu)$ is clearly a bijection between pairs of partitions $(\lambda, \mu)$ with $k$ distinct parts and partitions $\rho$ whose main diagonal (of the Ferrers diagram) has $k$ dots. Equivalently, $k$ is the largest integer $j$ for which $\rho_j \geq j$. Note that

$$|\rho| = |\lambda| + |\mu| - \ell(\lambda). \tag{8.19}$$

We now extend the above bijection to pairs $(P, Q)$ of reverse SSYT of the same shape. If $\lambda^i$ denotes the $i$th column of $P$ and $\mu^i$ the $i$th column of $Q$, then let $\pi(P, Q)$ be the array whose $i$th column is $\rho(\lambda^i, \mu^i)$. For instance, if

$$
P = \begin{matrix} 4 & 4 & 2 & 1 \\ 3 & 1 & 1 \\ 2 \end{matrix} \quad \text{and} \quad Q = \begin{matrix} 5 & 3 & 2 & 2 \\ 4 & 2 & 1 \\ 1 \end{matrix},
$$

then

$$
\pi(P, Q) = \begin{matrix} 4 & 4 & 2 & 1 \\ 4 & 2 & 2 & 1 \\ 4 & 2 \\ 2 \\ 2 \end{matrix} .
$$

It is easy to see that $\pi(P, Q)$ is a plane partition. Replace each row of $\pi(P, Q)$ by its conjugate to obtain another plane partition $\pi'(P, Q)$. With $\pi(P, Q)$ as above we obtain

$$
\pi'(P, Q) = \begin{matrix} 4 & 3 & 2 & 2 \\ 4 & 3 & 1 & 1 \\ 2 & 2 & 1 & 1 \\ 1 & 1 \\ 1 & 1 \end{matrix} .
$$

Write $|P|$ for the sum of the elements of $P$, and write $\max(P)$ for the largest element of $P$, and similarly for $Q$. When we merge $P$ and $Q$ into $\pi(P, Q)$, $\max(P)$ becomes the largest part of $\pi(P, Q)$. Thus when we conjugate each row, $\max(P)$ becomes the number $\mathrm{col}(\pi'(P, Q))$ of columns of $\pi'(P, Q)$ [why?]. Similarly, $\max(Q)$ becomes the number $\mathrm{row}(\pi'(P, Q))$ of rows of $\pi(P, Q)$ and of $\pi'(P, Q)$. In symbols,

$$
\begin{aligned}
\max P &= \mathrm{col}(\pi'(P, Q)) \\
\max Q &= \mathrm{row}(\pi'(P, Q)).
\end{aligned} \tag{8.20}
$$

Moreover, it follows from (8.19) that

$$|\pi'(P, Q)| = |\pi(P, Q)| = |P| + |Q| - v(P), \tag{8.21}$$

where $v(P)$ denotes the number of parts of $P$ (or of $Q$).

We now have all the ingredients necessary to prove the main result of this appendix.

**8.18 Theorem.** *Let* $pp_{rs}(n)$ *denote the number of plane partitions of* $n$ *with at most* $r$ *rows and at most* $s$ *columns. Then*

$$\sum_{n \geq 0} pp_{rs}(n)x^n = \prod_{i=1}^{r} \prod_{j=1}^{s} (1 - x^{i+j-1})^{-1}.$$

*Proof.* Let $A = (a_{ij})$ be an $r \times s$ $\mathbb{N}$-matrix. We can combine the bijections discussed above to obtain a plane partition $\pi(A)$ associated with $A$. Namely, first apply RSK to obtain $A \xrightarrow{\text{RSK}'} (P, Q)$, and then apply the merging process and row conjugation to obtain $\pi(A) = \pi'(P, Q)$. Since a column $\genfrac{}{}{0pt}{}{i}{j}$ of the two-line array $w_A$ occurs $a_{ij}$ times and results in an insertion of $j$ into $P$ and $i$ into $Q$, it follows that

$$|P| = \sum_{i,j} j a_{ij}$$

$$|Q| = \sum_{i,j} i a_{ij}$$

$$\max(P) = \max\{j : a_{ij} \neq 0\}$$

$$\max(Q) = \max\{i : a_{ij} \neq 0\}$$

Hence from (8.20) and (8.21), we see that the map $A \mapsto \pi(A)$ is a bijection from $r \times s$ $\mathbb{N}$-matrices $A$ to plane partitions with at most $r$ rows and at most $s$ columns. Moreover,

$$|\pi(A)| = |P| + |Q| - \nu(P)$$
$$= \sum_{i,j} (i + j - 1)a_{ij}.$$

Thus the enumeration of plane partitions is reduced to the much easier enumeration of $\mathbb{N}$-matrices. Specifically, we have

$$\sum_{n \geq 0} pp_{rs}(n)x^n = \sum_{\substack{\pi \\ \text{row}(\pi) \leq r \\ \text{col}(\pi) \leq s}} x^{|\pi|}$$

$$= \sum_{r \times s \ \mathbb{N}\text{-matrices } A} x^{\sum (i+j-1)a_{ij}}$$

$$= \prod_{i=1}^{r} \prod_{j=1}^{s} \left( \sum_{a_{ij} \geq 0} x^{\sum (i+j-1)a_{ij}} \right)$$

$$= \prod_{i=1}^{r} \prod_{j=1}^{s} (1 - x^{i+j-1})^{-1}.$$

$\square$

Write $pp_r(n)$ for the number of plane partitions of $n$ with at most $r$ rows. Letting $s \to \infty$ and then $r \to \infty$ in Theorem 8.18 produces the elegant generating functions of the next corollary.

**8.19 Corollary.** *We have*

$$\sum_{n \geq 0} pp_r(n) x^n = \prod_{i \geq 1} (1 - x^i)^{-\min(i,r)} \tag{8.22}$$

$$\sum_{n \geq 0} pp(n) x^n = \prod_{i \geq 1} (1 - x^i)^{-i}. \tag{8.23}$$

NOTE. Once one has seen the generating function

$$\frac{1}{(1-x)(1-x^2)(1-x^3)\cdots}$$

for one-dimensional (ordinary) partitions and the generating function

$$\frac{1}{(1-x)(1-x^2)^2(1-x^3)^3\cdots}$$

for two-dimensional (plane) partitions, it is quite natural to ask about higher-dimensional partitions. In particular, a *solid partition* of $n$ is a three-dimensional array $\pi = (\pi_{ijk})_{i,j,k \geq 1}$ of nonnegative integers, weakly decreasing in each of the three coordinate directions, and with elements summing to $n$. Let $sol(n)$ denote the number of solid partitions of $n$. It is easy to see that for any integer sequence $a_0 = 1$, $a_1, a_2, \ldots$, there are unique integers $b_1, b_2, \ldots$ for which

$$\sum_{n \geq 0} a_n x^n = \prod_{i \geq 1} (1 - x^i)^{-b_i}.$$

For the case $a_n = sol(n)$, we have

$$b_1 = 1, b_2 = 3, b_3 = 6, b_4 = 10, b_5 = 15,$$

which looks quite promising. Alas, the sequence of exponents continues

$$20, 26, 34, 46, 68, 97, 120, 112, 23, -186, -496, -735, -531, 779, \ldots.$$

The problem of enumerating solid partitions remains open and is considered most likely to be hopeless.

# Notes for Chapter 8

Standard Young tableaux (SYT) were first enumerated by MacMahon [89, p. 175] (see also [90, §103]). MacMahon formulated his result in terms of "generalized ballot sequences" or "lattice permutations" rather than SYT, but they are easily seen to be equivalent. He stated the result not in terms of the products of hook lengths as in Theorem 8.1, but as a more complicated product formula. The formulation in terms of hook lengths is due to Frame and appears first in the paper [45, Thm. 1] of Frame, Robinson, and Thrall; hence it is sometimes called the "Frame-Robinson-Thrall hook-length formula." (The actual definition of standard Young tableaux is due to Young [146, p. 258].)

Independently of MacMahon, Frobenius [48, eqn. (6)] obtained the same formula for the degree of the irreducible character $\chi^\lambda$ of $\mathfrak{S}_n$ as MacMahon obtained for the number of lattice permutations of type $\lambda$. Frobenius was apparently unaware of the combinatorial significance of deg $\chi^\lambda$, but Young showed in [146, pp. 260–261] that deg $\chi^\lambda$ was the number of SYT of shape $\lambda$, thereby giving an independent proof of MacMahon's result. (Young also provided his own proof of MacMahon's result in [146, Thm. II].)

A number of other proofs of the hook-length formula were subsequently found. Greene et al. [57] gave an elegant probabilistic proof. A proof of Hillman and Grassl [66] shows very clearly the role of hook lengths, though the proof is not completely bijective. A bijective version was later given by Krattenthaler [76]. Completely bijective proofs of the hook-length formula were first given by Franzblau and Zeilberger [46] and by Remmel [111]. An exceptionally elegant bijective proof was later found by Novelli et al. [97].

The use of the operators $U$ and $D$ to count walks in the Hasse diagram of Young's lattice was developed independently, in a more general context, by Fomin [43, 44] and Stanley [127, 129]. See also [130, §3.21] for a short exposition.

The RSK algorithm (known by a variety of other names, either "correspondence" or "algorithm" in connection with some subset of the names Robinson, Schensted, and Knuth) was first described, in a rather vague form, by Robinson [112, §5], as a tool in an attempted proof of a result now known as the "Littlewood–Richardson Rule." The RSK algorithm was later rediscovered by C.E. Schensted (see below), but no one actually analyzed Robinson's work until this was done by van Leeuwen [143, §7]. It is interesting to note that Robinson says in a footnote on page 754 that "I am indebted for this association I to Mr. D.E. Littlewood." Van Leeuwen's analysis makes it clear that "association I" gives the recording tableau $Q$ of the RSK algorithm $\pi \xrightarrow{\text{RSK}} (P, Q)$. Thus it might be correct to say that if $\pi \in \mathfrak{S}_n$ and $\pi \xrightarrow{\text{RSK}} (P, Q)$, then the definition of $P$ is due to Robinson, while the definition of $Q$ is due to Littlewood.

No further work related to Robinson's construction was done until Schensted published his seminal paper [115] in 1961. (For some information about the unusual life of Schensted, see [5].) Schensted's purpose was the enumeration of permutations in $\mathfrak{S}_n$ according to the length of their longest increasing and decreasing subsequences. According to Knuth [77, p. 726], the connection between the work of Robinson and that of Schensted was first pointed out by M.-P. Schützenberger, though as mentioned above the first person to describe this connection precisely was van Leeuwen.

Plane partitions were discovered by MacMahon in a series of papers which were not appreciated until much later. (See MacMahon's book [90, Sections IX and X] for an exposition of his results.) MacMahon's first paper dealing with plane partitions was [88]. In Article 43 of this paper he gives the definition of a plane partition (though not yet with that name). In Article 51 he conjectures that the generating function for plane partitions is the product

$$(1 - x)^{-1} (1 - x^2)^{-2} (1 - x^3)^{-3} (1 - x^4)^{-4} \cdots$$

(our (8.23)). In Article 52 he conjectures our (8.22) and Theorem 8.18, finally culminating in a conjectured generating function for plane partitions of $n$ with at most $r$ rows, at most $s$ columns, and with largest part at most $t$. (See Exercise 8.36.) MacMahon goes on in Articles 56–62 to prove his conjecture in the case of plane partitions with at most 2 rows and $s$ columns (the case $r = 2$ of our Theorem 8.18), mentioning on page 662 that an independent solution was obtained by A.R. Forsyth. (Though a publication reference is given to Forsyth's paper, apparently it never actually appeared.)

We will not attempt to describe MacMahon's subsequent work on plane partitions, except to say that the culmination of his work appears in [90, Art. 495], in which he proves his main conjecture from his first paper [88] on plane partitions, viz., our Exercise 8.36. MacMahon's proof is quite lengthy and indirect.

In 1972 Bender and Knuth [6] showed the connection between the theory of symmetric functions and the enumeration of plane partitions. They gave simple proofs based on the RSK algorithm of many results involving plane partitions, including the first bijective proof (the same proof that we give) of our Theorem 8.18. The process of merging two partitions with distinct parts into a single partition, discussed after Lemma 8.17, was first described by Frobenius [48] for a different purpose.

For further aspects of Young tableaux and the related topics of symmetric functions, representation theory of the symmetric group, Grassmann varieties, etc., see the expositions of Fulton [49], Sagan [114], and Stanley [131, Ch. 7].

# Exercises for Chapter 8

1. Draw all the standard Young tableaux of shape $(4, 2)$.
2. Using the hook-length formula, show that the number of SYT of shape $(n, n)$ is the Catalan number $C_n = \frac{1}{n+1} \binom{2n}{n}$.

3. How many maximal chains are in the poset $L(4, 4)$, where $L(m, n)$ is defined in Chapter 6? Express your answer in a form involving products and quotients of integers (no sums).

4. A *corner square* of a partition $\lambda$ is a square in the Young diagram of $\lambda$ whose removal results in the Young diagram of another partition (with the same upper-left corner). Let $c(\lambda)$ denote the number of corner squares (or distinct parts) of the partition $\lambda$. For instance, $c(5, 5, 4, 2, 2, 2, 1, 1) = 4$. (The distinct parts are $5, 4, 2, 1$.) Show that

$$\sum_{\lambda \vdash n} c(\lambda) = p(0) + p(1) + \cdots + p(n - 1),$$

   where $p(i)$ denotes the number of partitions of $i$ (with $p(0) = 1$). Try to give an elegant combinatorial proof.

5. Show that the number of odd hook lengths minus the number of even hook lengths of a partition $\lambda$ is a triangular number (a number of the form $k(k+1)/2$).

6. (moderately difficult) Show that the total number of SYT with $n$ entries and at most two rows is $\binom{n}{\lfloor n/2 \rfloor}$. Equivalently,

$$\sum_{i=0}^{\lfloor n/2 \rfloor} f^{(n-i,i)} = \binom{n}{\lfloor n/2 \rfloor}.$$

   Try to give an elegant combinatorial proof.

7. (difficult) (*) Let $f(n)$ be the number of partitions $\lambda$ of $2n$ whose Young diagram can be covered with $n$ nonoverlapping dominos (i.e., two squares with a common edge). For instance, the figure below shows a domino covering of the partition 43221.

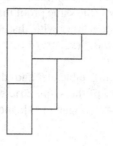

   Let

$$F(x) = \sum_{n \geq 0} f(n)x^n = 1 + 2x + 5x^2 + 10x^3 + 20x^4 + 36x^5 + \cdots .$$

Show that

$$F(x) = \prod_{n \geq 1}(1 - x^n)^{-2}.$$

8. (difficult) Let $\lambda$ be a partition. Let $m_k(\lambda)$ denote the number of parts of $\lambda$ that are equal to $k$, and let $\eta_k(\lambda)$ be the number of hooks of length $k$ of $\lambda$. Show that

$$\sum_{\lambda \vdash n} \eta_k(\lambda) = k \sum_{\lambda \vdash n} m_k(\lambda).$$

9. (moderately difficult) Let $\mu$ be a partition, and let $A_\mu$ be the infinite shape consisting of the quadrant $Q = \{(i, j) : i < 0, j > 0\}$ with the shape $\mu$ removed from the lower right-hand corner. Thus every square of $A_\mu$ has a finite hook and hence a hook length. For instance, when $\mu = (3, 1)$ we get the diagram

$$\vdots$$

|    | 10 | 9 | 8 | 6 | 5 | 3 |
|----|----|---|---|---|---|---|
|    | 9  | 8 | 7 | 5 | 4 | 2 |
| ···| 8  | 7 | 6 | 4 | 3 | 1 |
|    | 6  | 5 | 4 | 2 | 1 |   |
|    | 3  | 2 | 1 |   |   |   |

Show that the multiset of hook lengths of $A_\mu$ is equal to the union of the multiset of hook lengths of $Q$ (explicitly given by $\{1^1, 2^2, 3^3, \ldots\}$) and the multiset of hook lengths of $\mu$.

10. In how many ways can we begin with the empty partition $\emptyset$, then add $2n$ squares one at a time (always keeping a partition), then remove $n$ squares one at a time, then add $n$ squares one at a time, and finally remove $2n$ squares one at a time, ending up at $\emptyset$?

11. (difficult) Fix $n$. Show that the number of partitions $\lambda \vdash n$ for which $f^\lambda$ is odd is equal to $2^{k_1 + k_2 + \cdots}$, where $k_1 < k_2 < \cdots$ and $n = 2^{k_1} + 2^{k_2} + \cdots$ (the binary expansion of $n$). For instance, $75 = 2^0 + 2^1 + 2^3 + 2^6$, so the number of partitions $\lambda$ of 75 for which $f^\lambda$ is odd is $2^{6+3+1+0} = 1024$.

12. Let $U$ and $D$ be the linear transformations associated with Young's lattice. Write $D^2 U^2$ and $D^3 U^3$ in the form $\sum a_{ij} U^i D^j$.

13. Let $U$ and $D$ be the linear transformations associated with Young's lattice. Suppose that $f$ is some (noncommutative) polynomial in $U$ and $D$ satisfying $f(U, D) = 0$, e.g., $f(U, D) = DU - UD - I$. Let $i = \sqrt{-1}$. Show that $f(iD, iU) = 0$.

14. (*) Show that

$$U^n D^n = (UD - (n-1)I)(UD - (n-2)I) \cdots (UD - I)UD, \qquad (8.24)$$

where $U$ and $D$ are the linear transformations associated with Young's lattice (and $I$ is the identity transformation), and where both sides of (8.24) operate on the vector space $\mathbb{R}Y_j$ (for some fixed $j$).

15. (difficult) Give a bijective proof of Corollary 8.8, i.e., $\beta(2m, \emptyset) = 1 \cdot 3 \cdot 5 \cdots (2m-1)$. Your proof should be an analogue of the RSK algorithm. To start with, note that [why?] $1 \cdot 3 \cdot 5 \cdots (2m-1)$ is the number of *complete matchings* of $[2m]$, i.e., the number of graphs on the vertex set $[2m]$ with $m$ edges such that every vertex is incident to exactly one edge.

16. Fix a partition $\lambda \vdash n - 1$. Find a simple formula for the sum $t(\lambda) = \sum_{\mu > \lambda} f^\mu$ in terms of $f^\lambda$. The sum ranges over all partitions $\mu$ that cover $\lambda$ (i.e., $\mu > \lambda$ and nothing is in between, so $\mu \vdash n$) in Young's lattice $Y$. Give a simple proof using linear algebra rather than a combinatorial proof.

17. (a) (*) The *Bell number* $B(n)$ is defined to be the number of partitions of an $n$-element set $S$, i.e., the number of sets $\{B_1, \ldots, B_k\}$ where $B_i \neq \emptyset$, $B_i \cap B_j = \emptyset$ if $i \neq j$, and $\bigcup B_i = S$. Find a simple formula for the generating function

$$F(x) = \sum_{n \geq 0} B(n) \frac{x^n}{n!} = 1 + x + 2\frac{x^2}{2!} + 5\frac{x^3}{3!} + 15\frac{x^4}{4!} + \cdots .$$

(b) (moderately difficult) Let $f(n)$ be the number of ways to move from the empty partition $\emptyset$ to $\emptyset$ in $n$ steps, where each step consists of either (i) adding a box to the Young diagram, (ii) removing a box, or (iii) adding and then removing a box, always keeping the diagram of a partition (even in the middle of a step of type (iii)). For instance, $f(3) = 5$, corresponding to the five sequences

$$
\begin{array}{ccccc}
\emptyset & (1, \emptyset) & (1, \emptyset) & (1, \emptyset) & \\
\emptyset & (1, \emptyset) & 1 & \emptyset & \\
\emptyset & 1 & (2, 1) & \emptyset & . \\
\emptyset & 1 & (11, 1) & \emptyset & \\
\emptyset & 1 & \emptyset & (1, \emptyset) &
\end{array}
$$

Find (and prove) a formula for $f(n)$ in terms of Bell numbers.

18. (difficult) (*) For $n, k \geq 0$ let $\kappa(n \to n+k \to n)$ denote the number of *closed* walks in $Y$ that start at level $n$, go up $k$ steps to level $n+k$, and then go down $k$ steps to level $n$. Thus for instance $\kappa(n \to n+1 \to n)$ is the number of cover relations between levels $n$ and $n+1$. Show that

$$\sum_{n \geq 0} \kappa(n \to n+k \to n) q^n = k! \, (1-q)^{-k} F(Y, q).$$

Here $F(Y, q)$ is the rank-generating function of $Y$, which by Proposition 8.14 (letting $s \to \infty$) is given by

$$F(Y, q) = \prod_{i \geq 1}(1 - q^i)^{-1}.$$

19. Let $X$ denote the formal sum of all elements of Young's lattice $Y$. The operators $U$ and $D$ still act in the usual way on $X$, producing infinite linear combinations of elements of $Y$. For instance, the coefficient of the partition $(3, 1)$ in $DX$ is 3, coming from applying $D$ to $(4, 1)$, $(3, 2)$, and $(3, 1, 1)$.

    (a) Show that $DX = (U + I)X$, where as usual $I$ denotes the identity linear transformation.
    (b) Express the coefficient $s_n$ of $\emptyset$ (the empty partition) in $D^n X$ in terms of the numbers $f^\lambda$ for $\lambda \vdash n$. (For instance, $s_0 = s_1 = 1, s_2 = 2, s_3 = 4$.)
    (c) Show that

$$D^{n+1} X = (U D^n + D^n + n D^{n-1})X, \quad n \geq 0,$$

    where $D^{-1} = 0$, $D^0 = I$.
    (d) Find a simple recurrence relation satisfied by $s_n$.
    (e) Find a simple formula for the generating function

$$F(x) = \sum_{n \geq 0} s_n \frac{x^n}{n!}.$$

    (f) Show that $s_n$ is the number of involutions in $\mathfrak{S}_n$, i.e., the number of elements $\pi \in \mathfrak{S}_n$ satisfying $\pi^2 = \iota$.
    (g) (quite difficult) Show that if $\pi \in \mathfrak{S}_n$ and $\pi \xrightarrow{\text{RSK}} (P, Q)$, then $\pi^{-1} \xrightarrow{\text{RSK}} (Q, P)$.
    (h) Deduce your answer to (f) from (g).

20. (a) Consider the linear transformation $U_{n-1}D_n : \mathbb{R}Y_n \to \mathbb{R}Y_n$. Show that its eigenvalues are the integers $i$ with multiplicity $p(n - i) - p(n - i - 1)$, for $0 \leq i \leq n - 2$ and $i = n$.
    (b) (*) Use (a) to give another proof of Theorem 8.9.

21. (a) (moderately difficult) Let $Y_{[j-2,j]}$ denote the Hasse diagram of the restriction of Young's lattice $Y$ to the levels $j - 2$, $j - 1$, $j$. Let $p(n)$ denote the number of partitions of $n$, and write $\Delta p(n) = p(n) - p(n - 1)$. Show that the characteristic polynomial of the adjacency matrix of the graph $Y_{[j-2,j]}$ is given by

$$\pm x^{\Delta p(j)}(x^2 - 1)^{\Delta p(j-1)} \prod_{s=2}^{j}(x^3 - (2s - 1)x)^{\Delta p(j-1)},$$

    where the sign is $(-1)^{\#Y_{[j-2,j]}} = (-1)^{p(j-2)+p(j-1)+p(j)}$.

(b) (difficult) Extend to $Y_{[j-i,j]}$ for any $i \geq 0$. Express your answer in terms of the characteristic polynomial of matrices of the form

$$
\begin{bmatrix}
0 & a & 0 & & & 0 & 0 \\
1 & 0 & a+1 & & & 0 & 0 \\
0 & 1 & 0 & & \ddots & 0 & 0 \\
 & & & \ddots & & & \\
 & & & & \ddots & 0 & b \\
 & & & & & 1 & 0
\end{bmatrix}.
$$

22. (moderately difficult)

(a) Let $U$ and $D$ be operators (or just noncommutative variables) satisfying $DU - UD = I$. Show that for any power series $f(U) = \sum a_n U^n$ whose coefficients $a_n$ are real numbers, we have

$$
e^{Dt} f(U) = f(U + t) e^{Dt}.
$$

In particular,

$$
e^{Dt} e^{U} = e^{t+U} e^{Dt}. \tag{8.25}
$$

Here $t$ is a variable (indeterminate) commuting with $U$ and $D$. Regard both sides as power series in $t$ whose coefficients are (noncommutative) polynomials in $U$ and $D$. Thus for instance

$$
e^{Dt} e^{U} = \left( \sum_{m \geq 0} \frac{D^n t^n}{n!} \right) \left( \sum_{n \geq 0} \frac{U^n}{n!} \right)
$$

$$
= \sum_{m,n \geq 0} \frac{D^m U^n t^m}{m! \, n!}.
$$

(b) Show that $e^{(U+D)t} = e^{\frac{1}{2}t^2 + Ut} e^{Dt}$.

(c) Let $\delta_n$ be the total number of walks of length $n$ in Young's lattice $Y$ (i.e., in the Hasse diagram of $Y$) starting at $\emptyset$. For instance, $\delta_2 = 3$ corresponding to the walks $(\emptyset, 1, 2)$, $(\emptyset, 1, 11)$, and $(\emptyset, 1, \emptyset)$. Find a simple formula for the generating function $F(t) = \sum_{n \geq 0} \delta_n \frac{t^n}{n!}$.

23. Let $w$ be a *balanced* word in $U$ and $D$, i.e., the same number of $U$'s as $D$'s. For instance, $UUDUDDDU$ is balanced. Regard $U$ and $D$ as linear transformations on $\mathbb{R}Y$ in the usual way. A balanced word thus takes the space $\mathbb{R}Y_n$ to itself, where $Y_n$ is the $n$th level of Young's lattice $Y$. Show that the element $E_n = \sum_{\lambda \vdash n} f^\lambda \lambda \in \mathbb{R}Y_n$ is an eigenvector for $w$, and find the eigenvalue.

24. (*) Prove that any two balanced words (as defined in the previous exercise) commute.

25. Define a graded poset $Z$ inductively as follows. The bottom level $Z_0$ consists of a single element. Assume that we have constructed the poset up to level $n$. First "reflect" $Z_{n-1}$ through $Z_n$. More precisely, for each element $x \in Z_{n-1}$, let $x'$ be a new element of $Z_{n+1}$, with $x' > y$ (where $y \in Z_n$) if and only if $y > x$. Then for each element $y \in Z_n$, let $y'$ be a new element of $Z_{n+1}$ covering $y$ (and covering no other elements of $Z_n$). Figure 8.1 shows the poset $Z$ up to level 5. The cover relations obtained by the reflection construction are shown by solid lines, while those of the form $y' > y$ are shown by broken lines.

(a) Show that $\#Z_n = F_{n+1}$ (a Fibonacci number), so the rank-generating function of $Z$ is given by

$$F(Z, q) = \frac{1}{1 - q - q^2}.$$

(b) Define $U_i : \mathbb{R}Z_i \to \mathbb{R}Z_{i+1}$ and $D_i : \mathbb{R}Z_i \to \mathbb{R}Z_{i-1}$ exactly as we did for $Y$, namely, for $x \in Z_i$ we have

$$U_i(x) = \sum_{y > x} y$$

$$D_i(x) = \sum_{y < x} y.$$

Show that $D_{i+1}U_i - U_{i-1}D_i = I_i$. Thus all the results we have obtained for $Y$ based on this commutation relation also hold for $Z$! (For results involving $p(n)$, we need only replace $p(n)$ by $F_{n+1}$.)

26. (a) Suppose that $\pi \in \mathfrak{S}_n$ and $\pi \overset{\text{RSK}}{\longrightarrow} (P, Q)$. Let $f(\pi)$ be the largest integer $k$ for which $1, 2, \ldots, k$ all appear in the first row of $P$. Find a simple formula for the number of permutations $\pi \in \mathfrak{S}_n$ for which $f(\pi) = k$.

**Fig. 8.1** The poset $Z$ up to level 5

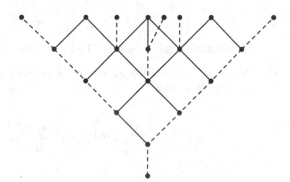

(b) Let $E(n)$ denote the expected value of $f(\pi)$ for $\pi \in \mathfrak{S}_n$, i.e.,

$$E(n) = \frac{1}{n!} \sum_{\pi \in \mathfrak{S}_n} f(\pi).$$

Find $\lim_{n \to \infty} E(n)$.

27. Suppose that $\pi \in \mathfrak{S}_n$ and $\pi \xrightarrow{\text{RSK}} (P, Q)$. Let $E_{12}(n)$ be the expected value of the $(1, 2)$-entry of $P$ (i.e., the second entry in the first row). Find $\lim_{n \to \infty} E_{12}(n)$.

28. (a) An *increasing subsequence* of a permutation $a_1 a_2 \cdots a_n \in \mathfrak{S}_n$ is a subsequence $a_{i_1} a_{i_2} \cdots a_{i_j}$ such that $a_{i_1} < a_{i_2} < \cdots < a_{i_j}$. For instance, 2367 is an increasing subsequence of the permutation 52386417. Suppose that the permutation $w \in \mathfrak{S}_n$ is sent into an SYT of shape $\lambda = (\lambda_1, \lambda_2, \dots)$ under the RSK algorithm. Show that $\lambda_1$ is the length of the longest increasing subsequence of $w$.

(b) (much harder) Define *decreasing subsequence* similarly to increasing subsequence. Show that $\lambda_1'$ (the number of parts of $\lambda$) is equal to the length of the longest decreasing subsequence of $\lambda$.

(c) Assuming (a) and (b), show that for $m, n \geq 1$, a permutation $w \in \mathfrak{S}_{mn+1}$ has an increasing subsequence of length $m + 1$ or a decreasing subsequence of length $n + 1$.

(d) How many permutations $w \in \mathfrak{S}_{mn}$ have longest increasing subsequence of length $m$ and longest decreasing subsequence of length $n$? (Use the hook length formula to obtain a simple explicit answer.)

29. Write down the 13 plane partitions of 4 and the 24 plane partitions of 5.

30. Prove the statement preceding Lemma 8.17 that in the bijection $A \xrightarrow{\text{RSK}'} (P, Q)$, equal elements of $Q$ are inserted from left to right.

31. Let $A$ be the $r \times s$ matrix of all 1's. Describe the plane partition $\pi'(A)$.

32. (a) Find the $\mathbb{N}$-matrix $A$ for which

$$\pi'(A) = \begin{matrix} 6 & 4 & 4 & 3 & 3 \\ 5 & 3 & 3 & 2 \\ 3 & 2 & 1 \end{matrix} \ .$$

(b) What message is conveyed by the nonzero entries of $A$?

33. (*) Let $f(n)$ denote the number of plane partitions $\pi = (\pi_{ij})$ of $n$ for which $\pi_{22} = 0$. Show that

$$\sum_{n \geq 0} f(n) x^n = \frac{\sum_{n \geq 0} (-1)^n x^{\binom{n+1}{2}}}{\prod_{i \geq 1} (1 - x^i)^2}.$$

34. (a) (quite difficult) Let $A$ be an $r \times s$ $\mathbb{N}$-matrix, and let $A \xrightarrow{\text{RSK}'} (P, Q)$. If $A^t$ denotes the transpose of $A$, then show that $A^t \xrightarrow{\text{RSK}'} (Q, P)$.

NOTE. This result is quite difficult to prove from first principles. If you can do Exercise 8.19(g), then the present exercise is a straightforward modification. In fact, it is possible to deduce the present exercise from Exercise 8.19(g).

(b) A plane partition $\pi = (\pi_{ij})$ is *symmetric* if $\pi_{ij} = \pi_{ji}$ for all $i$ and $j$. Let $s_r(n)$ denote the number of symmetric plane partitions of $n$ with at most $r$ rows. Assuming (a), show that

$$\sum_{n \geq 0} s_r(n) x^n = \prod_{i=1}^{r} \left(1 - x^{2i-1}\right)^{-1} \cdot \prod_{1 \leq i < j \leq r} \left(1 - x^{2(i+j-1)}\right)^{-1}.$$

(c) Let $s(n)$ denote the total number of symmetric plane partitions of $n$. Let $r \to \infty$ in (b) to deduce that

$$\sum_{n \geq 0} s(n) x^n = \prod_{i \geq 1} \frac{1}{(1 - x^{2i-1})(1 - x^{2i})^{\lfloor i/2 \rfloor}}.$$

(d) (very difficult; cannot be done using RSK) Let $s_{rt}(n)$ denote the number of symmetric plane partitions of $n$ with at most $r$ rows and with largest part at most $t$. Show that

$$\sum_{n \geq 0} s_{rt}(n) x^n = \prod_{1 \leq i < j \leq r} \prod_{k=1}^{t} \frac{1 - x^{(2-\delta_{ij})(i+j+k-1)}}{1 - x^{(2-\delta_{ij})(i+j+k-2)}}.$$

35. The *trace* of a plane partition $\pi = (\pi_{ij})$ is defined as $\text{tr}(\pi) = \sum_i \pi_{ii}$. Let $pp(n, k)$ denote the number of plane partitions of $n$ with trace $k$. Show that

$$\sum_{n \geq 0} \sum_{k \geq 0} pp(n, k) q^k x^n = \prod_{i \geq 1} (1 - q x^i)^{-i}.$$

36. (very difficult; cannot be done using RSK) Let $pp_{rst}(n)$ be the number of plane partitions of $n$ with at most $r$ rows, at most $s$ columns, and with largest part at most $t$. Show that

$$\sum_{n \geq 0} pp_{rst}(n) x^n = \prod_{i=1}^{r} \prod_{j=1}^{s} \prod_{k=1}^{t} \frac{1 - x^{i+j+k-1}}{1 - x^{i+j+k-2}}.$$

37. Let $f(n)$ denote the number of solid partitions of $n$ with largest part at most 1. Find the generating function $F(x) = \sum_{n \geq 0} f(n) x^n$.

# Chapter 9
# The Matrix-Tree Theorem

The Matrix-Tree Theorem is a formula for the number of spanning trees of a graph in terms of the determinant of a certain matrix. We begin with the necessary graph-theoretical background. Let $G$ be a finite graph, allowing multiple edges but not loops. (Loops could be allowed, but they turn out to be completely irrelevant.) Recall that $G$ is *connected* if there exists a walk between any two vertices of $G$. A *cycle* is a closed walk with no repeated vertices or edges, except for the first and last vertex. A *tree* is a connected graph with no cycles. In particular, a tree cannot have multiple edges, since a double edge is equivalent to a cycle of length two. The three nonisomorphic trees with five vertices are shown in Figure 9.1.

A basic theorem of graph theory (whose easy proof we leave as an exercise) is the following.

**9.1 Proposition.** *Let $G$ be a graph with $p$ vertices. The following conditions are equivalent:*

*(a)* *$G$ is a tree.*
*(b)* *$G$ is connected and has $p - 1$ edges.*
*(c)* *$G$ has no cycles and has $p - 1$ edges.*
*(d)* *There is a unique path (= walk with no repeated vertices) between any two vertices.*

A *spanning subgraph* of a graph $G$ is a graph $H$ with the same vertex set as $G$, and such that every edge of $H$ is an edge of $G$. If $G$ has $q$ edges, then the number of spanning subgraphs of $G$ is equal to $2^q$, since we can choose any subset of the edges of $G$ to be the set of edges of $H$. (Note that multiple edges between the same two vertices are regarded as *distinguishable*, in accordance with the definition of a graph in Chapter 1.) A spanning subgraph which is a tree is called a *spanning tree*. Clearly $G$ has a spanning tree if and only if it is connected [why?]. An important invariant of a graph $G$ is its number of spanning trees, called the *complexity* of $G$ and denoted $\kappa(G)$.

© Springer International Publishing AG, part of Springer Nature 2018
R. P. Stanley, *Algebraic Combinatorics*, Undergraduate Texts in Mathematics,
https://doi.org/10.1007/978-3-319-77173-1_9

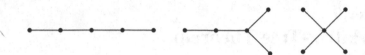

**Fig. 9.1** The three trees with five vertices

**9.2 Example.** Let $G$ be the graph illustrated below, with edges $a, b, c, d$, and $e$.

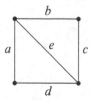

Then $G$ has eight spanning trees, namely, $abc, abd, acd, bcd, abe, ace, bde$, and $cde$ (where, e.g., $abc$ denotes the spanning subgraph with edge set $\{a, b, c\}$).

**9.3 Example.** Let $G = K_5$, the complete graph on five vertices. A simple counting argument shows that $K_5$ has 60 spanning trees isomorphic to the first tree in Figure 9.1, 60 isomorphic to the second tree, and 5 isomorphic to the third tree. Hence $\kappa(K_5) = 125$. It is even easier to verify that $\kappa(K_1) = 1$, $\kappa(K_2) = 1$, $\kappa(K_3) = 3$, and $\kappa(K_4) = 16$. Can the reader make a conjecture about the value of $\kappa(K_p)$ for any $p \geq 1$?

Our object is to obtain a "determinantal formula" for $\kappa(G)$. For this we need an important result from matrix theory, known as the *Binet–Cauchy theorem* or *Cauchy–Binet theorem* and which is often omitted from a beginning linear algebra course. Later (Theorem 10.4) we will prove a more general determinantal formula without the use of the Binet–Cauchy theorem. However, the use of the Binet–Cauchy theorem does afford some additional algebraic insight. The Binet–Cauchy theorem is a generalization of the familiar fact that if $A$ and $B$ are $n \times n$ matrices, then $\det AB = (\det A)(\det B)$, where $\det$ denotes determinant. We want to extend this formula to the case where $A$ and $B$ are rectangular matrices whose product is a square matrix (so that $\det AB$ is defined). In other words, $A$ will be an $m \times n$ matrix and $B$ an $n \times m$ matrix, for some $m, n \geq 1$.

We will use the following notation involving submatrices. Suppose $A = (a_{ij})$ is an $m \times n$ matrix, with $1 \leq i \leq m$, $1 \leq j \leq n$, and $m \leq n$. Given an $m$-element subset $S$ of $\{1, 2, \ldots, n\}$, let $A[S]$ denote the $m \times m$ submatrix of $A$ obtained by taking the columns indexed by the elements of $S$. In other words, if the elements of $S$ are given by $j_1 < j_2 < \cdots < j_m$, then $A[S] = (a_{i, j_k})$, where $1 \leq i \leq m$ and $1 \leq k \leq m$. For instance, if

$$
A = \begin{bmatrix}
1 & 2 & 3 & 4 & 5 \\
6 & 7 & 8 & 9 & 10 \\
11 & 12 & 13 & 14 & 15
\end{bmatrix}
$$

and $S = \{2, 3, 5\}$, then

$$A[S] = \begin{bmatrix} 2 & 3 & 5 \\ 7 & 8 & 10 \\ 12 & 13 & 15 \end{bmatrix}.$$

Similarly, let $B = (b_{ij})$ be an $n \times m$ matrix with $1 \leq i \leq n$, $1 \leq j \leq m$, and $m \leq n$. Let $S$ be an $m$-element subset of $\{1, 2, \ldots, n\}$ as above. Then $B[S]$ denotes the $m \times m$ matrix obtained by taking the *rows* of $B$ indexed by $S$. Note that $A^t[S] = A[S]^t$, where $^t$ denotes transpose.

**9.4 Theorem** (the Binet–Cauchy Theorem). *Let $A = (a_{ij})$ be an $m \times n$ matrix, with $1 \leq i \leq m$ and $1 \leq j \leq n$. Let $B = (b_{ij})$ be an $n \times m$ matrix with $1 \leq i \leq n$ and $1 \leq j \leq m$. (Thus $AB$ is an $m \times m$ matrix.) If $m > n$, then $\det(AB) = 0$. If $m \leq n$, then*

$$\det(AB) = \sum_S (\det A[S])(\det B[S]),$$

*where $S$ ranges over all $m$-element subsets of $\{1, 2, \ldots, n\}$.*

Before proceeding to the proof, let us give an example. We write $|a_{ij}|$ for the determinant of the matrix $(a_{ij})$. Suppose

$$A = \begin{bmatrix} a_1 & a_2 & a_3 \\ b_1 & b_2 & b_3 \end{bmatrix}, \quad B = \begin{bmatrix} c_1 & d_1 \\ c_2 & d_2 \\ c_3 & d_3 \end{bmatrix}.$$

Then

$$\det AB = \begin{vmatrix} a_1 & a_2 \\ b_1 & b_2 \end{vmatrix} \cdot \begin{vmatrix} c_1 & d_1 \\ c_2 & d_2 \end{vmatrix} + \begin{vmatrix} a_1 & a_3 \\ b_1 & b_3 \end{vmatrix} \cdot \begin{vmatrix} c_1 & d_1 \\ c_3 & d_3 \end{vmatrix} + \begin{vmatrix} a_2 & a_3 \\ b_2 & b_3 \end{vmatrix} \cdot \begin{vmatrix} c_2 & d_2 \\ c_3 & d_3 \end{vmatrix}.$$

sketch. First suppose $m > n$. Since from linear algebra we know that rank $AB \leq$ rank $A$ and that the rank of an $m \times n$ matrix cannot exceed $n$ (or $m$), we have that rank $AB \leq n < m$. But $AB$ is an $m \times m$ matrix, so $\det AB = 0$, as claimed.

Now assume $m \leq n$. We use notation such as $M_{rs}$ to denote an $r \times s$ matrix $M$. It is an immediate consequence of the definition of matrix multiplication (which the reader should check) that

$$\begin{bmatrix} R_{mm} & S_{mn} \\ T_{nm} & U_{nn} \end{bmatrix} \begin{bmatrix} V_{mn} & W_{mm} \\ X_{nn} & Y_{nm} \end{bmatrix} = \begin{bmatrix} RV + SX & RW + SY \\ TV + UX & TW + UY \end{bmatrix}. \tag{9.1}$$

In other words, we can multiply "block" matrices of suitable dimensions as if their entries were numbers. Note that the entries of the right-hand side of (9.1) all have well-defined dimensions (sizes), e.g., $RV + SX$ is an $m \times n$ matrix since both $RV$ and $SX$ are $m \times n$ matrices.

Now in (9.1) let $R = I_m$ (the $m \times m$ identity matrix), $S = A$, $T = O_{nm}$ (the $n \times m$ matrix of 0's), $U = I_n$, $V = A$, $W = O_{mm}$, $X = -I_n$, and $Y = B$. We get

$$\begin{bmatrix} I_m & A \\ O_{nm} & I_n \end{bmatrix} \begin{bmatrix} A & O_{mm} \\ -I_n & B \end{bmatrix} = \begin{bmatrix} O_{mn} & AB \\ -I_n & B \end{bmatrix}. \tag{9.2}$$

Take the determinant of both sides of (9.2). The first matrix on the left-hand side is upper triangular with 1's on the main diagonal. Hence its determinant is one. Since the determinant of a product of square matrices is the product of the determinants of the factors, we get

$$\begin{vmatrix} A & O_{mm} \\ -I_n & B \end{vmatrix} = \begin{vmatrix} O_{mn} & AB \\ -I_n & B \end{vmatrix}. \tag{9.3}$$

It is easy to see [why?] that the determinant on the right-hand side of (9.3) is equal to $\pm \det AB$. So consider the left-hand side. A nonzero term in the expansion of the determinant on the left-hand side is obtained by taking the product (with a certain sign) of $m + n$ nonzero entries, no two in the same row and column (so one in each row and each column). In particular, we must choose $m$ entries from the last $m$ columns. These entries belong to $m$ of the bottom $n$ rows [why?], say rows $m + s_1, m + s_2, \ldots, m + s_m$. Let $S = \{s_1, s_2, \ldots, s_m\} \subseteq \{1, 2, \ldots, n\}$. We must choose $n - m$ further entries from the last $n$ rows, and we have no choice but to choose the $-1$'s in those rows $m + i$ for which $i \notin S$. Thus every term in the expansion of the left-hand side of (9.3) uses exactly $n - m$ of the $-1$'s in the bottom left block $-I_n$.

What is the contribution to the expansion of the left-hand side of (9.3) from those terms which use exactly the $-1$'s from rows $m + i$ where $i \notin S$? We obtain this contribution by deleting all rows and columns to which these $-1$'s belong (in other words, delete row $m + i$ and column $i$ whenever $i \in \{1, 2, \ldots, n\} - S$), taking the determinant of the $2m \times 2m$ matrix $M_S$ that remains, and multiplying by an appropriate sign [why?]. But the matrix $M_S$ is in block-diagonal form, with the first block just the matrix $A[S]$ and the second block just $B[S]$. Hence $\det M_S = (\det A[S])(\det B[S])$ [why?]. Taking all possible subsets $S$ gives

$$\det AB = \sum_{\substack{S \subseteq \{1,2,\ldots,n\} \\ |S|=m}} \pm (\det A[S])(\det B[S]).$$

It is straightforward but somewhat tedious to verify that all the signs are $+$; we omit the details. This completes the proof.                                                    □

In Chapter 1 we defined the adjacency matrix $A(G)$ of a graph $G$ with vertex set $V = \{v_1, \ldots, v_p\}$ and edge set $E = \{e_1, \ldots, e_q\}$. We now define two related matrices. Continue to assume that $G$ has no loops. (This assumption is harmless since loops have no effect on $\kappa(G)$.)

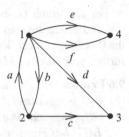

**9.5 Definition.** Let $G$ be as above. Give $G$ an *orientation* o, i.e., for every edge $e$ with vertices $u$, $v$, choose one of the ordered pairs $(u, v)$ or $(v, u)$. If we choose $(u, v)$, say, then we think of putting an arrow on $e$ pointing from $u$ to $v$; and we say that $e$ is directed from $u$ to $v$, that $u$ is the *initial vertex* and $v$ the *final vertex* of $e$, etc.

(a) The *incidence matrix* $M(G)$ of $G$ (with respect to the orientation o) is the $p \times q$ matrix whose $(i, j)$-entry $M_{ij}$ is given by

$$M_{ij} = \begin{cases} -1, & \text{if the edge } e_j \text{ has initial vertex } v_i \\ 1, & \text{if the edge } e_j \text{ has final vertex } v_i \\ 0, & \text{otherwise.} \end{cases}$$

(b) The *Laplacian matrix* $L(G)$ of $G$ is the $p \times p$ matrix whose $(i, j)$-entry $L_{ij}$ is given by

$$L_{ij} = \begin{cases} -m_{ij}, & \text{if } i \neq j \text{ and there are } m_{ij} \text{ edges between } v_i \text{ and } v_j \\ \deg(v_i), & \text{if } i = j, \end{cases}$$

where $\deg(v_i)$ is the number of edges incident to $v_i$. Note that $L(G)$ is symmetric and does not depend on the orientation o.

As an example, let $(G, o)$ be the oriented graph of Figure 9.2. Then

$$M(G) = \begin{bmatrix} 1 & -1 & 0 & -1 & -1 & -1 \\ -1 & 1 & -1 & 0 & 0 & 0 \\ 0 & 0 & 1 & 1 & 0 & 0 \\ 0 & 0 & 0 & 0 & 1 & 1 \end{bmatrix}$$

$$L(G) = \begin{bmatrix} 4 & -2 & -1 & -2 \\ -2 & 3 & -1 & 0 \\ -1 & -1 & 2 & 0 \\ -2 & 0 & 0 & 2 \end{bmatrix}.$$

For any graph $G$, every column of $M(G)$ contains one 1, one $-1$, and $q - 2$ 0's; and hence the sum of the entries in each column is 0. Thus all the rows sum to the 0 vector, a linear dependence relation which shows that rank($M(G)$) $< p$. Two further properties of $M(G)$ and $L(G)$ are given by the following lemma.

**9.6 Lemma.** *(a)* *We have $MM^t = L$.*
*(b)* *If $G$ is regular of degree $d$, then $L(G) = dI - A(G)$, where $A(G)$ denotes the adjacency matrix of $G$. Hence if $G$ (or $A(G)$) has eigenvalues $\lambda_1, \ldots, \lambda_p$, then $L(G)$ has eigenvalues $d - \lambda_1, \ldots, d - \lambda_p$.*

*Proof.* (a) This is immediate from the definition of matrix multiplication. Specifically, for $v_i, v_j \in V(G)$ we have

$$(MM^t)_{ij} = \sum_{e_k \in E(G)} M_{ik} M_{jk}.$$

If $i \neq j$, then in order for $M_{ik} M_{jk} \neq 0$, we must have that the edge $e_k$ connects the vertices $v_i$ and $v_j$. If this is the case, then one of $M_{ik}$ and $M_{jk}$ will be 1 and the other $-1$ [why?], so their product is always $-1$. Hence $(MM^t)_{ij} = -m_{ij}$, as claimed.

There remains the case $i = j$. Then $M_{ik} M_{ik}$ will be 1 if $e_k$ is an edge with $v_i$ as one of its vertices and will be 0 otherwise [why?]. So now we get $(MM^t)_{ii} = \deg(v_i)$, as claimed. This proves (a).
(b) Clear by (a), since the diagonal elements of $MM^t$ are all equal to $d$.     □

Now assume that $G$ is connected, and let $M_0(G)$ be $M(G)$ with its last row removed. Thus $M_0(G)$ has $p - 1$ rows and $q$ columns. Note that the number of rows is equal to the number of edges in a spanning tree of $G$. We call $M_0(G)$ the *reduced incidence matrix* of $G$. The next result tells us the determinants (up to sign) of all $(p - 1) \times (p - 1)$ submatrices $N$ of $M_0$. Such submatrices are obtained by choosing a set $X = \{e_{i_1}, \ldots, e_{i_{p-1}}\}$ of $p - 1$ edges of $G$, and taking all columns of $M_0$ indexed by the set $S = \{i_1, \ldots, i_{p-1}\}$. Thus this submatrix is just $M_0[S]$. For convenience we will not bother to distinguish between the set $S$ of indices with the corresponding set $X$ of edges.

**9.7 Lemma.** *Let $S$ be a set of $p - 1$ edges of $G$. If $S$ does not form the set of edges of a spanning tree, then $\det M_0[S] = 0$. If, on the other hand, $S$ is the set of edges of a spanning tree of $G$, then $\det M_0[S] = \pm 1$.*

*Proof.* If $S$ is not the set of edges of a spanning tree, then some subset $R$ of $S$ forms the edges of a cycle $C$ in $G$. Suppose that the cycle $C$ defined by $R$ has edges $f_1, \ldots, f_s$ in that order. Multiply the column of $M_0[S]$ indexed by $f_i$ by 1 if in going around $C$ we traverse $f_i$ in the direction of its arrow; otherwise multiply the column by $-1$. Then add these modified columns. It is easy to see (check a few small examples to convince yourself) that we get the 0 column. Hence the columns of $M_0[S]$ are linearly dependent, so $\det M_0[S] = 0$, as claimed.

Now suppose that $S$ is the set of edges of a spanning tree $T$. Let $e$ be an edge of $T$ which is connected to $v_p$ (the vertex which indexed the bottom row of $M$, i.e., the row removed to get $M_0$). The column of $M_0[S]$ indexed by $e$ contains exactly one nonzero entry [why?], which is $\pm 1$. Remove from $M_0[S]$ the row and column containing the nonzero entry of column $e$, obtaining a $(p-2) \times (p-2)$ matrix $M_0'$. Note that $\det M_0[S] = \pm \det M_0'$ [why?]. Let $T'$ be the tree obtained from $T$ by contracting the edge $e$ to a single vertex (so that $v_p$ and the remaining vertex of $e$ are merged into a single vertex $u$). Then $M_0'$ is just the matrix obtained from the incidence matrix $M(T')$ by removing the row indexed by $u$ [why?]. Hence by induction on the number $p$ of vertices (the case $p = 1$ being trivial), we have $\det M_0' = \pm 1$. Thus $\det M_0[S] = \pm 1$, and the proof follows. $\qquad\square$

NOTE. An alternative way of seeing that $\det M_0[S] = \pm 1$ when $S$ is the set of edges of a spanning tree $T$ is as follows. Let $u_1, u_2, \ldots, u_{p-1}$ be an ordering of the vertices $v_1, \ldots, v_{p-1}$ such that $u_i$ is an endpoint of the tree obtained from $T$ by removing vertices $u_1, \ldots, u_{i-1}$. (It is easy to see that such an ordering is possible.) Permute the rows of $M_0[S]$ so that the $i$th row is indexed by $u_i$. Then permute the columns in the order $e_1, \ldots, e_{p-1}$ so that $e_i$ is the unique edge adjacent to $u_i$ after $u_1, \ldots, u_{i-1}$ have been removed. Then we obtain a lower triangular matrix with $\pm 1$'s on the main diagonal, so the determinant is $\pm 1$.

We have now assembled all the ingredients for the main result of this chapter. Recall that $\kappa(G)$ denotes the number of spanning trees of $G$.

**9.8 Theorem** (the Matrix-Tree Theorem). *Let $G$ be a finite connected graph without loops, with laplacian matrix $L = L(G)$. Let $L_0$ denote $L$ with the last row and column removed (or with the $i$th row and column removed for any $i$). Then*

$$\det L_0 = \kappa(G).$$

*Proof.* Since $L = MM^t$ (Lemma 9.6(a)), it follows immediately that $L_0 = M_0 M_0^t$. Hence by the Binet–Cauchy theorem (Theorem 9.4), we have

$$\det L_0 = \sum_S (\det M_0[S])(\det M_0^t[S]), \qquad (9.4)$$

where $S$ ranges over all $(p-1)$-element subsets of $\{1, 2, \ldots, q\}$ (or equivalently, over all $(p-1)$-element subsets of the set of edges of $G$). Since in general $A^t[S] = A[S]^t$, (9.4) becomes

$$\det L_0 = \sum_S (\det M_0[S])^2. \qquad (9.5)$$

According to Lemma 9.7, $\det M_0[S]$ is $\pm 1$ if $S$ forms the set of edges of a spanning tree of $G$, and is 0 otherwise. Therefore the term indexed by $S$ in the sum on the right-hand side of (9.5) is 1 if $S$ forms the set of edges of a spanning tree of $G$, and is 0 otherwise. Hence the sum is equal to $\kappa(G)$, as desired. $\qquad\square$

The operation of removing a row and column from $L(G)$ may seem somewhat contrived. We would prefer a description of $\kappa(G)$ directly in terms of $L(G)$. Such a description will follow from the next lemma.

**9.9 Lemma.** *Let $M$ be a $p \times p$ matrix (with entries in a field) such that the sum of the entries in every row and column is 0. Let $M_0$ be the matrix obtained from $M$ by removing the last row and last column (or more generally, any row and any column). Then the coefficient of $x$ in the characteristic polynomial $\det(M - xI)$ of $M$ is equal to $-p \cdot \det(M_0)$. (Moreover, the constant term of $\det(M - xI)$ is 0.)*

*Proof.* The constant term of $\det(M - xI)$ is $\det M$, which is 0 since the rows of $M$ sum to 0.

For simplicity we prove the rest of the lemma only for removing the last row and column, though the proof works just as well for any row and column. Add all the rows of $M - xI$ except the last row to the last row. This doesn't affect the determinant and will change the entries of the last row all to $-x$ (since the rows of $M$ sum to 0). Factor out $-x$ from the last row, yielding a matrix $N(x)$ satisfying $\det(M - xI) = -x \det N(x)$. Hence the coefficient of $x$ in $\det(M - xI)$ is given by $-\det N(0)$. Now add all the columns of $N(0)$ except the last column to the last column. This does not affect $\det N(0)$. Because the columns of $M$ sum to 0, the last column of $N(0)$ becomes the column vector $[0, 0, \ldots, 0, p]^t$. Expanding the determinant by the last column shows that $\det N(0) = p \cdot \det M_0$, and the proof follows.                                                                                                                           $\square$

**9.10 Corollary.** *(a) Let $G$ be a connected (loopless) graph with $p$ vertices. Suppose that the eigenvalues of $L(G)$ are $\mu_1, \ldots, \mu_{p-1}, \mu_p$, with $\mu_p = 0$. Then*

$$\kappa(G) = \frac{1}{p}\mu_1\mu_2 \cdots \mu_{p-1}.$$

*(b) Suppose that $G$ is also regular of degree $d$, and that the eigenvalues of $A(G)$ are $\lambda_1, \ldots, \lambda_{p-1}, \lambda_p$, with $\lambda_p = d$. Then*

$$\kappa(G) = \frac{1}{p}(d - \lambda_1)(d - \lambda_2) \cdots (d - \lambda_{p-1}).$$

*Proof.* (a) We have

$$\det(L - xI) = (\mu_1 - x) \cdots (\mu_{p-1} - x)(\mu_p - x)$$

$$= -(\mu_1 - x)(\mu_2 - x) \cdots (\mu_{p-1} - x)x.$$

Hence the coefficient of $x$ is $-\mu_1\mu_2 \cdots \mu_{p-1}$. By Lemma 9.9, we get $-\mu_1\mu_2 \cdots \mu_{p-1} = -p \cdot \det(L_0)$. By Theorem 9.8 we have $\det(L_0) = \kappa(G)$, and the proof follows.

(b) Immediate from (a) and Lemma 9.6(b).                                                      $\square$

Let us look at a couple of examples of the use of the Matrix-Tree Theorem.

**9.11 Example.** Let $G = K_p$, the complete graph on $p$ vertices. Now $K_p$ is regular of degree $d = p - 1$, and by Proposition 1.5 its eigenvalues are $-1$ ($p - 1$ times) and $p - 1 = d$. Hence from Corollary 9.10 there follows

$$\kappa(K_p) = \frac{1}{p}((p - 1) - (-1))^{p-1} = p^{p-2}.$$

Naturally a combinatorial proof of such an elegant result is desirable. In the Appendix to this chapter we give three such proofs.

**9.12 Example.** Let $G = C_n$, the $n$-cube discussed in Chapter 2. Now $C_n$ is regular of degree $n$, and by Corollary 2.4 its eigenvalues are $n - 2i$ with multiplicity $\binom{n}{i}$ for $0 \leq i \leq n$. Hence from Corollary 9.10 there follows the amazing result

$$\kappa(C_n) = \frac{1}{2^n} \prod_{i=1}^{n} (2i)^{\binom{n}{i}}$$

$$= 2^{2^n - n - 1} \prod_{i=1}^{n} i^{\binom{n}{i}}.$$

A direct combinatorial proof (though not an explicit bijection) was found by O. Bernardi in 2012.

## Appendix: Three Elegant Combinatorial Proofs

In this appendix we give three elegant combinatorial proofs that the number of spanning trees of the complete graph $K_p$ is $p^{p-2}$ (Example 9.11). The proofs are given in chronological order of their discovery.

*First Proof* (Prüfer). Given a spanning tree $T$ of $K_p$, i.e., a tree on the vertex set $[p]$, remove the largest endpoint (leaf) $v$ and write down the vertex $a_1$ adjacent to $v$. Continue this procedure until only two vertices remain, obtaining a sequence $(a_1, \ldots, a_{p-2}) \in [p]^{p-2}$, called the *Prüfer sequence* of $T$. For the tree below, we first remove 11 and then record 8. Next remove 10 and record 1. Then remove 8 and record 4, etc., ending with the sequence $(8, 1, 4, 4, 1, 4, 9, 1, 9)$ and leaving the two vertices 1 and 9.

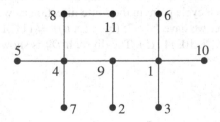

We claim that the map just defined from trees $T$ on $[p]$ to sequences $(a_1, \ldots, a_{p-2}) \in [p]^{p-2}$ is a bijection, thereby completing the proof since clearly $[p]^{p-2}$ has $p^{p-2}$ elements. The crucial observation is that the first vertex to be removed from $T$ is the largest vertex of $T$ missing from the sequence [why?—this takes a little thought]. This vertex is adjacent to $a_1$. For our example, we get that 11 was the first vertex removed, and that 11 is adjacent to 8. We can now proceed recursively. If $T_1$ denotes $T$ with the largest missing vertex removed, then the Prüfer sequence of $T_1$ is $(a_2, \ldots, a_{p-2})$. The first vertex to be removed from $T_1$ is the largest vertex of $T_1$ missing from $(a_2, \ldots, a_{p-2})$. This missing vertex is adjacent to $a_2$. For our example, this missing vertex is 10 (since 11 is not a vertex of $T_1$), which is adjacent to 1. Continuing in this way, we determine one new edge of $T$ at each step. At the end we have found $p - 2$ edges, and the remaining two unremoved vertices form the $(p - 1)$st edge.

*Second Proof* (Joyal). A *doubly rooted tree* is a tree $T$ with one vertex $u$ labelled $S$ (for "start") and one vertex $v$ (which may equal $u$) labelled $E$ ("end"). Let $t(p)$ be the number of trees $T$ on the vertex set $[p]$, and let $d(p)$ be the number of doubly rooted trees on $[p]$. Thus

$$d(p) = p^2 t(p), \tag{9.6}$$

since once we have chosen $T$ there are $p$ choices for $u$ and $p$ choices for $v$.

Let $T$ be a doubly-rooted tree. There is a unique path from $S$ to $E$, say with vertices $S = b_1, b_2, \ldots, b_k = E$ (in that order). The following diagram shows such a doubly-rooted tree.

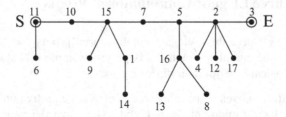

Let $a_1 < a_2 < \cdots < a_k$ be the increasing rearrangement of the numbers $b_1, b_2, \ldots, b_k$. Let $\pi$ be the permutation of the set $\{a_1, \ldots, a_k\}$ given by $\pi(a_i) = b_i$. Let $D_\pi$ be the digraph of $\pi$, that is, the vertex set of $D_\pi$ is $\{a_1, \ldots, a_k\}$, with a directed edge $a_i \to b_i$ for $1 \leq i \leq k$. Since any permutation $\pi$ of a finite set is a disjoint product of cycles, it follows that $D_\pi$ is a disjoint union of directed cycles (all edges of each cycle point in the same direction as we traverse the cycle). For the example above, we have $k = 7$, $(b_1, \ldots, b_7) = (11, 10, 15, 7, 5, 2, 3)$ and $(a_1, \ldots, a_7) = (2, 3, 5, 7, 10, 11, 15)$. The digraph $D_\pi$ is shown below.

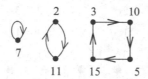

Now attach to each vertex $v$ of $D_\pi$ the same subgraph $T_v$ that was attached "below" $v$ in $T$ and direct the edges of $T_v$ toward $v$, obtaining a digraph $D_T$. For our example we get

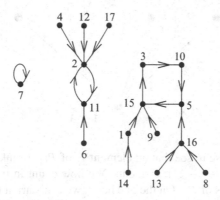

The graph $D_T$ has the crucial property that every vertex has outdegree one, that is, one arrow pointing out. In other words, $D_T$ is the graph of a function $f : [p] \to [p]$, with vertex set $[p]$ and edges $i \to f(i)$. Conversely, given a function $f : [p] \to [p]$, all the above steps can be reversed to obtain a unique doubly rooted tree $T$ for which $D_T$ is the graph of $f$. We have therefore found a bijection from doubly rooted trees on $[p]$ to functions $f : [p] \to [p]$. Since the number of such functions $f$ is $p^p$, it follows that $d(p) = p^p$. Then from (9.6) we get $t(p) = p^{p-2}$.

*Third Proof* (Pitman). A *forest* is a graph without cycles; thus every connected component is a tree. A *planted forest* is a forest $F$ for which every component $T$ has a distinguished vertex $r_T$ (called the *root* of $T$). Thus if a component $T$ has $k$ vertices, then there are $k$ ways to choose the root of $T$.

Let $P_p$ be the set of all planted forests on $[p]$. Let $uv$ be an edge of a forest $F \in P_p$ such that $u$ is closer than $v$ to the root $r$ of its component. Define $F$ to *cover* the planted forest $F'$ if $F'$ is obtained by removing the edge $uv$ from $F$, and rooting the new tree containing $v$ at $v$. This definition of cover defines the covering relation of a partial order on $P_p$. Under this partial order $P_p$ is graded of rank $p - 1$. The rank of a forest $F$ in $P_p$ is its number of edges. The following diagram shows the poset $P_3$, with the root of each tree being its top vertex.

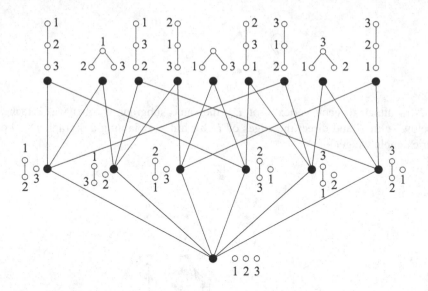

It is an easy exercise to see that an element $F$ of $P_p$ of rank $i$ covers $i$ elements and is covered by $(p - i - 1)p$ elements. We now count in two ways the number $M_p$ of maximal chains of $P_p$. On the one hand, we can start at the top. The number of maximal elements of $P_p$ is $p \cdot t(p)$, where $t(p)$ as above is the number of trees on the vertex set $[p]$, since there are $p$ ways to choose the root of such a tree. Once a maximal element $F$ is chosen, then there are $p - 1$ elements $F'$ that it covers, then $p - 2$ elements that $F'$ covers, etc., giving

$$M_p = p \cdot t(p)(p - 1)! = p!\, t(p). \tag{9.7}$$

On the other hand, we can start at the bottom. There is a unique element $F$ of rank zero (the planted forest with no edges), then $(p - 1)p$ elements $F'$ that cover $F$, then $(p - 2)p$ elements that cover $F'$, etc., giving

$$M_p = p^{p-1}(p - 1)!. \tag{9.8}$$

Comparing (9.7) and (9.8) gives $t(p) = p^{p-2}$.

Our third proof isn't an explicit bijection like the first two proofs. On the other hand, it has the virtue of not depending on the names of the vertices. Note that in the first two proofs it is necessary to know when one vertex is larger than another.

## Notes for Chapter 9

The concept of tree as a formal mathematical object goes back to G. Kirchhoff and K.G.C. von Staudt. Trees were first extensively investigated by A. Cayley, to whom the term "tree" is due. In particular, in [24] Cayley states the formula $\kappa(K_p) = p^{p-2}$ for the number of spanning trees of $K_p$, and he gives a vague idea of a combinatorial proof. Because of this paper, Cayley is often credited with the enumeration of labelled trees. Cayley pointed out, however, that an equivalent result had been proved earlier by Borchardt [11]. Moreover, this result appeared even earlier in a paper of Sylvester [134]. Undoubtedly Cayley and Sylvester could have furnished a complete, rigorous proof had they had the inclination to do so. The elegant combinatorial proofs given in the appendix are due to Prüfer [106], Joyal [71, Exam. 12, pp. 15–16], and Pitman [100].

    The Matrix-Tree Theorem (Theorem 9.8) was first proved by Borchardt [11] in 1860, though a similar result had earlier been published by Sylvester [134] in 1857. Cayley [23, p. 279] in fact in 1856 referred to the not-yet-published work of Sylvester. For further historical information on the Matrix-Tree theorem, see Moon [94, p. 42]. See Bernardi [8] for the combinatorial proof mentioned in Example 9.12.

## Exercises for Chapter 9

1. (*) Let $G_p$ be the complete graph $K_p$ with one edge removed. How many spanning trees does $G_p$ have?

2. Let $L = L(K_{rs})$ be the laplacian matrix of the complete bipartite graph $K_{rs}$.

   (a) Find a simple upper bound on rank$(L - rI)$. Deduce a lower bound on the number of eigenvalues of $L$ equal to $r$.
   (b) Assume $r \neq s$, and do the same as (a) for $s$ instead of $r$.
   (c) (*) Find the remaining eigenvalues of $L$.
   (d) Use (a)–(c) to compute $\kappa(K_{rs})$, the number of spanning trees of $K_{rs}$.
   (e) Give a combinatorial proof of the formula for $\kappa(K_{rs})$, by modifying either the proof of Prüfer or Joyal that $\kappa(K_p) = p^{p-2}$.

3. (a) (*) Let $1 \leq m \leq n$. Let $K_n - K_m$ denote the graph $K_n$ with all the edges of some subgraph $K_m$ removed. (In particular, $K_n - K_1 = K_n$.) Thus $K_n - K_m$ has $\binom{n}{2} - \binom{m}{2}$ edges. Use directly the Matrix-Tree Theorem to find the number of spanning trees of $K_n - K_m$.
   (b) Give another proof using Exercise 6.

4. Let $p \geq 5$, and let $G_p$ be the graph on the vertex set $\mathbb{Z}_p$ with edges $\{i, i+1\}$ and $\{i, i+2\}$, for $i \in \mathbb{Z}_p$. Thus $G_p$ has $2p$ edges. Show that $\kappa(G_p) = pF_p^2$, where $F_p$ is a Fibonacci number ($F_1 = F_2 = 1$, $F_p = F_{p-1} + F_{p-2}$ for $p \geq 3$).

5. Let $\overline{C}_n$ be the edge complement of the cube graph $C_n$, i.e., $\overline{C}_n$ has vertex set $\{0, 1\}^n$, with an edge $uv$ if $u$ and $v$ differ in at least two coordinates. Find a

formula for $\kappa(\overline{C}_n)$, the number of spanning trees of $\overline{C}_n$. Your answer should be expressed as a simple product.

6. Let $G$ be a finite graph on $p$ vertices with laplacian matrix $L(G)$. Let $G'$ be obtained from $G$ by adding a new vertex $v$ and connecting it to each vertex of $G$ (so we have $p$ new edges). Express $\kappa(G')$ (the number of spanning trees of $G'$) in terms of the eigenvalues $\mu_1, \ldots, \mu_p$ of $L(G)$.

7. (a) Let $G$ be a bipartite graph with vertex bipartition $(A, B)$. Suppose that $\deg v = a$ for all $v \in A$, and $\deg v = b$ for all $v \in B$. Let $A$ and $L$ denote the adjacency matrix and laplacian matrix of $G$, respectively. Show that if the eigenvalues of $L$ are $\lambda_1, \ldots, \lambda_p$, then the eigenvalues of $A^2$ are $(\lambda_1 - a)(\lambda_1 - b), \ldots, (\lambda_p - a)(\lambda_p - b)$.

   (b) (*) Find the number of spanning trees of the graph $C_{n,k}$ of Exercise 2.2.

8. (a) (*) Let $G$ be a finite loopless graph with $p$ vertices. Suppose that the eigenvalues of the Laplacian matrix $L(G)$ are $\theta_1, \ldots, \theta_{p-1}$ and $\theta_p = 0$. Let $J$ be the $p \times p$ matrix of all 1's, and let $\alpha \in \mathbb{R}$. Show that the eigenvalues of $L + \alpha J$ are $\theta_1, \ldots, \theta_{p-1}, \alpha p$.

   (b) Let $G \cup K_p$ be the graph obtained from $G$ by adding one new edge between every pair of distinct vertices. Express the number of spanning trees of $G \cup K_p$ in terms of $\theta_1, \ldots, \theta_{p-1}$.

   (c) Suppose that $G$ is simple, and let $\overline{G}$ be the *complementary graph*, i.e., $G$ and $\overline{G}$ have the same vertex set, and two distinct vertices are adjacent in $G$ if and only if they are not adjacent in $\overline{G}$. Express the number of spanning trees of $\overline{G}$ in terms of $\theta_1, \ldots, \theta_{p-1}$.

   (d) (*) Let $G$ be simple with $p$ vertices, and define the polynomial

$$P(G, x) = \sum_F x^{c(F)-1},$$

   where $F$ ranges over all spanning planted forests of $G$, and where $c(F)$ is the number of components of $F$. Show that

$$P(\overline{G}, x) = (-1)^{p-1} P(G, -x - p).$$

9. (*) Let $V$ be the subset of $\mathbb{Z} \times \mathbb{Z}$ on or inside some simple closed polygonal curve whose vertices belong to $\mathbb{Z} \times \mathbb{Z}$, such that every line segment that makes up the curve is parallel to either the $x$-axis or $y$-axis. Draw an edge $e$ between any two points of $V$ at distance one apart, provided $e$ lies on or inside the boundary curve. We obtain a planar graph $G$, an example being

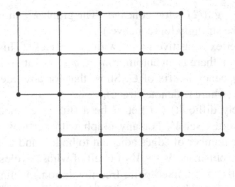

Let $G'$ be the dual graph $G^*$ with the "outside" vertex deleted. (The vertices of $G'$ are the interior regions of $G$. For each edge $e$ of $G$, say with regions $R$ and $R'$ on the two sides of $e$, there is an edge of $G'$ between $R$ and $R'$. See Section 11.4 for more information on duals to planar graphs.) For the above example, $G'$ is given by

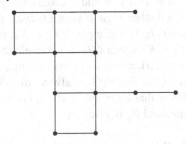

Let $\lambda_1, \ldots, \lambda_p$ denote the eigenvalues of $G'$ (i.e., of the adjacency matrix $A(G')$). Show that

$$\kappa(G) = \prod_{i=1}^{p} (4 - \lambda_i).$$

10. (a) (*) Let $L$ be the laplacian matrix of a graph $G$ on a $p$-element vertex set $V$. For a $k$-element subset $S$ of the vertices, let $L[S, S]$ be the $k \times k$ submatrix of $L$ whose rows and columns are indexed by the vertices in $S$. Show that $\det L[S, S]$ is equal to the number of planted spanning forests (as defined in the third proof of the Appendix to this chapter) of $G$ whose set of roots is $V - S$.

(b) Deduce from (a) that the coefficient of $x^j$ in the characteristic polynomial $\det(L - xI)$ is equal to $(-1)^j$ times the number $f_j(G)$ of planted spanning forests of $G$ with $j$ components.

(c) Deduce from Theorem 5.12 that the sequence $f_1(G), f_2(G), \ldots, f_p(G)$ is strongly log-concave.

(d) (extremely difficult) Let $g_k(G)$ denote the number of spanning forests (but not planted) of $G$ with $j$ components. Show that the sequence $g_1(G)$,

$g_2(G), \ldots, g_p(G)$ is log-concave. (The graph $K_3$ shows that this sequence need not be strongly log-concave.)

11. Let $G$ be a vertex-transitive graph with $p$ vertices. (This means that for any two vertices $u$, $v$ there is an automorphism $\varphi$ of $G$ satisfying $\varphi(u) = v$.) Let $A$ denote the adjacency matrix of $G$. Show that for any integer $n \geq 1$, $\text{tr}(A^n)$ is divisible by $p$ (where tr denotes trace).

12. (a) (moderately difficult) (*) Let $G$ be a (finite) connected graph on a $2m$-element vertex set $V$. For any graph with vertices $u$ and $v$, let $\mu(u, v)$ denote the number of edges adjacent to both $u$ and $v$. Suppose that there is an automorphism $\varphi: V \rightarrow V$ of $G$ all of whose cycles have length two. (In other words, $\varphi$ is a fixed-point free involution.) Define the *quotient graph* $G/\varphi$ as follows. The vertices of $G/\varphi$ are the orbits of $\varphi$. Thus $G/\varphi$ has $m$ vertices. Write $[v]$ for the orbit containing vertex $v$, so $[v] = \{v, \varphi(v)\}$. Set

$$\mu([u], [v]) = \mu(u, v) + \mu(\varphi(u), v).$$

For instance, if $G$ is a 4-cycle and $\varphi$ takes each vertex to its antipode, then $G/\varphi$ consists of a double edge. If $G$ is a 6-cycle and $\varphi$ takes each vertex to its antipode, then $G/\varphi$ is a triangle. If $G = K_4$, then $G/\varphi$ is a double edge (for any $\varphi$). If $G = K_{3,3}$ then $G/\varphi$ is a 3-cycle for any $\varphi$. Show that $2\kappa(G)$ is divisible by $\kappa(G/\varphi)$, where $\kappa$ denotes the number of spanning trees.

   (b) (difficult) Show in fact that $\kappa(G)$ is always divisible by $\kappa(G/\varphi)$.

13. (a) (difficult) (*) Show that the number $s(n, q)$ of invertible $n \times n$ symmetric matrices over the field $\mathbb{F}_q$ is given by

$$s(n, q) = \begin{cases} q^{m(m-1)}(q - 1)(q^3 - 1) \cdots (q^{2m-1} - 1), & n = 2m - 1 \\ q^{m(m+1)}(q - 1)(q^3 - 1) \cdots (q^{2m-1} - 1), & n = 2m. \end{cases}$$

   (b) Find a formula for the number $f(p)$ of simple graphs on the vertex set $[p]$ with an odd number of spanning trees.

14. (*) Let $S$ be a $k$-element subset of $[p]$. Show that the number $f_S(p)$ of planted forests on the vertex set $[p]$ with exactly $k$ components, whose set of roots is $S$, is given by $f_S(p) = kn^{n-k-1}$. Deduce that the total number $f_k(p)$ of planted forests on $[p]$ with $k$ components is given by

$$f_k(p) = k\binom{p}{k}p^{p-k-1} = \binom{p-1}{k-1}p^{p-k}.$$

# Chapter 10
# Eulerian Digraphs and Oriented Trees

A famous problem which goes back to Euler asks for what graphs $G$ is there a closed walk which uses every edge exactly once. (There is also a version for non-closed walks.) Such a walk is called an *Eulerian tour* (also known as an *Eulerian cycle*). A graph which has an Eulerian tour is called an *Eulerian graph*. Euler's famous theorem (the first real theorem of graph theory) states that a graph $G$ without isolated vertices (which clearly would be irrelevant) is Eulerian if and only if it is connected and every vertex has even degree. Here we will be concerned with the analogous theorem for directed graphs. We want to know not just whether an Eulerian tour exists, but also how many there are. We will prove an elegant determinantal formula for this number closely related to the Matrix-Tree Theorem. For the case of undirected graphs no analogous formula is known, explaining why we consider only the directed case.

A (finite) *directed graph* or *digraph* $D$ consists of a vertex set $V = \{v_1, \ldots, v_p\}$ and edge set $E = \{e_1, \ldots, e_q\}$, together with a function $\varphi : E \to V \times V$ (the set of ordered pairs $(u, v)$ of elements of $V$). If $\varphi(e) = (u, v)$, then we think of $e$ as an arrow from $u$ to $v$. We then call $u$ the *initial vertex* and $v$ the *final vertex* of $e$, denoted init$(e)$ and fin$(e)$, respectively. (These concepts arose in the definition of an orientation in Definition 9.5.)

A walk in a digraph $D$ is defined analogously to the undirected case. Namely, a *walk* in $D$ of *length* $\ell$ from a vertex $u$ to a vertex $v$ is a sequence $v_1, e_1, e_2, \ldots, v_\ell, e_\ell, v_{\ell+1}$ such that:

- each $v_i$ is a vertex of $D$
- each $e_j$ is an edge of $D$ with initial vertex $v_i$ and final vertex $v_{i+1}$
- init$(e_i) = v_i$ and fin$(e_i) = v_{i+1}$, for $1 \le i \le \ell$
- $v_1 = u$ and $v_{\ell+1} = v$.

There are two notions of connected in a digraph $D$. We say that $D$ is *strongly connected* if there is a walk (as defined above) between any two vertices. We also say that $D$ is *weakly connected*, or just *connected*, if $D$ is connected as an undirected

R. P. Stanley, *Algebraic Combinatorics*, Undergraduate Texts in Mathematics,
https://doi.org/10.1007/978-3-319-77173-1_10

graph, i.e., regard each edge as an *undirected* edge. Thus a strongly connected digraph is (weakly) connected.

A *tour* in $D$ is a sequence $e_1, e_2, \ldots, e_r$ of *distinct* edges such that the final vertex of $e_i$ is the initial vertex of $e_{i+1}$ for all $1 \le i \le r - 1$, and the final vertex of $e_r$ is the initial vertex of $e_1$. A tour is *Eulerian* if every edge of $D$ occurs at least once (and hence exactly once). A digraph which has no isolated vertices and contains an Eulerian tour is called an *Eulerian digraph*. Clearly an Eulerian digraph is strongly connected. The *outdegree* of a vertex $v$, denoted outdeg($v$), is the number of edges of $D$ with initial vertex $v$. Similarly the *indegree* of $v$, denoted indeg($v$), is the number of edges of $D$ with final vertex $v$. A loop (edge $e$ for which $\varphi(e) = (v, v)$) contributes one to both the indegree and outdegree. A digraph is *balanced* if indeg($v$) = outdeg($v$) for all vertices $v$.

**10.1 Theorem.** *A digraph $D$ without isolated vertices is Eulerian if and only if it is connected (strongly or weakly) and balanced.*

*Proof.* Assume $D$ is Eulerian, and let $e_1, \ldots, e_q$ be an Eulerian tour. As we move along the tour, whenever we enter a vertex $v$ we must exit it, except at the very end we enter the final vertex $v$ of $e_q$ without exiting it. However, at the beginning we exited $v$ without having entered it. Hence every vertex is entered as often as it is exited and so must have the same outdegree as indegree. Therefore $D$ is balanced, and as noted above $D$ is clearly strongly (and therefore weakly) connected.

Now assume that $D$ is balanced and weakly connected. We may assume that $D$ has at least one edge. We first claim that for any edge $e$ of $D$, $D$ has a tour for which $e = e_1$. If $e_1$ is a loop we are done. Otherwise we have entered the vertex fin($e_1$) for the first time, so since $D$ is balanced there is some exit edge $e_2$. Either fin($e_2$) = init($e_1$) and we are done, or else we have entered the vertex fin($e_2$) once more than we have exited it. Since $D$ is balanced there is new edge $e_3$ with fin($e_2$) = init($e_3$). Continuing in this way, either we complete a tour or else we have entered the current vertex once more than we have exited it, in which case we can exit along a new edge. Since $D$ has finitely many edges, eventually we must complete a tour. Thus $D$ does have a tour which uses $e_1$.

Now let $e_1, \ldots, e_r$ be a tour $C$ of maximum length. We must show that $r = q$, the number of edges of $D$. Assume to the contrary that $r < q$. Since in moving along $C$ every vertex is entered as often as it is exited (with init($e_1$) exited at the beginning and entered at the end), when we remove the edges of $C$ from $D$ we obtain a digraph $H$ which is still balanced, though it need not be connected. However, since $D$ is connected, at least one connected component $H_1$ of $H$ contains at least one edge and has a vertex $v$ in common with $C$ [why?]. Since $H_1$ is balanced, there is an edge $e$ of $H_1$ with initial vertex $v$. The argument of the previous paragraph shows that $H_1$ has a tour $C'$ of positive length beginning with the edge $e$. But then when moving along $C$, when we reach $v$ we can take the "detour" $C'$ before continuing with $C$. This gives a tour of length longer than $r$, a contradiction. Hence $r = q$, and the theorem is proved.                                                                                   □

Our primary goal is to count the number of Eulerian tours of a connected balanced digraph. A key concept in doing so is that of an oriented tree. An *oriented tree* with root $v$ is a (finite) digraph $T$ with $v$ as one of its vertices, such that there is a unique directed path from any vertex $u$ to $v$. In other words, there is a unique sequence of edges $e_1, \ldots, e_r$ such that (a) $\mathrm{init}(e_1) = u$, (b) $\mathrm{fin}(e_r) = v$, and (c) $\mathrm{fin}(e_i) = \mathrm{init}(e_{i+1})$ for $1 \leq i \leq r - 1$. (All three of these conditions are considered to be vacuously true if the sequence $e_1, \ldots, e_r$ is empty.) It's easy to see that this means that the underlying undirected graph (i.e., "erase" all the arrows from the edges of $T$) is a tree, and that all arrows in $T$ "point toward" $v$. From now on, an oriented subtree of $D$ will always mean a subdigraph of $D$ that is an oriented tree with the full vertex set $V$, or in other words, a *spanning* oriented subtree of $D$. There is a surprising connection between Eulerian tours and oriented trees, given by the next result.

**10.2 Theorem.** *Let $D$ be a connected balanced digraph with vertex set $V$. Fix an edge $e$ of $D$, and let $v = \mathrm{init}(e)$. Let $\tau(D, v)$ denote the number of oriented (spanning) subtrees of $D$ with root $v$, and let $\epsilon(D, e)$ denote the number of Eulerian tours of $D$ starting with the edge $e$. Then*

$$\epsilon(D, e) = \tau(D, v) \prod_{u \in V} (\mathrm{outdeg}(u) - 1)!. \tag{10.1}$$

*Proof.* Let $e = e_1, e_2, \ldots, e_q$ be an Eulerian tour $\mathcal{E}$ in $D$. For each vertex $u \neq v$, let $e(u)$ be the "last exit" from $u$ in the tour, i.e., let $e(u) = e_j$ where $\mathrm{init}(e_j) = u$ and $\mathrm{init}(e_k) \neq u$ for any $k > j$.

*Claim #1.* The vertices of $D$, together with the edges $e(u)$ for all vertices $u \neq v$, form an oriented subtree of $D$ with root $v$.

*Proof of Claim #1.* This is a straightforward verification. Let $T$ be the spanning subgraph of $D$ with edges $e(u)$, $u \neq v$. Thus if $\#V = p$, then $T$ has $p$ vertices and $p - 1$ edges [why?]. There are three items to check to insure that $T$ is an oriented tree with root $v$:

(a) $T$ does not have two edges $f$ and $f'$ satisfying $\mathrm{init}(f) = \mathrm{init}(f')$. This is clear since both $f$ and $f'$ can't be last exits from the same vertex.
(b) $T$ does not have an edge $f$ with $\mathrm{init}(f) = v$. This is clear since by definition the edges of $T$ consist only of last exits from vertices other than $v$, so no edge of $T$ can exit from $v$.
(c) $T$ does not have a (directed) cycle $C$. For suppose $C$ were such a cycle. Let $f$ be that edge of $C$ which occurs after all the other edges of $C$ in the Eulerian tour $\mathcal{E}$. Let $f'$ be the edge of $C$ satisfying $\mathrm{fin}(f) = \mathrm{init}(f') (= u$, say). We can't have $u = v$ by (b). Thus when we enter $u$ via $f$, we must exit $u$. We can't exit $u$ via $f'$ since $f$ occurs after $f'$ in $\mathcal{E}$. Hence $f'$ is not the last exit from $u$, contradicting the definition of $T$.

It's easy to see that conditions (a)–(c) imply that $T$ is an oriented tree with root $v$, proving the claim.

*Claim #2.* We claim that the following converse to Claim #1 is true. Given a connected balanced digraph $D$ and a vertex $v$, let $T$ be an oriented (spanning) subtree of $D$ with root $v$. Then we can construct an Eulerian tour $\mathcal{E}$ as follows. Choose an edge $e_1$ with init$(e_1) = v$. Then continue to choose any edge possible to continue the tour, except we never choose an edge $f$ of $T$ unless we have to, i.e., unless it's the only remaining edge exiting the vertex at which we stand. Then we never get stuck until all edges are used, so we have constructed an Eulerian tour $\mathcal{E}$. Moreover, the set of last exits of $\mathcal{E}$ from vertices $u \neq v$ of $D$ coincides with the set of edges of the oriented tree $T$.

*Proof of Claim #2.* Since $D$ is balanced, the only way to get stuck is to end up at $v$ with no further exits available, but with an edge still unused. Suppose this is the case. At least one unused edge must be a last exit edge, i.e., an edge of $T$ [why?]. Let $u$ be a vertex of $T$ closest to $v$ in $T$ such that the unique edge $f$ of $T$ with init$(f) = u$ is not in the tour. Let $y = \text{fin}(f)$. Suppose $y \neq v$. Since we enter $y$ as often as we leave it, we don't use the last exit from $y$. Thus $y = v$. But then we can leave $v$, a contradiction. This proves Claim #2.

We have shown that every Eulerian tour $\mathcal{E}$ beginning with the edge $e$ has associated with it a "last exit" oriented subtree $T = T(\mathcal{E})$ with root $v = \text{init}(e)$. Conversely, given an oriented subtree $T$ with root $v$, we can obtain all Eulerian tours $\mathcal{E}$ beginning with $e$ and satisfying $T = T(\mathcal{E})$ by choosing for each vertex $u \neq v$ the order in which the edges from $u$, except the edge of $T$, appear in $\mathcal{E}$; as well as choosing the order in which all the edges from $v$ except for $e$ appear in $\mathcal{E}$. Thus for each vertex $u$ we have (outdeg$(u) - 1$)! choices, so for each $T$ we have $\prod_u (\text{outdeg}(u) - 1)!$ choices. Since there are $\tau(D, v)$ choices for $T$, the proof is complete.                                                                                    $\square$

**10.3 Corollary.** *Let $D$ be a connected balanced digraph, and let $v$ be a vertex of $D$. Then the number $\tau(D, v)$ of oriented subtrees with root $v$ is independent of $v$.*

*Proof.* Let $e$ be an edge with initial vertex $v$. By (10.1), we need to show that the number $\epsilon(D, e)$ of Eulerian tours beginning with $e$ is independent of $e$. But $e_1 e_2 \cdots e_q$ is an Eulerian tour if and only if $e_i e_{i+1} \cdots e_q e_1 e_2 \cdots e_{i-1}$ is also an Eulerian tour, and the proof follows [why?].                                              $\square$

What we obviously need to do next is find a formula for $\tau(D, v)$. This result turns out to be very similar to the Matrix-Tree Theorem, and indeed we will show (Example 10.6) that the Matrix-Tree Theorem is a simple corollary to Theorem 10.4.

**10.4 Theorem.** *Let $D$ be a digraph with vertex set $V = \{v_1, \ldots, v_p\}$ and with $l_i$ loops at vertex $v_i$. Let $L(D)$ be the $p \times p$ matrix defined by*

$$L_{ij} = \begin{cases} -m_{ij}, & \text{if } i \neq j \text{ and there are } m_{ij} \text{ edges with} \\ & \text{initial vertex } v_i \text{ and final vertex } v_j \\ \\ \text{outdeg}(v_i) - l_i, & \text{if } i = j. \end{cases}$$

*(Thus L is the directed analogue of the laplacian matrix of an undirected graph.)*
*Let $L_0$ denote L with the last row and column deleted. Then*

$$\det L_0 = \tau(D, v_p). \qquad (10.2)$$

NOTE. *If we remove the ith row and column from L instead of the last row and column, then (10.2) still holds with $v_p$ replaced with $v_i$.*

*Sketch.* We first prove the case where $D$ is not connected. In this case $L_0$ has a set of rows summing to 0, so $\det L_0 = 0$. Since a disconnected graph has no spanning trees, we also have $\tau(D, v_p) = 0$, proving the theorem when $D$ is not connected.

Now assume that $D$ is connected. Induction on $q$, the number of edges of $D$. The fewest number of edges which $D$ can have is $p - 1$ (since $D$ is connected). Suppose then that $D$ has $p - 1$ edges, so that as an undirected graph $D$ is a tree. If $D$ is not an oriented tree with root $v_p$, then some vertex $v_i \neq v_p$ of $D$ has outdegree 0 [why?]. Then $L_0$ has a zero row, so $\det L_0 = 0 = \tau(D, v_p)$. If on the other hand $D$ is an oriented tree with root $v_p$, then an argument like that used to prove Lemma 9.7 (in the case when $S$ is the set of edges of a spanning tree) shows that $\det L_0 = 1 = \tau(D, v_p)$.

Now assume that $D$ has $q > p - 1$ edges, and assume the theorem for digraphs with at most $q - 1$ edges. We may assume that no edge $f$ of $D$ has initial vertex $v_p$, since such an edge belongs to no oriented tree with root $v_p$ and also makes no contribution to $L_0$. It then follows, since $D$ has at least $p$ edges, that there exists a vertex $u \neq v_p$ of $D$ of outdegree at least two. Let $e$ be an edge with init$(e) = u$. Let $D_1$ be $D$ with the edge $e$ removed. Let $D_2$ be $D$ with all edges $e'$ removed such that init$(e) = $ init$(e')$ and $e' \neq e$. (Note that $D_2$ is strictly smaller than $D$ since outdeg$(u) \geq 2$.) By induction or because $D_1$ and/or $D_2$ is not connected, we have $\det L_0(D_1) = \tau(D_1, v_p)$ and $\det L_0(D_2) = \tau(D_2, v_p)$. Clearly $\tau(D, v_p) = \tau(D_1, v_p) + \tau(D_2, v_p)$, since in an oriented tree $T$ with root $v_p$, there is exactly one edge whose initial vertex coincides with that of $e$. On the other hand, it follows immediately from the multilinearity of the determinant [why?] that

$$\det L_0(D) = \det L_0(D_1) + \det L_0(D_2).$$

From this the proof follows by induction. □

**10.5 Corollary.** *Let $D$ be a connected balanced digraph with vertex set $V = \{v_1, \ldots, v_p\}$. Let $e$ be an edge of $D$. Then the number $\epsilon(D, e)$ of Eulerian tours of $D$ with first edge $e$ is given by*

$$\epsilon(D, e) = (\det L_0(D)) \prod_{u \in V} (\text{outdeg}(u) - 1)!.$$

*Equivalently (since D is balanced, so Lemma 9.9 applies), if $L(D)$ has eigenvalues $\mu_1, \ldots, \mu_p$ with $\mu_p = 0$, then*

$$\epsilon(D, e) = \frac{1}{p}\mu_1 \cdots \mu_{p-1} \prod_{u \in V} (\text{outdeg}(u) - 1)!.$$

*Proof.* Combine Theorems 10.2 and 10.4. □

**10.6 Example** (the Matrix-Tree Theorem revisited). Let $G = (V, E, \varphi)$ be a connected loopless undirected graph. Let $\widehat{G} = (V, \widehat{E}, \widehat{\varphi})$ be the digraph obtained from $G$ by replacing each edge $e$ of $G$, where $\varphi(e) = \{u, v\}$, with a pair $e'$ and $e''$ of directed edges satisfying $\widehat{\varphi}(e') = (u, v)$ and $\widehat{\varphi}(e'') = (v, u)$. Clearly $\widehat{G}$ is balanced and connected. Choose a vertex $v$ of $G$. There is an obvious one-to-one correspondence between spanning trees $T$ of $G$ and oriented spanning trees $\widehat{T}$ of $\widehat{G}$ with root $v$, namely, direct each edge of $T$ toward $v$. Moreover, $L(G) = L(\widehat{G})$ [why?]. Hence the Matrix-Tree Theorem is an immediate consequence of Theorem 10.4.

**10.7 Example** (the efficient mail carrier). A mail carrier has an itinerary of city blocks to which he (or she) must deliver mail. He wants to accomplish this by walking along each block twice, once in each direction, thus passing along houses on each side of the street. He also wants to end up where he started, which is where his car is parked. The blocks form the edges of a graph $G$, whose vertices are the intersections. The mail carrier wants simply to walk along an Eulerian tour in the digraph $\widehat{G}$ of the previous example. Making the plausible assumption that the graph is connected, not only does an Eulerian tour always exist, but also we can tell the mail carrier how many there are. Thus he will know how many different routes he can take to avoid boredom. For instance, suppose $G$ is the $3 \times 3$ grid illustrated below.

This graph has 192 spanning trees. Hence the number of mail carrier routes beginning with a fixed edge (in a given direction) is $192 \cdot 1!^4 \, 2!^4 \, 3! = 18{,}432$. The total number of routes is thus 18,432 times twice the number of edges [why?], viz., $18{,}432 \times 24 = 442{,}368$. Assuming the mail carrier delivered mail 250 days a year, it would be 1769 years before he would have to repeat a route!

**10.8 Example** (binary de Bruijn sequences). A *binary sequence* is just a sequence of 0's and 1's. A *binary de Bruijn sequence* of degree $n$ is a binary sequence $A = a_1 a_2 \cdots a_{2^n}$ such that every binary sequence $b_1 \cdots b_n$ of length $n$ occurs exactly once as a "circular factor" of $A$, i.e., as a sequence $a_i a_{i+1} \cdots a_{i+n-1}$, where the subscripts are taken modulo $2^n$ if necessary. For instance, some circular factors of the sequence $abcdefg$ are $a$, $bcde$, $fgab$, and $defga$. Note that there are exactly $2^n$ binary sequences of length $n$, so the only possible length of a binary de Bruijn sequence of degree $n$ is $2^n$ [why?]. Clearly any cyclic shift $a_i a_{i+1} \cdots a_{2^n} a_1 a_2 \cdots a_{i-1}$ of a binary de Bruijn sequence $a_1 a_2 \cdots a_{2^n}$ is also a binary de Bruijn sequence, and we call two such sequences *equivalent*. This relation of equivalence is obviously an equivalence relation, and every equivalence class contains exactly one sequence beginning with $n$ 0's [why?]. Up to equivalence, there is one binary de Bruijn sequence of degree two, namely, 0011. It's easy to check that there are two inequivalent binary de Bruijn sequences of degree three, namely, 00010111 and 00011101. However, it's not clear at this point whether binary de Bruijn sequences exist for all $n$. By a clever application of Theorems 10.2 and 10.4, we will not only show that such sequences exist for all positive integers $n$, but we will also count the number of them. It turns out that there are *lots* of them. For instance, the number of inequivalent binary de Bruijn sequences of degree eight is equal to

$$132922799578491587290380706028034457 6,$$

as the reader can easily check by writing down all these sequences. De Bruijn sequences have a number of interesting applications to the design of switching networks and related topics.

Our method of enumerating binary de Bruijn sequences will be to set up a correspondence between them and Eulerian tours in a certain directed graph $D_n$, the *de Bruijn graph* of degree $n$. The graph $D_n$ has $2^{n-1}$ vertices, which we will take to consist of the $2^{n-1}$ binary sequences of length $n - 1$. The edges are indexed by binary sequences $a_1 a_2 \cdots a_n$ with initial vertex $a_1 a_2 \cdots a_{n-1}$ and final vertex $a_2 a_3 \cdots a_n$. Thus every vertex has indegree two and outdegree two [why?], so $D_n$ is balanced. The number of edges of $D_n$ is $2^n$. Moreover, it's easy to see that $D_n$ is connected (see Lemma 10.9). The graphs $D_3$ and $D_4$ look as follows:

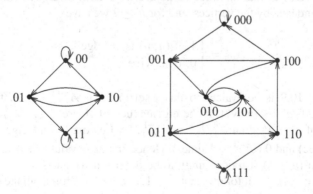

Suppose that $\mathcal{E} = e_1 e_2 \cdots e_{2^n}$ is an Eulerian tour in $D_n$. If $\mathrm{fin}(e_i)$ is the binary sequence $a_{i,1} a_{i,2} \cdots a_{i,n-1}$, then replace $e_i$ in $\mathcal{E}$ by the last bit $a_{i,n-1}$. For instance, the Eulerian tour (where we simply write the vertices)

$$000, 000, 001, 010, 101, 011, 111, 111, 110, 101, 010, 100, 001, 011, 110, 100, 000$$

corresponds to the sequence 0101111010011000 (the last bits of the vertices above, excluding the first vertex 000). It is easy to see that the resulting sequence $\beta(\mathcal{E}) = a_{1,n-1} a_{2,n-1} \cdots a_{2^n,n-1}$ is a binary de Bruijn sequence, and conversely every binary de Bruijn sequence arises in this way. In particular, since $D_n$ is balanced and connected there exists at least one binary de Bruijn sequence. In order to count the total number of such sequences, we need to compute $\det L_0(D_n)$. One way to do this is by a clever but messy sequence of elementary row and column operations which transforms the determinant into triangular form. We will give instead an elegant computation of the eigenvalues of $L(D_n)$ based on the following simple lemma.

**10.9 Lemma.** *Let $u$ and $v$ be any two vertices of $D_n$. Then there is a unique (directed) path from $u$ to $v$ of length $n - 1$.*

*Proof.* Suppose $u = a_1 a_2 \cdots a_{n-1}$ and $v = b_1 b_2 \cdots b_{n-1}$. Then the unique path of length $n - 1$ from $u$ to $v$ has vertices

$$a_1 a_2 \cdots a_{n-1}, a_2 a_3 \cdots a_{n-1} b_1, a_3 a_4 \cdots a_{n-1} b_1 b_2, \ldots,$$

$$a_{n-1} b_1 \cdots b_{n-2}, b_1 b_2 \cdots b_{n-1}.$$

$\square$

**10.10 Theorem.** *The eigenvalues of $L(D_n)$ are 0 (with multiplicity one) and 2 (with multiplicity $2^{n-1} - 1$).*

*Proof.* Let $A(D_n)$ denote the directed adjacency matrix of $D_n$, i.e., the rows and columns are indexed by the vertices, and for $n \geq 2$ we have

$$A_{uv} = \begin{cases} 1, & \text{if } (u, v) \text{ is an edge} \\ 0, & \text{otherwise.} \end{cases}$$

Now Lemma 10.9 is equivalent to the assertion that $A^{n-1} = J$, the $2^{n-1} \times 2^{n-1}$ matrix of all 1's [why?]. If the eigenvalues of $A$ are $\lambda_1, \ldots \lambda_{2^{n-1}}$, then the eigenvalues of $J = A^{n-1}$ are $\lambda_1^{n-1}, \ldots, \lambda_{2^{n-1}}^{n-1}$. By Lemma 1.4, the eigenvalues of $J$ are $2^{n-1}$ (once) and 0 ($2^{n-1} - 1$ times). Hence the eigenvalues of $A$ are $2\zeta$ (once, where $\zeta$ is an $(n - 1)$-st root of unity to be determined), and 0 ($2^{n-1} - 1$ times). Since the trace of $A$ is 2, it follows that $\zeta = 1$, and we have found all the eigenvalues of $A$.

Now $L(D_n) = 2I - A(D_n)$ [why?]. Hence the eigenvalues of $L$ are $2 - \lambda_1, \ldots, 2 - \lambda_{2^n-1}$, and the proof follows from the above determination of $\lambda_1, \ldots, \lambda_{2^n-1}$.                                                                                        □

**10.11 Corollary.** *The number $B_0(n)$ of binary de Bruijn sequences of degree $n$ beginning with $n$ 0's is equal to $2^{2^{n-1}-n}$. The total number $B(n)$ of binary de Bruijn sequences of degree $n$ is equal to $2^{2^{n-1}}$.*

*Proof.* By the above discussion, $B_0(n)$ is the number of Eulerian tours in $D_n$ whose first edge is the loop at vertex $00 \cdots 0$. Moreover, the outdegree of every vertex of $D_n$ is two. Hence by Corollary 10.5 and Theorem 10.10 we have

$$B_0(n) = \frac{1}{2^{n-1}} 2^{2^{n-1}-1} = 2^{2^{n-1}-n}.$$

Finally, $B(n)$ is obtained from $B_0(n)$ by multiplying by the number $2^n$ of edges, and the proof follows.                                                                                                □

Note that the total number of binary sequences of length $2^n$ is $N = 2^{2^n}$. By the previous corollary, the number of these which are de Bruijn sequences is just $\sqrt{N}$. This suggests the following problem, which remained open until 2009. Let $\mathcal{A}_n$ be the set of all binary sequences of length $2^n$. Let $\mathcal{B}_n$ be the set of binary de Bruijn sequences of degree $n$. Find an explicit bijection

$$\psi : \mathcal{B}_n \times \mathcal{B}_n \to \mathcal{A}_n,                                          \qquad (10.3)$$

thereby giving a combinatorial proof of Corollary 10.11.

# Notes for Chapter 10

The characterization of Eulerian digraphs given by Theorem 10.1 is a result of Good [52], while the fundamental connection between oriented subtrees and Eulerian tours in a balanced digraph that was used to prove Theorem 10.2 was shown by van Aardenne-Ehrenfest and de Bruijn [142, Thm. 5a]. This result is sometimes called the BEST Theorem, after de Bruijn, van Aardenne-Ehrenfest, Smith, and Tutte. However, Smith and Tutte were not involved in the original discovery. (In [118] Smith and Tutte give a determinantal formula for the number of Eulerian tours in a special class of balanced digraphs. Van Aardenne-Ehrenfest and de Bruijn refer to the paper of Smith and Tutte in a footnote added in proof.) The determinantal formula for the number of oriented subtrees of a directed graph (Theorem 10.4) is due to Tutte [139, Thm. 3.6].

De Bruijn sequences are named from the paper [30] of de Bruijn, where they are enumerated in the binary case. However, it was discovered by Stanley in 1975 that this work had been done earlier by Flye Sainte-Marie [42] in 1894, as reported by de Bruijn [32]. The generalization to $d$-ary de Bruijn sequences (Exercise 10.2) is due to van Ardenne-Ehrenfest and de Bruijn [142]. Some recent work in this area appears in a special issue [132] of *Discrete Mathematics*. Some entertaining applications to magic are given by Diaconis and Graham [35, Chs. 2–4]. The bijection $\psi$ of (10.3) is due to Bidkhori and Kishore [9].

## Exercises for Chapter 10

1. Choose positive integers $a_1, \ldots, a_{p-1}$. Let $D = D(a_1, \ldots, a_{p-1})$ be the digraph defined as follows. The vertices of $D$ are $v_1, \ldots, v_p$. For each $1 \le i \le p - 1$, there are $a_i$ edges from $x_i$ to $x_{i+1}$ and $a_i$ edges from $x_{i+1}$ to $x_i$. For instance, $D(1, 3, 2)$ looks like

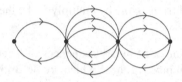

   (a) Find by a direct argument (no determinants) the number $\tau(D, v)$ of oriented subtrees with a given root $v$.
   (b) Find the number $\epsilon(D, e)$ of Eulerian tours of $D$ whose first edge is $e$.

2. Let $d > 1$. A *$d$-ary sequence* is a sequence whose terms belong to $\{0, 1, \ldots, d - 1\}$. A *$d$-ary de Bruijn sequence* of degree $n$ is a $d$-ary sequence $A = a_1 a_2 \cdots a_{d^n}$ such that every $d$-ary sequence $b_1 b_2 \cdots b_n$ of length $n$ occurs exactly once as a circular factor of $A$. Find the number of $d$-ary de Bruijn sequences of length $n$ which begin with $n$ 0's.

3. (a) Let $D$ be a digraph and $v$ a vertex of $D$. For $k \ge 1$, let $D^{[k]}$ denote the digraph obtained from $D$ by replacing each directed edge $e$ with $k$ edges with the same initial and final vertices as $e$. Express $\tau(D^{[k]}, v)$ in terms of $\tau(D, v)$.
   (b) Suppose also that $D$ is balanced and connected, with vertex outdegrees $d_1, \ldots, d_p$. Let $e$ be an edge of $D$, and let $e'$ be an edge of $D^{[k]}$. Express $\epsilon(D^{[k]}, e')$ in terms of $\epsilon(D, e)$ and the numbers $d_1, \ldots, d_p$.

4. (a) (*) Let $n, d, k \ge 1$. Show that the number of $d$-ary sequences $a_1 a_2 \cdots a_{kd^n+n-1}$ such that every $d$-ary sequence $b_1 b_2 \cdots b_n$ occurs exactly $k$ times as a *linear* factor $a_j a_{j+1} \cdots a_{j+n-1}$ (so $1 \le j \le kd^n$) is equal to

$$\left( \frac{(kd)!}{k!^d} \right)^{d^{n-1}}.$$

(b) (more difficult) (*) Show that the number of $d$-ary sequences $a_1 a_2 \cdots a_{kd^n}$ such that every $d$-ary sequence $b_1 b_2 \cdots b_n$ occurs exactly $k$ times as a *circular* factor is equal to

$$\frac{1}{k} \sum_{r | k} \phi(k/r) \left( \frac{(rd)!}{r!^d} \right)^{d^{n-1}}.$$

5. Let $G$ be a regular loopless (undirected) graph of degree $d$ with $p$ vertices and $q$ edges.

(a) Find a simple relation between $p$, $q$, and $d$.

(b) (*) Express the largest eigenvalue of the adjacency matrix $A$ of $G$ in terms of $p$, $q$, and $d$.

(c) Suppose also that $G$ has no multiple edges. Express the number of closed walks in $G$ of length two in terms of $p$, $q$, and $d$.

(d) Suppose that $G$ has no multiple edges and that the number of closed walks in $G$ of length $\ell$ is given by

$$6^\ell + 2 \cdot (-3)^\ell.$$

Find the number $\kappa(G)$ of spanning trees of $G$. (Don't forget that $A$ may have some eigenvalues equal to 0.) Give a purely numerical answer, not involving $p$, $q$, or $d$.

(e) Let $G$ be as in (d). How many closed walks in $G$ walk along each edge of $G$ exactly once in each direction? Give a purely numerical answer.

6. (a) (difficult) Let $f(p)$ be the number of loopless connected digraphs $D$ on the vertex set $[p]$ such that $D$ has exactly one Eulerian tour (up to cyclic shift). For instance, $f(3) = 5$; two such digraphs are triangles, and three consist of two 2-cycles with a common vertex. Show that

$$f(p) = (p+2)(p+3) \cdots (2p-1), \quad p \geq 2.$$

(b) (somewhat more difficult) Suppose now that loops are allowed, and let $g(p)$ denote the resulting number of digraphs. For instance, $g(1) = 2$, the two graphs being a single vertex and a single vertex with a loop. Show that

$$\sum_{p \geq 1} g(p) \frac{x^p}{(p-1)!} = \frac{(1-x)^2 - (1+x)\sqrt{1 - 6x + x^2}}{4x}.$$

7. Suppose that the connected digraph $D$ has $p$ vertices, each of outdegree $d$ and indegree $d$. Let $D'$ be the graph obtained from $D$ by doubling each edge, i.e., replacing each edge $u \to v$ with two such edges. Express $\epsilon(D', e')$ (the number of Eulerian tours of $D'$ beginning with the edge $e'$ of $D'$) in terms of $\epsilon(D, e)$.

8. Let $D$ be a digraph with $p$ vertices, and let $\ell$ be a fixed positive integer. Suppose that for every pair $u$, $v$ of vertices of $D$, there is a unique (directed) walk of length $\ell$ from $u$ to $v$.

   (a) (*) What are the eigenvalues of the (directed) adjacency matrix $A(D)$?
   (b) How many loops $(v, v)$ does $D$ have?
   (c) (*) Show that $D$ is connected and balanced.
   (d) Show that all vertices have the same indegree $d$ and same outdegree, which by (c) is also $d$. Find a simple formula relating $p$, $d$, and $\ell$.
   (e) How many Eulerian tours does $D$ have?
   (f) (*) (open–ended) What more can be said about $D$? Show that $D$ need not be a de Bruijn graph (the graphs used to solve #2).

9. (a) Let $n \geq 3$. Show that there does not exist a sequence $a_1, a_2, \ldots, a_{n!}$ such that the $n!$ circular factors $a_i, a_{i+1}, \ldots, a_{i+n-1}$ (subscripts taken modulo $n!$ if necessary) are the $n!$ permutations of $[n]$.
   (b) Show that for all $n \geq 1$ there does exist a sequence $a_1, a_2, \ldots, a_{n!}$ such that the $n!$ circular factors $a_i, a_{i+1}, \ldots, a_{i+n-2}$ consist of the first $n - 1$ terms $b_1, \ldots, b_{n-1}$ of all permutations $b_1, b_2, \ldots, b_n$ of $[n]$. Such sequences are called *universal cycles for* $\mathfrak{S}_n$. When $n = 3$, an example of such a universal cycle is 123213.
   (c) When $n = 3$, find the number of universal cycles beginning with 123.
   (d) (unsolved) Find the number $U_n$ of universal cycles for $\mathfrak{S}_n$ beginning with $1, 2, \ldots, n$. It is known that

$$U_4 = 2^7 \cdot 3$$
$$U_5 = 2^{33} \cdot 3^8 \cdot 5^3$$
$$U_6 = 2^{190} \cdot 3^{49} \cdot 5^{33}$$
$$U_7 = 2^{1217} \cdot 3^{123} \cdot 5^{119} \cdot 7^5 \cdot 11^{28} \cdot 43^{35} \cdot 73^{20} \cdot 79^{21} \cdot 109^{35}.$$

Moreover, $U_9$ is divisible by $p^{168}$, where $p = 59229013196333$ is prime. Most likely there is not a "nice" formula for $U_n$. Some, but not all, of the factorization of $U_n$ into lots of factors can be explained using the representation theory of $\mathfrak{S}_n$, a topic beyond the scope of this text.

# Chapter 11
# Cycles, Bonds, and Electrical Networks

## 11.1 The Cycle Space and Bond Space

In this chapter we will deal with some interesting linear algebra related to the structure of a directed graph. Let $D = (V, E)$ be a digraph. A function $f: E \to \mathbb{R}$ is called a *circulation* or *flow* if for every vertex $v \in V$, we have

$$\sum_{\substack{e \in E \\ \text{init}(e)=v}} f(e) = \sum_{\substack{e \in E \\ \text{fin}(e)=v}} f(e). \tag{11.1}$$

Thus if we think of the edges as pipes and $f$ as measuring the flow (quantity per unit of time) of some commodity (such as oil) through the pipes in the specified direction (so that a negative value of $f(e)$ means a flow of $|f(e)|$ in the direction opposite the direction of $e$), then (11.1) simply says that the amount flowing into each vertex equals the amount flowing out. In other words, the flow is *conservative*. The figure below illustrates a circulation in a digraph $D$.

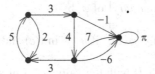

Let $\mathcal{C} = \mathcal{C}_D$ denote the set of all circulations on $D$. Clearly if $f, g \in \mathcal{C}$ and $\alpha, \beta \in \mathbb{R}$ then $\alpha f + \beta g \in \mathcal{C}$. Hence $\mathcal{C}$ is a (real) vector space, called the *cycle space* of $D$. Thus if $q = \#E$, then $\mathcal{C}_D$ is a subspace of the $q$-dimensional vector space $\mathbb{R}^E$ of all functions $f: E \to \mathbb{R}$.

© Springer International Publishing AG, part of Springer Nature 2018
R. P. Stanley, *Algebraic Combinatorics*, Undergraduate Texts in Mathematics,
https://doi.org/10.1007/978-3-319-77173-1_11

What do circulations have to do with something "circulating," and what does the cycle space have to do with actual cycles? To see this, define a *circuit* or *elementary cycle* in $D$ to be a set of edges of a closed walk, *ignoring the direction of the arrows*, with no repeated vertices except the first and last. Suppose that a circuit $C$ has been assigned an orientation (direction of travel) $\mathfrak{o}$. (Note that this meaning of orientation is not the same as that appearing in Definition 9.5.)

Define a function $f_C \colon E \to \mathbb{R}$ (which also depends on the orientation $\mathfrak{o}$, though we suppress it from the notation) by

$$f_C(e) = \begin{cases} \phantom{-}1, & \text{if } e \in C \text{ and } e \text{ agrees with } \mathfrak{o} \\ -1, & \text{if } e \in C \text{ and } e \text{ is opposite to } \mathfrak{o} \\ \phantom{-}0, & \text{otherwise.} \end{cases}$$

It is easy to see that $f_C$ is a circulation. Later we will see that the circulations $f_C$ span the cycle space $\mathcal{C}$, explaining the terminology "circulation" and "cycle space." The figure below shows a circuit $C$ with an orientation $\mathfrak{o}$, and the corresponding circulation $f_C$.

Given a function $\varphi \colon V \to \mathbb{R}$, called a *potential* on $D$, define a new function $\delta\varphi \colon E \to \mathbb{R}$, called the *coboundary*[1] of $\varphi$, by

$$\delta\varphi(e) = \varphi(v) - \varphi(u), \quad \text{if } u = \operatorname{init}(e) \text{ and } v = \operatorname{fin}(e).$$

Figure 11.1 shows a digraph $D$ with the value $\varphi(v)$ of some function $\varphi \colon V \to \mathbb{R}$ indicated at each vertex $v$, and the corresponding values $\delta\varphi(e)$ shown at each edge $e$.

One should regard $\delta$ as an operator which takes an element $\varphi$ of the vector space $\mathbb{R}^V$ of all functions $V \to \mathbb{R}$ and produces an element of the vector space $\mathbb{R}^E$ of all functions $E \to \mathbb{R}$. It is immediate from the definition of $\delta$ that $\delta$ is *linear*, i.e.,

$$\delta(a\varphi_1 + b\varphi_2) = a \cdot \delta\varphi_1 + b \cdot \delta\varphi_2,$$

---

[1]The term "coboundary" arises from algebraic topology, but we will not explain the connection here.

**Fig. 11.1** A function
(potential) and its coboundary

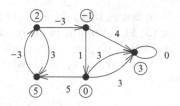

for all $\varphi_1, \varphi_2 \in \mathbb{R}^V$ and $a, b \in \mathbb{R}$. Thus $\delta$ is simply a certain linear transformation $\delta \colon \mathbb{R}^V \to \mathbb{R}^E$ between two finite-dimensional vector spaces.

A function $g \colon E \to \mathbb{R}$ is called a *potential difference* on $D$ if $g = \delta\varphi$ for some $\varphi \colon V \to \mathbb{R}$. (Later we will see the connection with electrical networks that accounts for the terminology "potential difference.") Let $\mathcal{B} = \mathcal{B}_D$ be the set of all potential differences on $D$. Thus $\mathcal{B}$ is just the image of the linear transformation $\delta$ and is hence a real vector space, called the *bond space* of $D$.

Let us explain the reason behind the terminology "bond space." A *bond* in a digraph $D$ is a set $B$ of edges such that (a) removing $B$ from $D$ disconnects some (undirected) component of $D$ (i.e., removing $B$ creates a digraph which has more connected components, as an undirected graph, than $D$), and (b) no proper subset of $B$ has this property. A subset of edges satisfying (a) is called a *cutset*, so a bond is just a minimal cutset. Suppose, for example, that $D$ is given as follows (with no arrows drawn since they are irrelevant to the definition of bond):

Then the bonds are the six subsets $ab, de, acd, bce, ace,$ and $bcd$.

Let $B$ be a bond. Suppose $B$ disconnects the component $(V', E')$ into two pieces (a bond always disconnects some component into exactly two pieces [why?]) with vertex set $S$ in one piece and $\bar{S}$ in the other. Thus $S \cup \bar{S} = V'$ and $S \cap \bar{S} = \emptyset$. Define

$$[S, \bar{S}] = \{e \in E \colon \text{exactly one vertex of } e \text{ lies in } S \text{ and one lies in } \bar{S}\}.$$

Clearly $B = [S, \bar{S}]$. It is often convenient to use the notation $[S, \bar{S}]$ for a bond.

Given a bond $B = [S, \bar{S}]$ of $D$, define a function $g_B \colon E \to \mathbb{R}$ by

$$g_B(e) = \begin{cases} 1, & \text{if init}(e) \in \bar{S}, \text{fin}(e) \in S \\ -1, & \text{if init}(e) \in S, \text{fin}(e) \in \bar{S} \\ 0, & \text{otherwise.} \end{cases}$$

Note that $g_B$ really depends not just on $B$, but also on whether we write $B$ as $[S, \bar{S}]$ or $[\bar{S}, S]$. Writing $B$ in the reverse way simply changes the sign of $g_B$. Whenever we deal with $g_B$ we will assume that some choice $B = [S, \bar{S}]$ has been made.

Now note that $g_B = \delta\varphi$, where

$$\varphi(v) = \begin{cases} 1, & \text{if } v \in S \\ 0, & \text{if } v \notin S. \end{cases}$$

Hence $g_B \in \mathcal{B}$, the bond space of $D$. We will later see that $\mathcal{B}$ is in fact spanned by the functions $g_B$, explaining the terminology "bond space."

**11.1 Example.** In the digraph below, open (white) vertices indicate an element of $S$ and closed (black) vertices an element of $\bar{S}$ for a certain bond $B = [S, \bar{S}]$. The elements of $B$ are drawn with solid lines. The edges are labelled by the values of $g_B$, and the vertices by the function $\varphi$ for which $g_B = \delta\varphi$.

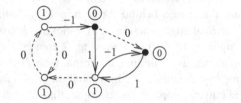

Recall that in Definition 9.5 we defined the incidence matrix $M(G)$ of a loopless undirected graph $G$ with respect to an orientation $o$. We may just as well think of $G$ together with its orientation $o$ as a directed graph. We also will allow loops. Thus if $D = (V, E)$ is any (finite) digraph, define the *incidence matrix* $M = M(D)$ to be the $p \times q$ matrix whose rows are indexed by $V$ and columns by $E$, as follows. The entry in row $v \in V$ and column $e \in E$ is denoted $m_v(e)$ and is given by

$$m_v(e) = \begin{cases} -1, & \text{if } v = \text{init}(e) \text{ and } e \text{ is not a loop} \\ 1, & \text{if } v = \text{fin}(e) \text{ and } e \text{ is not a loop} \\ 0, & \text{otherwise}. \end{cases}$$

For instance, if $D$ is given by

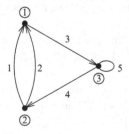

then

$$M(D) = \begin{bmatrix} 1 & 1 & -1 & 0 & 0 \\ -1 & -1 & 0 & 1 & 0 \\ 0 & 0 & 1 & -1 & 0 \end{bmatrix}.$$

**11.2 Theorem.** *The row space of $M(D)$ is the bond space $\mathcal{B}_D$. Equivalently, the functions $m_v \colon E \to \mathbb{R}$, where $v$ ranges over all vertices of $D$, span $\mathcal{B}_D$.*

*Proof.* Let $g = \delta\varphi$ be a potential difference on $D$, so

$$g(e) = \varphi(\mathrm{fin}(e)) - \varphi(\mathrm{init}(e))$$

$$= \sum_{v \in V} \varphi(v) m_v(e).$$

Thus $g = \sum_{v \in V} \varphi(v) m_v$, so $g$ belongs to the row space of $M$.

Conversely, if $g = \sum_{v \in V} \psi(v) m_v$ is in the row space of $M$, where $\psi \colon V \to \mathbb{R}$, then $g = \delta\psi \in \mathcal{B}$.                                                                 $\square$

We now define a scalar product (or inner product) on the space $\mathbb{R}^E$ by

$$\langle f, g \rangle = \sum_{e \in E} f(e)g(e),$$

for any $f, g \in \mathbb{R}^E$. If we think of the numbers $f(e)$ and $g(e)$ as the coordinates of $f$ and $g$ with respect to the basis $E$, then $\langle f, g \rangle$ is just the usual dot product of $f$ and $g$. Because we have a scalar product, we have a notion of what it means for $f$ and $g$ to be *orthogonal*, viz., $\langle f, g \rangle = 0$. If $\mathcal{V}$ is any subspace of $\mathbb{R}^E$, then define the *orthogonal complement* $\mathcal{V}^{\perp}$ of $\mathcal{V}$ by

$$\mathcal{V}^{\perp} = \{ f \in \mathbb{R}^E : \langle f, g \rangle = 0 \text{ for all } g \in \mathcal{V} \}.$$

Recall from linear algebra that

$$\dim \mathcal{V} + \dim \mathcal{V}^{\perp} = \dim \mathbb{R}^E = \#E. \tag{11.2}$$

Furthermore, $\left( \mathcal{V}^{\perp} \right)^{\perp} = \mathcal{V}$. Let us also note that since we are working over $\mathbb{R}$, we have $\mathcal{V} \cap \mathcal{V}^{\perp} = \{0\}$. Thus $\mathbb{R}^E = \mathcal{V} \oplus \mathcal{V}^{\perp}$ (direct sum).

Intuitively there is a kind of "duality" between elementary cycles and bonds. Cycles "hold vertices together," while bonds "tear them apart." The precise statement of this duality is given by the next result.

**11.3 Theorem.** *The cycle and bond spaces of $D$ are related by $\mathcal{C} = \mathcal{B}^{\perp}$. (Equivalently, $\mathcal{B} = \mathcal{C}^{\perp}$.)*

*Proof.* Let $f : E \to \mathbb{R}$. Then $f$ is a circulation if and only if

$$\sum_{e \in E} m_v(e) f(e) = 0$$

for all $v \in V$ [why?]. But this is exactly the condition that $f \in \mathcal{B}^\perp$.

$\square$

## 11.2   Bases for the Cycle Space and Bond Space

We want to examine the incidence matrix $M(D)$ in more detail. In particular, we would like to determine which rows and columns of $M(D)$ are linearly independent, and which span the row and column spaces. As a corollary, we will determine the dimension of the spaces $\mathcal{B}$ and $\mathcal{C}$. We begin by defining the *support* $\|f\|$ of $f : E \to \mathbb{R}$ to be the set of edges $e \in E$ for which $f(e) \neq 0$.

**11.4 Lemma.** *If $0 \neq f \in \mathcal{C}$, then $\|f\|$ contains an undirected circuit.*

*Proof.* If not, then $\|f\|$ has a vertex of degree one [why?], which is clearly impossible.                                    $\square$

**11.5 Lemma.** *If $0 \neq g \in \mathcal{B}$, then $\|g\|$ contains a bond.*

*Proof.* Let $0 \neq g \in \mathcal{B}$, so $g = \delta\varphi$ for some $\varphi : V \to \mathbb{R}$. Choose a vertex $v$ which is incident to an edge of $\|g\|$, and set

$$U = \{u \in V : \varphi(u) = \varphi(v)\}.$$

Let $\bar{U} = V - U$. Note that $\bar{U} \neq \emptyset$, since otherwise $\varphi$ is constant so $g = 0$. Since $g(e) \neq 0$ for all $e \in [U, \bar{U}]$ [why?], we have that $\|g\|$ contains the cutset $[U, \bar{U}]$. Since a bond is by definition a minimal cutset, it follows that $\|g\|$ contains a bond.

$\square$

A matrix $B$ is called a *basis matrix* of $\mathcal{B}$ if the rows of $B$ form a basis for $\mathcal{B}$. Similarly define a basis matrix $C$ of $\mathcal{C}$.

Recall the notation of Theorem 9.4: let $A$ be a matrix with at least as many columns as rows, whose columns are indexed by the elements of a set $T$. If $S \subseteq T$, then $A[S]$ denotes the submatrix of $A$ consisting of the columns indexed by the elements of $S$. In particular, $A[e]$ (short for $A[\{e\}]$) denotes the column of $A$ indexed by $e$. We come to our first significant result about bases for the vector spaces $\mathcal{B}$ and $\mathcal{C}$.

**11.6 Theorem.** *Let $B$ be a basis matrix of $\mathcal{B}$, and $C$ a basis matrix of $\mathcal{C}$. (Thus the columns of $B$ and $C$ are indexed by the edges $e \in E$ of $D$.) Let $S \subseteq E$, Then:*

(i) *The columns of $B[S]$ are linearly independent if and only if $S$ is acyclic (i.e., contains no circuit as an undirected graph).*

(ii) *The columns of $C[S]$ are linearly independent if and only if $S$ contains no bond.*

*Proof.* The columns of $B[S]$ are linearly dependent if and only if there exists a function $f : E \to \mathbb{R}$ such that

$$f(e) \neq 0 \text{ for some } e \in S$$

$$f(e) = 0 \text{ for all } e \notin S$$

$$\sum_{e \in E} f(e) B[e] = \mathbf{0}, \text{ the column vector of } 0\text{'s.} \tag{11.3}$$

The last condition is equivalent to $\langle f, m_v \rangle = 0$ for all $v \in V$, i.e., $f$ is a circulation. Thus the columns of $B[S]$ are linearly dependent if and only if there exists a nonzero circulation $f$ such that $\|f\| \subseteq S$. By Lemma 11.4, $\|f\|$ (and therefore $S$) contains a circuit. Conversely, if $S$ contains a circuit $C$ then $0 \neq f_C \in C$ and $\|f_C\| = C \subseteq S$, so $f_C$ defines a linear dependence relation (11.3) among the columns. Hence the columns of $B[S]$ are linearly independent if and only if $S$ is acyclic, proving (i). (Part (i) can also be deduced from Lemma 9.7.)

The proof of (ii) is similar and is left as an exercise. $\qquad \square$

**11.7 Corollary.** *Let $D = (V, E)$ be a digraph with $p$ vertices, $q$ edges, and $k$ connected components (as an undirected graph). Then*

$$\dim \mathcal{B} = p - k$$

$$\dim \mathcal{C} = q - p + k.$$

*Proof.* For any matrix $X$, the rank of $X$ is equal to the maximum number of linearly independent columns. Now let $B$ be a basis matrix of $\mathcal{B}$. By Theorem 11.6(i), the rank of $B$ is then the maximum size (number of elements) of an acyclic subset of $E$. In each connected component $D_i$ of $D$, the largest acyclic subsets are the spanning trees, whose number of edges is $p(D_i) - 1$, where $p(D_i)$ is the number of vertices of $D_i$. Hence

$$\text{rank } B = \sum_{i=1}^{k} (p(D_i) - 1)$$

$$= p - k.$$

Since $\dim \mathcal{B} + \dim \mathcal{C} = \dim \mathbb{R}^E = q$ by (11.2) and Theorem 11.3, we have

$$\dim \mathcal{C} = q - (p - k) = q - p + k.$$

(It is also possible to determine $\dim \mathcal{C}$ by a direct argument similar to our determination of $\dim \mathcal{B}$.) $\qquad \square$

The number $q - p + k$ (which should be thought of as the number of independent cycles in $D$) is called the *cyclomatic number* of $D$ (or of its undirected version $G$, since the direction of the edges has no effect).

Our next goal is to describe explicit bases of $C$ and $B$. Recall that a *forest* is an undirected graph without circuits, or equivalently, a disjoint union of trees. We extend the definition of forest to directed graphs by ignoring the arrows, i.e., a directed graph is a forest if it has no circuits as an undirected graph. Equivalently [why?], $\dim C = 0$.

Pick a maximal forest $T$ of $D = (V, E)$. Thus $T$ restricted to each component of $D$ is a spanning tree. If $e$ is an edge of $D$ not in $T$, then it is easy to see that $T \cup e$ contains a unique circuit $C_e$.

**11.8 Theorem.** *Let $T$ be as above. Then the set $S$ of circulations $f_{C_e}$, as $e$ ranges over all edges of $D$ not in $T$, is a basis for the cycle space $C$.*

*Proof.* The circulations $f_{C_e}$ are linearly independent, since for each $e \in E(D) - E(T)$ only $f_{C_e}$ doesn't vanish on $e$. Moreover,

$$\#S = \#E(D) - \#E(T) = q - p + k = \dim C,$$

so $S$ is a basis.                                                                                    □

**11.9 Example.** Let $D$ be the digraph shown below, with the edges $a, b, c$ of $T$ shown by dotted lines.

Orient each circuit $C_t$ in the direction of the added edge, i.e., $f_{C_t}(t) = 1$. Then the basis matrix $C$ of $C$ corresponding to the basis $f_{C_d}, f_{C_e}, f_{C_f}$ is given by

$$C = \begin{bmatrix} 0 & -1 & -1 & 1 & 0 & 0 \\ -1 & -1 & -1 & 0 & 1 & 0 \\ 0 & 0 & -1 & 0 & 0 & 1 \end{bmatrix}. \tag{11.4}$$

We next want to find a basis for the bond space $B$ analogous to that of Theorem 11.8.

**11.10 Lemma.** *Let $T$ be a maximal forest of $D = (V, E)$. Let $T^* = D - E(T)$ (the digraph obtained from $D$ by removing the edges of $T$), called a* cotree *if $D$ is connected. Let $e$ be an edge of $T$. Then $E(T^*) \cup e$ contains a unique bond.*

*Proof.* Removing $E(T^*)$ from $D$ leaves a maximal forest $T$, so removing one further edge $e$ disconnects some component of $D$. Hence $E(T^*) \cup e$ contains a bond $B$. It remains to show that $B$ is unique. Removing $e$ from $T$ breaks some component of $T$ into two connected graphs $T_1$ and $T_2$ with vertex sets $S$ and $\bar{S}$. It follows [why?] that we must have $B = [S, \bar{S}]$, so $B$ is unique.                                  □

Let $T$ be a maximal forest of the digraph $D$, and let $e$ be an edge of $T$. By the previous lemma, $E(T^*) \cup e$ contains a unique bond $B_e$. Let $g_{B_e}$ be the corresponding element of the bond space $\mathcal{B}$, chosen for definiteness so that $g_{B_e}(e) = 1$.

**11.11 Theorem.** *The set of functions $g_{B_e}$, as $e$ ranges over all edges of $T$, is a basis for the bond space $\mathcal{B}$.*

*Proof.* The functions $g_{B_e}$ are linearly independent, since only $g_{B_e}$ is nonzero on $e \in E(T)$. Since

$$\#E(T) = p - k = \dim \mathcal{B},$$

it follows that the $g_{B_e}$'s are a basis for $\mathcal{B}$.                           □

**11.12 Example.** Let $D$ and $T$ be as in the previous diagram. Thus a basis for $\mathcal{B}$ is given by the functions $g_{B_a}, g_{B_b}, g_{B_c}$. The corresponding basis matrix is given by

$$\boldsymbol{B} = \begin{bmatrix} 1 & 0 & 0 & 0 & 1 & 0 \\ 0 & 1 & 0 & 1 & 1 & 0 \\ 0 & 0 & 1 & 1 & 1 & 1 \end{bmatrix}.$$

Note that the rows of $\boldsymbol{B}$ are orthogonal to the rows of the matrix $\boldsymbol{C}$ of (11.4), in accordance with Theorem 11.3. Equivalently, $\boldsymbol{B}\boldsymbol{C}^t = \boldsymbol{0}$, the $3 \times 3$ zero matrix. (In general, $\boldsymbol{B}\boldsymbol{C}^t$ will have $q - p + k$ rows and $p - k$ columns. Here it is just a coincidence that these two numbers are equal.)

The basis matrices $\boldsymbol{C}_T$ and $\boldsymbol{B}_T$ of $\mathcal{C}$ and $\mathcal{B}$ obtained from a maximal forest $T$ have an important property. A real $m \times n$ matrix $A$ with $m \leq n$ is said to be *unimodular* if every $m \times m$ submatrix has determinant 0, 1, or $-1$. For instance, the adjacency matrix $\boldsymbol{M}(D)$ of a digraph $D$ is unimodular, as proved in Lemma 9.7 (by showing that the expansion of the determinant of a full submatrix has at most one nonzero term).

**11.13 Theorem.** *Let $T$ be a maximal forest of $D$. Then the basis matrices $\boldsymbol{C}_T$ of $\mathcal{C}$ and $\boldsymbol{B}_T$ of $\mathcal{B}$ are unimodular.*

*Proof.* First consider the case $\boldsymbol{C}_T$. Let $\boldsymbol{P}$ be a full submatrix of $\boldsymbol{C}$ (so $\boldsymbol{P}$ has $q - p + k$ rows and columns). Assume $\det \boldsymbol{P} \neq 0$. We need to show $\det \boldsymbol{P} = \pm 1$. Since $\det \boldsymbol{P} \neq 0$, it follows from Theorem 11.6(ii) that $\boldsymbol{P} = \boldsymbol{C}_T[T_1^*]$ for the complement $T_1^*$ of some maximal forest $T_1$. Note that the rows of the matrix $\boldsymbol{C}_T[T_1^*]$ are indexed by $T^*$ and the columns by $T_1^*$. Similarly the rows of the basis matrix $\boldsymbol{C}_{T_1}$ are indexed by $T_1^*$ and the columns by $E$ (the set of all edges of $D$). Hence it makes sense to define the matrix product

$$Z = C_T[T_1^*]C_{T_1},$$

a matrix whose rows are indexed by $T^*$ and columns by $E$.

Note that the matrix $Z$ is a basis matrix for the cycle space $\mathcal{C}$ since its rows are linear combinations of the rows of the basis matrix $C_{T_1}$, and it has full rank since the matrix $C_T[T_1^*]$ is invertible. Now $C_{T_1}[T_1^*] = I_{T_1^*}$ (the identity matrix indexed by $T_1^*$), so $Z[T_1^*] = C_T[T_1^*]$. Thus $Z$ agrees with the basis matrix $C_T$ in columns $T_1^*$. Hence the rows of $Z - C_T$ are circulations supported on a subset of $T_1$. Since $T_1$ is acyclic, it follows from Lemma 11.4 that the only such circulation is identically 0, so $Z = C_T$.

We have just shown that

$$C_T[T_1^*]C_{T_1} = C_T.$$

Restricting both sides to $T^*$, we obtain

$$C_T[T_1^*]C_{T_1}[T^*] = C_T[T^*] = I_{T^*}.$$

Taking determinants yields

$$\det(C_T[T_1^*])\det(C_{T_1}[T^*]) = 1.$$

Since all the matrices we have been considering have integer entries, the above determinants are integers. Hence

$$\det C_T[T_1^*] = \pm 1,$$

as was to be proved.

A similar proof works for $B_T$.                                                                    □

## 11.3   Electrical Networks

We will give a brief indication of the connection between the above discussion and the theory of electrical networks. Let $D$ be a digraph, which for convenience we assume is *connected* and *loopless*. Suppose that at each edge $e$ there is a voltage (potential difference) $V_e$ from init $e$ to fin $e$, and a current $I_e$ in the direction of $e$ (so a negative current $I_e$ indicates a current of $|I_e|$ in the direction opposite to $e$). Think of $V$ and $I$ as functions on the edges, i.e., as elements of the vector space $\mathbb{R}^E$. There are three fundamental laws relating the quantities $V_e$ and $I_e$.

**Kirchhoff's First Law.** $I \in \mathcal{C}_D$. *In other words, the current flowing into a vertex equals the current flowing out. In symbols,*

$$\sum_{\substack{e \\ \text{init } e=v}} I_e = \sum_{\substack{e \\ \text{fin } e=v}} I_e,$$

*for all vertices $v \in V$.*

**Kirchhoff's Second Law.** $V \in \mathcal{C}_D^{\perp} = \mathcal{B}$. *In other words, the sum of the voltages around any circuit (called loops by electrical engineers), taking into account orientations, is 0.*

**Ohm's Law.** *If edge $e$ has resistance $R_e > 0$, then $V_e = I_e R_e$.*

The central problem of electrical network theory dealing with the above three laws[2] is the following: which of the $3q$ quantities $V_e, I_e, R_e$ need to be specified to uniquely determine all the others, and how can we find or stipulate the solution in a fast and elegant way? We will be concerned here only with a special case, perhaps the most important special case in practical applications. Namely, suppose we apply a voltage $V_q \neq 0$ at edge $e_q$, with resistances $R_1, \ldots, R_{q-1}$ at the other edges $e_1, \ldots, e_{q-1}$. Let $V_i, I_i$ be the voltage and current at edge $e_i$. We would like to express each $V_i$ and $I_i$ in terms of $V_q$ and $R_1, \ldots, R_{q-1}$. By "physical intuition" there should be a unique solution, since we can actually build a network meeting the specifications of the problem. Note that if we have quantities $V_i, I_i, R_i$ satisfying the three network laws above, then for any scalar $\alpha$ the quantities $\alpha V_i, \alpha I_i, R_i$ are also a solution. This means that we might as well assume that $V_q = 1$, since we can always multiply all voltages and currents afterwards by whatever value we want $V_q$ to be.

When we apply a voltage $V_q$ the current will flow along $e_q$ from the lower potential (say at vertex $u$) to the higher (at vertex $v$). Since by convention the direction of current flow is from the higher potential to the lower, if we orient $e_q$ from $u$ to $v$, then $V_q < 0$ and $I_q > 0$. Thus $V_q/I_q < 0$ so we should define the *total resistance* $R(D)$ of the network $D$, together with the distinguished edge $e$, by $R(D) = -V_q/I_q$. If we replace the entire network $D$ except for the edge $e_q$ by a single edge from $u$ to $v$, then the resistance of this edge will be $R(D)$.

As an illustration of a simple method of computing the total resistance of a network, the following diagram illustrates the notion of a *series connection* $D_1 + D_2$ and a *parallel connection* $D_1 \parallel D_2$ of two networks $D_1$ and $D_2$ with a distinguished edge $e$ at which a voltage is applied.

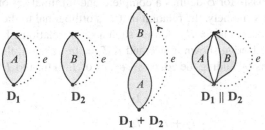

$$D_1 \qquad D_2 \qquad \qquad D_1 \parallel D_2$$
$$D_1 + D_2$$

---

[2]Of course the situation becomes much more complicated when one introduces *dynamic* network elements like capacitors, alternating current, etc.

It is well-known and easy to deduce from the three network laws that

$$R(D_1 + D_2) = R(D_1) + R(D_2)$$

$$\frac{1}{R(D_1 \parallel D_2)} = \frac{1}{R(D_1)} + \frac{1}{R(D_2)}. \qquad (11.5)$$

A network that is built up from a single edge by a sequence of series and parallel connections is called a *series–parallel network*. An example is the following, with the distinguished edge $e$ shown by a broken line from bottom to top.

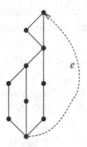

The simplest network which is not a series-parallel network is called the *Wheatstone bridge* and is illustrated below. (The direction of the arrows has been chosen arbitrarily.) We will use this network as our main example in the discussion that follows.

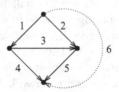

We now return to an arbitrary connected loopless digraph $D$, with currents $I_i$, voltages $V_i$, and resistances $R_i$ at the edges $e_i$. Recall that we are fixing $V_q = 1$ and $R_1, \ldots, R_{q-1}$. Let $T$ be a spanning tree of $D$. Since $I$ is a current if and only if it is orthogonal to the bond space $\mathcal{B}$ (Theorem 11.3 and Kirchhoff's First Law), it follows that any basis for $\mathcal{B}$ defines a complete and minimal set of linear relations satisfied by the $I_i$'s (namely, the relation that $I$ is orthogonal to the basis elements). In particular, the basis matrix $\boldsymbol{B}_T$ defines such a set of relations. For example, if $D$ is the Wheatstone bridge shown above and if $T = \{e_1, e_2, e_5\}$, then we obtain the following relations by adding the edges $e_1, e_2, e_5$ of $T$ in turn to $T^*$:

$$I_1 - I_3 - I_4 = 0$$

$$I_2 + I_3 + I_4 + I_6 = 0 \qquad (11.6)$$

$$I_4 + I_5 + I_6 = 0.$$

These three ($= p - 1$) equations give all the relations satisfied by the $I_i$'s alone, and the equations are linearly independent.

Similarly if $V$ is a voltage then it is orthogonal to the cycle space $\mathcal{C}$. Thus any basis for $\mathcal{C}$ defines a complete and minimal set of linear relations satisfied by the $V_i$'s (namely, the relation that $V$ is orthogonal to the basis elements). In particular, the basis matrix $C_T$ defines such a set of relations. Continuing our example, we obtain the following relations by adding the edges $e_3, e_4, e_6$ of $T^*$ in turn to $T$.

$$V_1 - V_2 + V_3 = 0$$
$$V_1 - V_2 + V_4 - V_5 = 0 \tag{11.7}$$
$$V_2 + V_5 = 1,$$

These three ($= q - p + k$) equations give all the relations satisfied by the $V_i$'s alone, and the equations are linearly independent.

In addition, Ohm's Law gives the $q - 1$ equations $V_i = R_i I_i$, $1 \le i \le q - 1$. We have a total of $(p - k) + (q - p + k) + (q - 1) = 2q - 1$ equations in the $2q - 1$ unknowns $I_i$ ($1 \le i \le q$) and $V_i$ ($1 \le i \le q - 1$). Moreover, it is easy to see that these $2q - 1$ equations are linearly independent, using the fact that we already know that just the equations involving the $I_i$'s alone are linearly independent, and similarly the $V_i$'s. Hence this system of $2q - 1$ equations in $2q - 1$ unknowns has a unique solution. We have now reduced the problem to straightforward linear algebra. However, it is possible to describe the solution explicitly. We will be content here with giving a formula just for the total resistance $R(D) = -V_q/I_q = 1/I_q$. (Recall that we take $V_q < 0$.)

Write the $2q - 1$ equations in the form of a $(2q - 1) \times 2q$ matrix $K$. The columns of the matrix are indexed by $I_1, I_2, \ldots, I_q, V_1, V_2 \ldots, V_q$. The last column $V_q$ of the matrix keeps track of the constant terms of the equations. The rows of $K$ are given first by the equations among the $I_i$'s, then the $V_i$'s, and finally Ohm's Law. For our example of the Wheatstone bridge, we obtain the matrix

$$K =
\begin{array}{|cccccc|cccccc|}
\hline
I_1 & I_2 & I_3 & I_4 & I_5 & I_6 & V_1 & V_2 & V_3 & V_4 & V_5 & V_6 \\
\hline
1 & 0 & -1 & -1 & 0 & 0 & 0 & 0 & 0 & 0 & 0 & 0 \\
0 & 1 & 1 & 1 & 0 & 1 & 0 & 0 & 0 & 0 & 0 & 0 \\
0 & 0 & 0 & 1 & 1 & 1 & 0 & 0 & 0 & 0 & 0 & 0 \\
0 & 0 & 0 & 0 & 0 & 0 & 1 & -1 & 1 & 0 & 0 & 0 \\
0 & 0 & 0 & 0 & 0 & 0 & 1 & -1 & 0 & 1 & -1 & 0 \\
0 & 0 & 0 & 0 & 0 & 0 & 0 & -1 & 0 & 0 & -1 & -1 \\
\hline
R_1 & 0 & 0 & 0 & 0 & 0 & -1 & 0 & 0 & 0 & 0 & 0 \\
0 & R_2 & 0 & 0 & 0 & 0 & 0 & -1 & 0 & 0 & 0 & 0 \\
0 & 0 & R_3 & 0 & 0 & 0 & 0 & 0 & -1 & 0 & 0 & 0 \\
0 & 0 & 0 & R_4 & 0 & 0 & 0 & 0 & 0 & -1 & 0 & 0 \\
0 & 0 & 0 & 0 & R_5 & 0 & 0 & 0 & 0 & 0 & -1 & 0 \\
\hline
\end{array}$$

We want to solve for $I_q$ by Cramer's rule. Call the submatrix consisting of all but the last column $X$. Let $Y$ be the result of replacing the $I_q$ column of $X$ by the last column of $K$. Cramer's rule then asserts that

$$I_q = \frac{\det Y}{\det X}.$$

We evaluate $\det X$ by taking a Laplace expansion along the first $p - 1$ rows. In other words,

$$\det X = \sum_{S} \pm \det(X[[p - 1], S]) \cdot \det(X[[p - 1]^c, \bar{S}]), \tag{11.8}$$

where (a) $S$ indexes all $(p - 1)$-element subsets of the columns, (b) $X[[p - 1], S]$ denotes the submatrix of $X$ consisting of entries in the first $p - 1$ rows and in the columns $S$, and (c) $X[[p - 1]^c, \bar{S}]$ denotes the submatrix of $X$ consisting of entries in the last $2q - p$ rows and in the columns other than $S$. In order for $\det(X[[p-1], S]) \neq 0$, we must choose $S = \{I_{i_1}, \ldots, I_{i_{p-1}}\}$, where $\{e_{i_1}, \ldots, e_{i_{p-1}}\}$ is a spanning tree $T_1$ (by Theorem 11.6(i)). In this case, $\det(X[[p-1], S]) = \pm 1$ by Theorem 11.13. If $I_q \notin S$, then the $I_q$ column of $X[[p - 1]^c, \bar{S}]$ will be zero. Hence to get a nonzero term in (11.8), we must have $e_q \in S$. The matrix $X[[p - 1]^c, \bar{S}]$ will have one nonzero entry in each of the first $q - p + 1$ columns, namely, the resistances $R_j$ where $e_j$ is not an edge of $T_1$. This accounts for $q - p + 1$ entries from the last $q - 1$ rows of $X[[p - 1]^c, \bar{S}]$. The remaining $p - 2$ of the last $q - 1$ rows have available only one nonzero entry each, namely, a $-1$ in the columns indexed by $V_j$ where $e_j$ is an edge of $T_1$ other than $e_q$. Hence we need to choose $q - p + 1$ remaining entries from rows $p$ through $q$ and columns indexed by $V_j$ for $e_j$ not an edge of $T_1$. By Theorems 11.6(ii) and 11.13, this remaining submatrix has determinant $\pm 1$. It follows that

$$\det(X[[p - 1], S]) \cdot \det(X[[p - 1]^c, \bar{S}]) = \pm \prod_{e_j \notin E(T_1)} R_j.$$

Hence by (11.8), we get

$$\det X = \sum_{T_1} \pm \left( \prod_{e_j \notin E(T_1)} R_j \right), \tag{11.9}$$

where $T_1$ ranges over all spanning trees of $D$ containing $e_q$. A careful analysis of the signs (omitted here) shows that all signs in (11.9) are positive, so we finally arrive at the remarkable formula

$$\det X = \sum_{\substack{\text{spanning trees } T_1 \\ \text{containing } e_q}} \prod_{e_j \notin E(T_1)} R_j.$$

For example, if $D$ is the Wheatstone bridge as above, and if we abbreviate $R_1 = a$, $R_2 = b$, $R_3 = c$, $R_4 = d$, $R_5 = e$, then

$$\det X = abc + abd + abe + ace + ade + bcd + bde + cde.$$

Now suppose we replace column $I_q$ in $X$ by column $V_q$ in the matrix $K$, obtaining the matrix $Y$. There is a unique nonzero entry in the new column, so it must be chosen in any nonzero term in the expansion of $\det Y$. The argument now goes just as it did for $\det X$, except we have to choose $S$ to correspond to a spanning tree $T_1$ that *doesn't* contain $e_q$. We therefore obtain

$$\det Y = \sum_{\substack{\text{spanning trees } T_1 \\ \text{not containing } e_q}} \prod_{\substack{e_j \notin E(T_1) \\ e_j \neq e_q}} R_j.$$

For example, for the Wheatstone bridge we get

$$\det Y = ac + ad + ae + bc + bd + be + cd + ce.$$

Recall that $I_q = \det(Y)/\det(X)$ and that the total resistance of the network is $1/I_q$. Putting everything together gives our main result on electrical networks.

**11.14 Theorem.** *In the situation described above, the total resistance of the network is given by*

$$R(D) = \frac{1}{I_q} = \frac{\displaystyle\sum_{\substack{\text{spanning trees } T_1 \\ \text{containing } e_q}} \prod_{e_j \notin E(T_1)} R_j}{\displaystyle\sum_{\substack{\text{spanning trees } T_1 \\ \text{not containing } e_q}} \prod_{\substack{e_j \notin E(T_1) \\ e_j \neq e_q}} R_j}.$$

**11.15 Corollary.** *If the resistances $R_1, \ldots, R_{q-1}$ are all equal to one, then the total resistance of the network is given by*

$$R(D) = \frac{1}{I_q} = \frac{\text{number of spanning trees containing } e_q}{\text{number of spanning trees not containing } e_q}.$$

In particular, if $R_1 = \cdots = R_{q-1} = 1$, then the total resistance, when reduced to lowest terms $a/b$, has the curious property that the number $\kappa(D)$ of spanning trees of $D$ is divisible by $a + b$.

## 11.4 Planar Graphs (Sketch)

A graph $G$ is *planar* if it can be drawn in the plane $\mathbb{R}^2$ without crossing edges. A drawing of $G$ in this way is called a *planar embedding*. An example of a planar embedding is shown in Figure 11.2. In this section we state the basic results on the bond and cycle spaces of a planar graph. The proofs are relatively straightforward and are omitted.

If the vertices and edges of a planar embedding of $G$ are removed from $\mathbb{R}^2$, then we obtain a disjoint union of open sets, called the *faces* (or *regions*) of $G$. (More precisely, these open sets are the faces of the planar embedding of $G$. Often we will not bother to distinguish between a planar graph and a planar embedding if no confusion should result.) Let $R = R(G)$ be the set of faces of $G$, and as usual $V(G)$ and $E(G)$ denote the set of vertices and edges of $G$, respectively.

NOTE. If $G$ is a simple (no loops or multiple edges) planar embedding, then it can be shown that there exists a planar embedding of the same graph with edges as straight lines and with faces (regarding as the sequence of vertices and edges obtained by walking around the boundaries of the faces) preserved.

The *dual* $G^*$ of the planar embedded graph $G$ has vertex set $R(G)$ and edge set $E^*(G) = \{e^* : e \in E(G)\}$. If $e$ is an edge of $G$, then let $r$ and $r'$ be the faces on its two sides. (Possibly $r = r'$; there are five such edges in Figure 11.2.) Then define $e^*$ to connect $r$ and $r'$. We can always draw $G^*$ to be planar, letting $e$ and $e^*$ intersect once. If $G$ is connected then every face of $G^*$ contains exactly one (nonisolated) vertex of $G$ and $G^{**} \cong G$. For any planar embedded graph $G$, the dual $G^*$ is connected. Then $G \cong G^{**}$ if and only if $G$ is connected. In general, we always have

**Fig. 11.2** A planar embedding

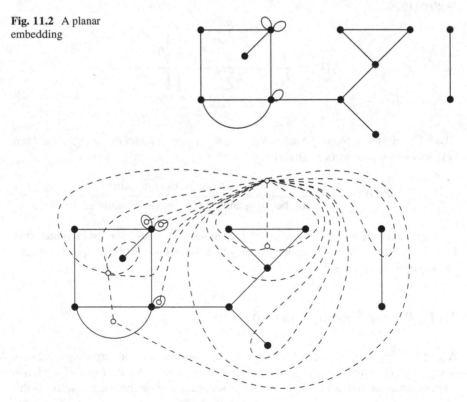

**Fig. 11.3** A planar embedding and its dual

$G^* \cong G^{***}$. Figure 11.3 shows the dual $G^*$ to the graph $G$ of Figure 11.2, with the vertices of $G^*$ drawn as open circles and the edges as broken lines.

**11.16 Example.** Let $G$ consist of two disjoint edges. Then $G^*$ has one vertex and two loops, while $G^{**}$ is a three-vertex path. The unbounded face of $G^*$ contains two vertices of $G$, and $G^{**} \not\cong G$.

Orient the edges of the planar graph $G$ in any way to get a digraph $D$. Let $r$ be an interior (i.e., bounded) face of $D$. An *outside edge* of $r$ is an edge $e$ such that $r$ lies on one side of the edge, and a *different* face lies on the other side. The outside edges of any interior face $r$ define a circulation (shown as solid edges in the diagram below), and these circulations (as $r$ ranges over all interior faces of $D$) form a basis for the cycle space $\mathcal{C}_G$ of $G$.

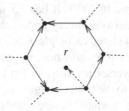

Given the orientation $D$ of $G$, orient the edges of $G^*$ as follows: as we walk along $e$ in the direction of its orientation, $e^*$ points to our *right*.

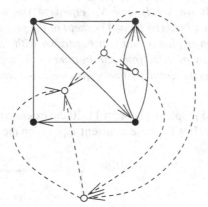

**11.17 Theorem.** *Let* $f : E(G) \to \mathbb{R}$. *Define* $f^* : E(G^*) \to \mathbb{R}$ *by* $f^*(e^*) = f(e)$. *Then*

$$f \in \mathcal{B}_G \iff f^* \in \mathcal{C}_{G^*}$$
$$f \in \mathcal{C}_G \iff f^* \in \mathcal{B}_{G^*}.$$

**11.18 Proposition.** *The set $S$ is the set of edges of a spanning tree $T$ of $G$ if and only if $S^* = \{e^* : e \in S\}$ is the set of edges of a cotree $T^*$ of $G^*$.*

**11.19 Corollary.** $\kappa(G) = \kappa(G^*)$

For nonplanar graphs there is still a notion of a "dual" object, but it is no longer a graph but rather something called a *matroid*. Matroid theory is a flourishing subject which may be regarded as a combinatorial abstraction of linear algebra.

## 11.5  Squaring the Square

A *squared rectangle* is a rectangle partitioned into finitely many (but more than one) squares. A squared rectangle is *perfect* if all the squares are of different sizes. The earliest perfect squared rectangle was found in 1936; its size is $33 \times 32$ and consists of nine squares, as shown in Figure 11.4.

The question then arose: does there exist a perfect squared square? A single example was found by Sprague in 1939; it has 55 squares. Then Brooks, Smith, Stone, and Tutte developed a network theory approach which we now explain.

The *Smith diagram D* of a squared rectangle $R$ is a directed graph whose vertices are the horizontal line segments of $R$ and whose edges are the squares of $R$, directed from top to bottom. The top vertex (corresponding to the top edge of $R$) and the bottom vertex (corresponding to the bottom edge) are called *poles*. Label each edge by the side length of the square to which it corresponds. Figure 11.5 shows the Smith diagram of the (perfect) squared rectangle in Figure 11.4.

The following result concerning Smith diagrams is straightforward to verify.

**11.20 Theorem.** *(a) If we set $I_e$ and $V_e$ equal to the label of edge e, then Kirchhoff's two laws hold (so $R_e = 1$) except at the poles.*

*(b) The Smith diagram is planar and can be drawn without separation of poles. Joining the poles by an edge from the bottom to the top gives a 3-connected graph, i.e., a connected graph that remains connected when one or two vertices are removed.*

Call the 3-connected graph of Theorem 11.20 the *extended* Smith diagram of the $a \times b$ squared rectangle. If we impose a current $I_{e_1} = b$ on the new edge $e_1$ (directed

**Fig. 11.4** A squared rectangle

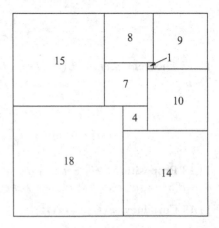

**Fig. 11.5** A Smith diagram

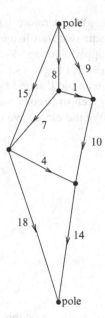

from bottom to top) between poles, and a voltage $V_{e_1} = -a$, then Kirchhoff's two laws hold at *all* vertices. The diagram below shows the extended Smith diagram corresponding to Figure 11.5, with the new edge $e_1$ labelled by the current $I_{e_1}$.

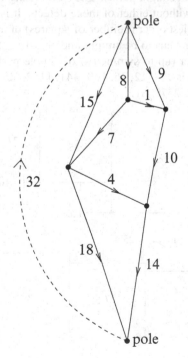

We therefore have a recipe for searching for perfect squared rectangles and squares: start listing all three-connected planar graphs. Then choose an edge $e_1$ to apply a voltage $V_1$. Put a resistance $R_e = 1$ at the remaining edges $e$. Solve for $I_e$ ($= V_e$) to get a squared rectangle, and hope that one of these will be a square. One example $\Gamma$ found by Brooks et al. was a $112 \times 75$ rectangle with 14 squares. It was given to Brooks' mother as a jigsaw puzzle, and she found a different solution $\Delta$! We therefore have found a squared square (though not perfect):

|                |           |
|:--------------:|:---------:|
| $\Delta$       | 75 x 75   |
| 112 x 112      | $\Gamma$  |

Building on this idea, Brooks et al. finally found two $422 \times 593$ perfect rectangles with thirteen squares, all 26 squares being of different sizes. Putting them together as above gives a perfect squared square. This example has two defects: (a) it contains a smaller perfect squared rectangle (and is therefore not *simple*), and (b) it contains a "cross" (four squares meeting a point). They eventually found a perfect squared square with 69 squares without either of these defects. It is now known (thanks to computers) that the smallest order (number of squares) of a perfect squared square is 21. It is unique and happens to be simple and crossfree. See the figure below. It is known that the number (up to symmetry) of simple perfect squared squares of order $n$ for $21 \le n \le 35$ is 1, 8, 12, 26, 160, 441, 1152, 3001, 7901, 20566, 54541, 144161, 378197, 990981, 2578081.

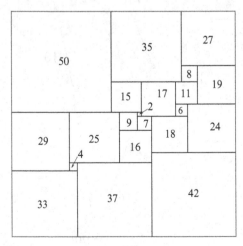

## Notes for Chapter 11

The theory of cycle spaces and bond spaces developed here had its origins with the pioneering work of Kirchhoff [75] in 1847. The proof given here of Theorem 11.13 is due to Tutte [140] in 1965. A nice account of the history of squaring the square due to Tutte appears in a *Scientific American* column by Martin Gardner [51]. See also [141] for another article by Tutte. A further survey article on this topic is by Kazarinoff and Weitzenkamp [73]. For some interesting connections between random walks and electrical networks, see Doyle and Snell [36].

## Exercises for Chapter 11

1. (a) Let $C_n$ be the graph of the $n$-cube. Find the dimension of the bond space and cycle space of $C_n$. Does there exist a circulation (with respect to some orientation of $C_n$) supported on three edges?

   (b) Show that the cycle space $\mathcal{C}_{C_n}$ (with respect to some orientation of $C_n$) is spanned by circulations $f_C$, where $C$ is a circuit of length four.

2. We have defined *real-valued* flows $f : E \to \mathbb{R}$, but we could just as easily allow the values to be in any abelian group $A$, i.e., $f : E \to A$ such that (11.1) is satisfied.

   (a) A *bridge* or *isthmus* in a graph $G$ (directed or undirected) is an edge $e$ whose removal disconnects the connected component (an undirected graph in the case where $G$ is directed) to which it belongs. For instance, $G$ is a forest if and only if every edge is a bridge. If $e$ is an isthmus of the digraph $D$ and $f : E \to A$ is a flow, then show that $f(e) = 0$.

   (b) Let $G$ be an undirected (finite) graph and $o$ an orientation. For a positive integer $n$, define $C_G(n)$ to be the number of flows $f : G \to \mathbb{Z}_n$ (with respect to the orientation $o$, where $\mathbb{Z}_n$ denotes the integers modulo $n$) which never take the value 0. Show that $C_G(n)$ depends only on $G$ and $n$, not on $o$.

   (c) Let $G$ be a finite bridgeless graph, i.e., a graph with no bridges. Show that $C_G(n)$ is a polynomial function of $n$ whose degree is the dimension of the cycle space of $G$. (If $G$ has an bridge, then $C_G(n) = 0$ by part (a) of this exercise.)

   (d) Let $G_p$ denote the cycle of length $p \geq 3$. Find $C_{G_p}(n)$.

   (e) (*) Let $d = \deg C_G(n)$. Show that for $0 \leq i \leq d$, the coefficient of $n^i$ in the polynomial $(-1)^d C_G(-n)$ is positive.

   (f) (difficult) Let $K_p$ denote the complete graph with $p$ vertices. Show that

$$\sum_{p\geq 0}(-1)^{\binom{p-1}{2}}C_{K_p}(n)\frac{x^p}{p!} = 1 - \left(\sum_{k\geq 0}(1-n)^{\binom{k}{2}}(-n)^{-k}\frac{x^k}{k!}\right)^n$$

$$= x - (n-1)\frac{x^3}{3!}$$

$$-(n-1)(n-2)(n-3)\frac{x^4}{4!}$$

$$+(n-1)(n^5 - 9n^4 + 36n^3 - 79n^2 + 96n - 51)\frac{x^5}{5!} + \cdots.$$

(g) Find some positive integer $j$ with the following property: for any finite bridgeless graph $G$ we have $C_G(j) > 0$.

(h) (very difficult) Show that if $G$ is finite bridgeless graph, then $C_G(6) > 0$.

(i) (unsolved) Show that if $G$ is finite bridgeless graph, then $C_G(5) > 0$. This inequality is a famous conjecture of Tutte, known as the *nowhere-zero 5-flow conjecture*.

3. What digraphs have the property that every nonempty set of edges is a cutset?

4. What is the size of the largest set of edges of the complete graph $K_p$ that doesn't contain a bond? How many such sets are there?

5. (*) The cycle space $\mathcal{C}_D$ and bond space $\mathcal{B}_D$ of a finite digraph $D$ were defined over $\mathbb{R}$. However, the definition works over any field $K$, even those of positive characteristic. Show that the dimensions of these two spaces remain the same for any $K$, i.e., $\dim \mathcal{C}_D = q - p + k$ and $\dim \mathcal{B}_D = p - k$.

6. (a) Let $K_p$ be the complete graph on the vertex set $[p]$. Suppose that a voltage $V_e = 1$ is applied to an edge $e$, and all other edges have a resistance of 1. Without doing any computation, find the current $I_f$ along any edge $f$ disjoint from $e$ (i.e., $e$ and $f$ have no common vertices).

(b) (*) Using (a), find the total resistance of the network of (a).

(c) Check that your answer to (b) agrees with Corollary 11.15.

7. (a) A graph $G$ is *edge-transitive* if its automorphism group $\text{Aut}(G)$ is transitive on the edges, i.e., for any two edges $e, e'$ of $G$, there is an automorphism $\phi$ which takes $e$ to $e'$. For instance, the cube graph $C_n$ is edge-transitive. Is an edge-transitive graph also vertex-transitive? What about conversely? If we consider only simple graphs (no loops or multiple edges), does that affect the answers?

(b) Suppose that $G$ is edge-transitive and has $p$ vertices and $q$ edges. A one volt battery is placed on one edge $e$, and all other edges have a resistance of one ohm. Express the total resistance $R_e = -V_e/I_e$ of the network in terms of $p$ and $q$.

8. Let $D$ be a loopless connected digraph with $q$ edges. Let $T$ be a spanning tree of $D$. Let $C$ be the basis matrix for the cycle space $\mathcal{C}$ of $D$ obtained from $T$, and similarly $B$ for the bond space (as described in Theorems 11.8 and 11.11).

(a) (*) Show that $\det CC^t = \det BB^t = \kappa(D)$, the number of spanning trees of $D$ (ignoring the directions of the edges).

(b) (*) Let

$$Z = \begin{bmatrix} C \\ B \end{bmatrix},$$

a $q \times q$ matrix. Show that $\det Z = \pm\kappa(D)$.

9. (difficult) Let $M$ be an $m \times n$ real unimodular matrix such that every two columns of $M$ are linearly independent. Show that $n \le \binom{m}{2}$

10. Let $D$ be a planar electrical network with edges $e_1, \ldots, e_q$. Place resistances $R_i = 1$ at $e_i$, $1 \le i \le q-1$, and place a voltage $V_q = 1$ at $e_q$. Let $D^*$ be the dual network, with the same resistances $R_i$ at $e_i^*$ and voltage $V_q^*$ at $e_q^*$. What is the connection between the total resistances $R(D)$ and $R(D^*)$?

11. Let $D$ be the extended Smith diagram of a squared rectangle, with the current $I$ and voltage $V$ as defined in the text. What is the "geometric" significance of the fact that $\langle I, V \rangle = 0$?

12. Let $D$ be the extended Smith diagram of an $a \times b$ squared rectangle. Show that the number of spanning trees of $D$ is divisible by $a + b$.

# Chapter 12
# A Glimpse of Combinatorial Commutative Algebra

## 12.1 Simplicial Complexes

In this chapter we will discuss a profound connection between commutative rings and some combinatorial properties of simplicial complexes. The deepest and most interesting results in this area require a background in algebraic topology and homological algebra beyond the scope of this book. However, we will be able to prove a highly nontrivial combinatorial result that relies on commutative algebra (i.e., the theory of commutative rings and modules over them) in an essential way. This result is our Theorem 12.25, the characterization of $f$-vectors of shellable simplicial complexes. Of course we must first define these terms and then set up the necessary machinery.

Let $V = \{x_1, \ldots, x_n\}$ be a finite set, called a *vertex set*. An *abstract simplicial complex* on $V$, or just *simplicial complex* for short, is a collection $\Delta$ of subsets of $V$ satisfying the following two conditions (of which the second is the significant one):

1. $\{x_i\} \in \Delta$ for $1 \le i \le n$
2. If $F \in \Delta$ and $G \subseteq F$, then $G \in \Delta$.

An element $F$ of $\Delta$ is called a *face*. A maximal face $F$, i.e., a face that is not contained in any larger face, is called a *facet*. The *dimension* of $F$ is $\#F - 1$. In particular, the empty set $\emptyset$ is a face of dimension $-1$, unless $\Delta = \emptyset$ (see the remark below concerning empty simplicial complexes and empty faces). An $i$-dimensional face is called an $i$-*face*. Soon we will see the geometric reason for our definition of dimension.

**12.1 Remark.** There is a small subtlety about the definition of simplicial complex that can lead to confusion. Namely, one must distinguish between the empty simplicial complex $\Delta = \emptyset$ which has no faces whatsoever, and the simplicial complex $\Delta = \{\emptyset\}$ whose only face is the empty set $\emptyset$.

© Springer International Publishing AG, part of Springer Nature 2018    187
R. P. Stanley, *Algebraic Combinatorics*, Undergraduate Texts in Mathematics,
https://doi.org/10.1007/978-3-319-77173-1_12

If $\Gamma$ is any finite collection of finite sets, then $\langle \Gamma \rangle$ denotes the smallest simplicial complex containing the elements of $\Gamma$. Thus

$$\langle \Gamma \rangle = \{F \ : \ F \subseteq G \text{ for some } G \in \Gamma\}.$$

In presenting examples we will often abbreviate a set such as $\{1, 2, 3\}$ as simply 123. Thus for instance $\langle 123, 14, 24 \rangle$ denotes the simplicial complex with faces

$$\emptyset, 1, 2, 3, 4, 12, 13, 23, 14, 24, 123.$$

It is worthwhile to understand simplicial complexes geometrically, though such understanding is not really germane to our main results here. Let us first review some basic definitions. A *convex set* in $\mathbb{R}^d$ is a subset $S$ of $\mathbb{R}^d$ such that if $u, v \in S$, then the line segment joining $u$ and $v$ is also in $S$. Equivalently, $\lambda u + (1 - \lambda)v \in S$ for all real numbers $0 \le \lambda \le 1$. Clearly the intersection of convex sets is convex. The *convex hull* of any subset $S$ of $\mathbb{R}^d$, denoted conv($S$), is defined to be the intersection of all convex sets containing $S$. It is therefore the smallest convex set in $\mathbb{R}^d$ containing $S$.

A set $\{v_0, v_1, \ldots, v_j\} \subset \mathbb{R}^d$ is *affinely independent* if the following condition holds: if $\alpha_0, \alpha_1, \ldots, \alpha_j$ are real numbers for which $\sum \alpha_i v_i = 0$ and $\sum \alpha_i = 0$, then $\alpha_0 = \alpha_1 = \cdots = \alpha_j = 0$. Equivalently, define an *affine subspace* of $\mathbb{R}^d$ to be the translate of a linear subspace, i.e., a set

$$A = \{v \in \mathbb{R}^d \ : \ v \cdot y^{(1)} = \alpha_1, \ldots, v \cdot y^{(k)} = \alpha_k\},$$

where $y^{(1)}, \ldots, y^{(k)} \in \mathbb{R}^d$ (with each $y^{(i)} \ne 0$) and $\alpha_1, \ldots, \alpha_k \in \mathbb{R}$ are fixed, and where $v \cdot y$ denotes the usual dot product in $\mathbb{R}^d$. The *dimension* of $A$ is the dimension of the linear subspace

$$\{v \in \mathbb{R}^d \ : \ v \cdot y^{(1)} = 0, \ldots, v \cdot y^{(k)} = 0\}.$$

The *affine span* of a subset $S$ of $\mathbb{R}^d$, denoted aff($S$), is the intersection of all affine subspaces containing $S$. It is easy to see that aff($S$) is itself an affine subspace. It is then true that a set of $k + 1$ points of $\mathbb{R}^d$ is affinely independent if and only if its affine span has dimension $k$, the maximum possible. In particular, the largest number of points of an affinely independent subset of $\mathbb{R}^d$ is $d + 1$.

A *simplex* (plural *simplices*) $\sigma$ in $\mathbb{R}^d$ is the convex hull of an affinely independent subset of $\mathbb{R}^d$. The *dimension* of a simplex $\sigma$ is the dimension of its affine span. Equivalently, if $\sigma$ is the convex hull of $j + 1$ affinely independent points, then $\dim \sigma = j$. If $S$ is affinely independent and $\sigma = \text{conv}(S)$, then a *face* of $\sigma$ is a set conv($T$) for some $T \subseteq S$. In particular, taking $T = \emptyset$ shows that $\emptyset$ is a face of $\sigma$. A face $\tau$ of dimension zero (i.e., $\tau$ is a single point) is called a *vertex* of $\Gamma$. If $\dim \sigma = j$, then $\sigma$ has $\binom{j+1}{i+1}$ $i$-dimensional faces. For instance, a zero-dimensional simplex is a point, a one-dimensional simplex is a line segment, a two-dimensional simplex is a triangle, a three-dimensional simplex is a tetrahedron, etc.

A (finite) *geometric simplicial complex* is a finite set $\Gamma$ of simplices in $\mathbb{R}^d$ such that the following two conditions hold:

1. If $\sigma \in \Gamma$ and $\tau$ is a face of $\sigma$, then $\tau \in \Gamma$.
2. If $\sigma, \tau \in \Gamma$, then $\sigma \cap \tau$ is a common face (possibly empty) of $\sigma$ and $\tau$.

We sometimes identify $\Gamma$ with the union $\bigcup_{\sigma \in \Gamma} \sigma$ of its simplices. In this situation $\Gamma$ is just a subset of $\mathbb{R}^d$, but it is understood that it has been described as a union of certain simplices.

There is an obvious abstract simplicial complex $\Delta$ that we can associate with a geometric simplicial complex. Namely, the vertex set $V$ of $\Delta$ consists of the vertices of $\Gamma$, and a set $F$ of vertices of $\Delta$ is a face of $\Delta$ if $F$ is the set of vertices of some simplex $\sigma \in \Gamma$. We then say that $\Gamma$ (regarded as a union of its simplices) is a *geometric realization* of $\Delta$, denoted $\Gamma = |\Delta|$. Note that if $F$ is a face of $\Delta$ with $k + 1$ vertices, then it corresponds to a $k$-dimensional simplex in $\Gamma$, explaining why we defined $\dim F = \#F - 1$.

NOTE. In some situations it is useful for $\Delta$ to have a unique (canonical) geometric realization. We can do this as follows. Suppose that $\Delta$ has $n$ vertices $v_1, \ldots, v_n$. Let $\delta_i$ be the $i$th unit coordinate vector in $\mathbb{R}^n$. For each face $F = \{v_{i_1}, \ldots, v_{i_k}\} \in \Delta$, define the simplex $\sigma_F = \mathrm{conv}(\delta_{i_1}, \ldots, \delta_{i_k})$. The linear independence of the $\delta_i$'s guarantees that $\sigma_F$ is indeed a simplex and that $\sigma_F \cap \sigma_G = \sigma_{F \cap G}$. Hence the set $\Gamma = \{\sigma_F : F \in \Delta\}$ is a geometric realization of $\Delta$, so we could define $\Gamma$ as *the* geometric realization of $\Delta$ (unique once we have labelled the vertices $v_1, \ldots, v_n$). However, for our purposes we don't need this uniqueness.

**12.2 Remark** (for those with some knowledge of topology). The geometric realization $|\Delta|$ is a topological space $X$ (a topological subspace of some $\mathbb{R}^d$). We say that $\Delta$ is a *triangulation* of $X$.

**12.3 Example.** Let $\Delta = \langle 123, 234, 235, 36, 56, 57, 8 \rangle$. A geometric realization of $\Delta$ is shown in Figure 12.1, projected from three dimensions. Note that since three triangles share the edge 23, any geometric realization in $\mathbb{R}^d$ requires $d \geq 3$. It is a result of Karl Menger, though irrelevant for us, that any $d$-dimensional simplicial complex can be realized in $\mathbb{R}^{2d+1}$, and that this result is best possible, i.e., the dimension $2d + 1$ cannot in general be decreased. In fact, the simplicial complex

**Fig. 12.1** A geometric realization

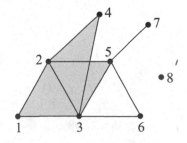

whose facets are all the $(d+1)$-element subsets of a $(2d+3)$-element set cannot be realized in $\mathbb{R}^{2d}$. For example, when $d = 1$ we get that the complete graph $K_5$ cannot be embedded in the plane (without crossing edges), a famous result in graph theory known to Euler at least implicitly, since he showed in 1750 that $f_1 \leq 3f_0 - 6$ for any planar graph (where $f_i$ is defined below). The first person to realize explicitly that $K_5$ is not planar seems to be A. F. Möbius in 1840, who stated the result in the form of a puzzle.

**12.4 Example.** Let $V = \{1, \bar{1}, 2, \bar{2}, 3, \bar{3}\}$ and

$$\Delta = \langle 123, \bar{1}23, 1\bar{2}3, 12\bar{3}, \bar{1}\bar{2}3, \bar{1}2\bar{3}, 1\bar{2}\bar{3}, \bar{1}\bar{2}\bar{3} \rangle.$$

Then the boundary of an octahedron is a geometric realization of $\Delta$. See Figure 12.2. Thus we can also say, following Remark 12.2, that $\Delta$ is a triangulation of the 2-sphere (two-dimensional sphere).

We now come to the combinatorial information about simplicial complexes that is our primary interest in this chapter. For $i \geq -1$, let $f_i$ be the number of $i$-dimensional faces of $\Delta$. Thus $f_{-1} = 1$ unless $\Delta = \emptyset$, and $f_0 = \#V$, the number of vertices of $\Delta$. If $\dim \Delta = d - 1$, then the vector

$$f(\Delta) = (f_0, f_1, \ldots, f_{d-1})$$

is called the $f$-*vector* of $\Delta$. Thus the simplicial complex $\Delta$ of Figure 12.1 has $f$-vector $(8, 10, 3)$, while that of Figure 12.2 has $f$-vector $(6, 12, 8)$.

An important general problem is to characterize the $f$-vector of various classes of simplicial complexes. The first class to come to mind is *all* simplicial complexes.

**Fig. 12.2** The boundary of an octahedron

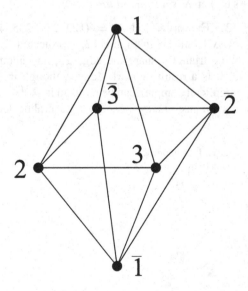

In other words, what vectors $(f_0, f_1, \ldots, f_{d-1})$ of positive integers are $f$-vectors of $(d - 1)$-dimensional simplicial complexes? Although this result is not directly related to the upcoming connection with commutative algebra, we will discuss it because of its general interest and its analogy to the upcoming Theorem 12.28.

We first make some strange-looking definitions and then explain their connection with $f$-vectors.

**12.5 Proposition.** *Given positive integers n and j, there exist unique integers*

$$n_j > n_{j-1} > \cdots > n_1 \geq 0$$

*such that*

$$n = \binom{n_j}{j} + \binom{n_{j-1}}{j-1} + \cdots + \binom{n_1}{1}. \tag{12.1}$$

*Proof.* The proof is based on the following simple combinatorial identity. Let $1 \leq i \leq m$. Then

$$\binom{m}{i} + \binom{m-1}{i-1} + \cdots + \binom{m-i+1}{1} + 1 = \binom{m+1}{i}. \tag{12.2}$$

This identity can easily be proved by induction on $i$, for instance. It also has a simple combinatorial interpretation. Namely, the right-hand side is the number of $i$-element subsets $S$ of the set $[m + 1] = \{1, 2, \ldots, m + 1\}$. The number of such subsets for which the least missing element is $s+1$ is equal to $\binom{m-s}{i-s}$. Summing over all $0 \leq s \leq i$ completes the proof of (12.2).

We now prove the proposition by induction on $j$. For $j = 1$ we have $n = \binom{n}{1}$, while $n \neq \binom{m}{1}$ for $m \neq n$. Hence the proposition is true for $j = 1$.

Assume the proposition for $j - 1$. Given $n$, $j$, define $m_j$ to be the largest integer for which $n \geq \binom{m_j}{j}$. Hence if the proposition is true for $n$ and $j$, then $n_j \leq m_j$. But by (12.2),

$$\binom{m_j - 1}{j} + \binom{m_j - 2}{j-1} + \cdots + \binom{m_j - j}{1} = \binom{m_j}{j} - 1 < n.$$

Since the above sum is the largest possible number of the form (12.1) beginning with $\binom{m_j - 1}{j}$, we have $n_j \geq m_j$. Hence $n_j = m_j$. By induction there is a unique way to write

$$n - \binom{n_j}{j} = \binom{n_{j-1}}{j-1} + \binom{n_{j-2}}{j-2} + \cdots + \binom{n_1}{1},$$

where $n_{j-1} > n_{j-2} > \cdots > n_1 \geq 0$. Thus we need only check that $n_{j-1} < n_j$. If on the contrary $n_{j-1} \geq n_j$, then

$$\binom{n_j}{j} + \binom{n_{j-1}}{j-1} \geq \binom{n_j}{j} + \binom{n_j}{j-1} = \binom{n_j + 1}{j},$$

contradicting the maximality of $n_j$ and completing the proof. $\qquad\square$

The representation of $n$ in the form (12.1) is called the $j$-*binomial expansion* of $n$. Given this formula, define

$$n^{(j)} = \binom{n_j}{j+1} + \binom{n_{j-1}}{j} + \cdots + \binom{n_1}{2}.$$

In other words, add 1 to the bottom of all the binomial coefficients in the $j$-binomial expansion of $n$. For instance, $51 = \binom{7}{4} + \binom{5}{3} + \binom{4}{2} + \binom{0}{1}$, so $51^{(4)} = \binom{7}{5} + \binom{5}{4} + \binom{4}{3} + \binom{0}{2} = 30$. For notational simplicity we sometimes suppress the binomial coefficients equal to 0, e.g., $51 = \binom{7}{4} + \binom{5}{3} + \binom{4}{2}$. Note that a binomial coefficient $\binom{n_i}{i} = 0$ in the $j$-binomial expansion of $n$ if and only if $n_i = i - 1$, in which case $n_r = r - 1$ for all $1 \leq r \leq i$.

We can now state a famous theorem of Schützenberger and Kruskal–Katona, often called the *Kruskal–Katona theorem*.

**12.6 Theorem.** *A vector* $(f_0, f_1, \ldots, f_{d-1}) \in \mathbb{P}^d$ *is the $f$-vector of a $((d-1)$- dimensional) simplicial complex if and only if*

$$f_{i+1} \leq f_i^{(i+1)}, \quad 0 \leq i \leq d - 2. \tag{12.3}$$

As an example, the fact that $51^{(4)} = 30$ means that in any simplicial complex with $f_3 = 51$ we must have $f_4 \leq 30$, and that this result is best possible. Theorem 12.6 says qualitatively the intuitively clear result that given $f_i$, the number $f_{i+1}$ cannot be too big. However, the precise quantitative result given by this theorem is by no means intuitively obvious. Let us try to provide some intuition and at the same time convey some idea of the proof.

Let $\alpha = (a_1, \ldots, a_j)$ and $\beta = (b_1, \ldots, b_j)$ be two sequences of nonnegative integers of the same length $j$. We say that $\alpha$ is less than $\beta$ in *reverse lexicographic order* (or *reverse lex order* for short), denoted $\alpha \overset{R}{<} \beta$, if for some $0 \leq i \leq j - 1$ we have

$$a_j = b_j, \ a_{j-1} = b_{j-1}, \ldots, a_{j-i+1} = b_{j-i+1}, \text{ and } a_{j-i} < b_{j-i}. \tag{12.4}$$

Equivalently, if we regard the nonnegative integers as the letters of an alphabet in their usual order, then the reverse sequences $(a_j, \ldots, a_1)$ and $(b_j, \ldots, b_1)$ are in dictionary (lexicographic) order. If $S$ and $T$ are two $j$-element subsets of $\mathbb{N}$, then we say that $S \overset{R}{<} T$ if $S' \overset{R}{<} T'$, where $A'$ denotes the sequence of elements of the set $A$ written in increasing order. If we abbreviate a set like $\{2, 4, 7\}$ as 247, then the one-element subsets of $\mathbb{N}$ in reverse lex order are

$$0 \overset{R}{<} 1 \overset{R}{<} 2 \overset{R}{<} 3 \overset{R}{<} 4 \overset{R}{<} 5 \overset{R}{<} 6 \overset{R}{<} \cdots .$$

The two-element subsets are

$$01 \overset{R}{<} 02 \overset{R}{<} 12 \overset{R}{<} 03 \overset{R}{<} 13 \overset{R}{<} 23 \overset{R}{<} 04 \overset{R}{<} \cdots .$$

The three-element subsets are

$$012 \overset{R}{<} 013 \overset{R}{<} 023 \overset{R}{<} 123 \overset{R}{<} 014 \overset{R}{<} 024 \overset{R}{<} 124 \overset{R}{<} 034$$
$$\overset{R}{<} 134 \overset{R}{<} 234 \overset{R}{<} 015 \overset{R}{<} \cdots .$$

The next result explains the connection between the $j$-binomial expansion and reverse lex order on $j$-element subsets of $\mathbb{N}$.

**12.7 Theorem.** *Let $S_0, S_1, \ldots$ be the sequence of $j$-element subsets of $\mathbb{N}$ in reverse lex order. Suppose that $S_n = \{a_1, \ldots, a_j\}$ with $a_1 < \cdots < a_j$. Then*

$$n = \binom{a_j}{j} + \binom{a_{j-1}}{j-1} + \cdots + \binom{a_1}{1},$$

*and this formula gives the $j$-binomial expansion of $n$.*

Before beginning the proof, here is an example. What is the 1985th term (calling the first term $S_0$ the 0th term) $S_{1985}$ of the reverse lex order on four-element subsets of $\mathbb{N}$? We have

$$1985 = \binom{16}{4} + \binom{11}{3}.$$

Hence $S_{1985} = \{16, 11, 1, 0\}$.

*Proof.* It suffices to show that the number of $j$-element subsets of $\mathbb{N}$ that are smaller than $S_n$ in reverse lex order is $\binom{a_j}{j} + \binom{a_{j-1}}{j-1} + \cdots + \binom{a_1}{1}$. We claim that for each $1 \leq k \leq n$ the number of such subsets that agree with $S_n$ in their largest $j - k$ elements but differ in their $(j - k + 1)$st largest one is $\binom{a_k}{k}$. In fact, these are just the union of the $k$-element subsets of $\{0, 1, \ldots, a_k - 1\}$ with $\{a_{k+1}, a_{k+2}, \ldots, a_j\}$, so the proof follows.                                                                              $\square$

Note that this argument assumes nothing about $j$-binomial expansions. Since for each $n$ there is just one $S_n$, the above proof in fact yields a new proof of Proposition 12.5.

Now suppose that $f = (f_0, \ldots, f_{d-1}) \in \mathbb{P}^d$. Define a collection $\Gamma_f$ of subsets of $\mathbb{N}$ to consist of the empty set $\emptyset$ together with the first $f_i$ of the $(i + 1)$-element subsets of $\mathbb{N}$ in reverse lex order. For example, if $f = (6, 8, 5, 2)$, then (writing as usual $\{1, 2, 3\} = 123$, etc.)

$$\Gamma_f = \{\emptyset, 0, 1, 2, 3, 4, 5, 01, 02, 12, 03, 13, 23, 04, 14,$$

$$012, 013, 023, 123, 014, 0123, 0124\}.$$

Note that for this example, $\Gamma_f$ is not a simplicial complex.

**12.8 Theorem.** *The set $\Gamma_f$ is a simplicial complex if and only if $f_{i+1} \le f_i^{(i+1)}$ for* $0 \le i \le d - 2$.

*Proof.* Let us use the notation $[0, m] = \{0, 1, \ldots, m\}$ and for any set $S$,

$$\binom{S}{k} = \{T \subseteq S : \#T = k\}.$$

Let $f_i = \binom{n_{i+1}}{i+1} + \binom{n_i}{i} + \cdots + \binom{n_1}{1}$ be the $(i + 1)$-binomial expansion of $f_i$. By the definition of reverse lex order, we see that the set $X$ of the first $f_i$ $(i + 1)$-elements of $\mathbb{N}$ in reverse lex order is given by

$$X = \binom{[0, n_{i+1} - 1]}{i + 1} \bigcup \left( \{n_{i+1}\} \cup \binom{[0, n_i - 1]}{i} \right)$$

$$\bigcup \left( \{n_{i+1}, n_i\} \cup \binom{[0, n_{i-1} - 1]}{i - 1} \right) \bigcup \cdots .$$

The set of $(i + 2)$-elements subsets $F$ of $\mathbb{N}$ all of whose $(i + 1)$-element subsets belong to $X$ is given by

$$X = \binom{[0, n_{i+1} - 1]}{i + 2} \bigcup \left( \{n_{i+1}\} \cup \binom{[0, n_i - 1]}{i + 1} \right)$$

$$\bigcup \left( \{n_{i+1}, n_i\} \cup \binom{[0, n_{i-1} - 1]}{i} \right) \bigcup \cdots .$$

These are just the first $f_i^{(i+1)}$ $(i + 2)$-element subsets of $\mathbb{N}$ in reverse lex order, and the proof follows.                                                                                    $\square$

Theorem 12.8 establishes the "if" direction of the Kruskal–Katona theorem (Theorem 12.6), i.e., condition (12.3) is sufficient for the existence of a simplicial complex with $f$-vector $f = (f_0, f_1, \ldots, f_{d-1})$. We have in fact constructed a "canonical" simplicial complex $\Gamma_f$ with this $f$-vector. Such a simplicial complex is called *compressed*.

The difficult part of the Kruskal–Katona theorem is the "only if" direction. We need to show that every simplicial complex $\Delta$ has the same $f$-vector as some compressed simplicial complex $\Gamma_f$. This is proved by transforming $\Delta$ to $\Gamma_f$ by a sequence of steps preserving the simplicial complex property and preserving the $f$-vector. It is not necessary to understand this argument (or in fact even the statement of the Kruskal–Katona theorem) in order to understand the main result of this chapter (Theorem 12.25) and its proof, so we will omit it here. See the "Notes for Chapter 13" below for a reference to a readable proof.

**12.9 Example.** Is $f = (5, 7, 5)$ an $f$-vector? Of course we could simply check whether the Kruskal–Katona conditions (12.3) hold. Alternatively, we can construct $\Gamma_f$ and check whether it is a simplicial complex. In fact,

$$\Gamma_f = \{\emptyset, 0, 1, 2, 3, 4, 01, 02, 12, 03, 13, 23, 04, 012, 013, 023, 123, 014\}.$$

This is *not* a simplicial complex since 14 is a subset of $014 \in \Gamma_f$, but $14 \notin \Gamma_f$. Hence $(5, 7, 5)$ is not an $f$-vector. In fact, we have $7 = \binom{4}{2} + \binom{1}{1}$ and $7^{(2)} = \binom{4}{3} + \binom{1}{2} = 4 < 5$.

We next want to characterize the $f$-vectors of a certain class of simplicial complexes, called *shellable* simplicial complexes. The result will be very similar to the Kruskal–Katona theorem, but the proof is vastly different. It will use tools from commutative algebra. First we will define shellable simplicial complexes and state the characterization of their $f$-vectors. We will then develop the algebraic tools necessary for the proof. Finally we will discuss a connection with an analogue of the Kruskal–Katona theorem.

We say that a simplicial complex is *pure* if every facet (maximal face) has the same dimension. For instance, the simplicial complex of Figure 12.1 is not pure; it has facets of dimensions zero, one and two. A *subcomplex* $\Delta'$ of a simplicial complex $\Delta$ is a subset of $\Delta$ that is itself a simplicial complex. (We don't require that $\Delta'$ has the same vertex set as $\Delta$.)

**12.10 Definition.** A $(d - 1)$-dimensional simplicial complex $\Delta$ is *shellable* if $\Delta$ is pure and there exists an ordering $F_1, F_2, \ldots, F_t$ of its facets (so $t = f_{d-1}$) such that the following property holds. For $0 \leq j \leq t$ let $\Delta_j = \langle F_1, \ldots, F_j \rangle$, the subcomplex of $\Delta$ generated by $F_1, \ldots, F_j$. In particular, $\Delta_0 = \emptyset$. Now let $1 \leq j \leq t$. Then we require that the set of faces of $F_j$ (i.e., the set of all subsets of $F_j$) has a *unique minimal element* $G_j$ not belonging to $\Delta_{j-1}$. Call the sequence $F_1, \ldots, F_t$ a *shelling order* or just *shelling* of $\Delta$, and call $G_j$ the *restriction* of $F_j$ (with respect to the shelling $F_1, \ldots, F_t$).

NOTE. Let $\Delta$ be a pure $(d - 1)$-dimensional simplicial complex. It is easy to see (Exercise 2) that a facet ordering $F_1, \ldots F_t$ is a shelling if and only if for all $2 \leq i \leq t$, the subcomplex $\langle F_1, \ldots, F_{i-1} \rangle \cap \langle F_i \rangle$ (i.e., the set of faces of $F_i$ that already belong to $\langle F_1, \ldots, F_{i-1} \rangle$) is a pure simplicial complex of dimension $d - 2$. More informally, $F_i$ attaches along some nonempty union of its facets.

It takes some time looking at examples to develop a feeling for shellings. First note that since $\Delta_0 = \emptyset$, the empty set is the unique minimal element of $F_1$ not in $\Delta_0$. Thus we always have $G_1 = \emptyset$.

**12.11 Example.** (a) Consider the one-dimensional simplicial complex $\Delta$ of Figure 12.3a. The ordering 1,2,3 of the facets is a shelling order, with $G_1 = \emptyset$, $G_2 = \{c\}$ (abbreviated as $c$), $G_3 = d$. For instance, when we attach facet 2 to facet 1, we create the two new faces $c$ and $bc$. The unique minimal element (with respect to inclusion) of the two sets $c$ and $bc$ is $G_2 = c$. Another shelling

order is 2,1,3, with $G_1 = \emptyset$, $G_2 = a$, $G_3 = d$. In fact, there are exactly four
shelling orders: 123, 213, 231, 312. For instance, 132 is *not* a shelling order.
When we adjoin 3 to 1 we create the new faces $c, d, cd$, and now we have *two*
minimal elements $c$ and $d$.

(b) One shelling order of the simplicial complex of Figure 12.3b is 1,2,3,4, with
$G_1 = \emptyset$, $G_2 = c$, $G_3 = d$, $G_4 = ad$.

(c) As essentially explained in (a) above, the simplicial complex $\Delta_c$ of Figure 12.3c
is not shellable.

(d) The simplicial complex $\Delta_d$ of Figure 12.3d is also not shellable. Otherwise by
symmetry we can assume 1, 2 is a shelling order. But when we adjoin facet 2
to 1, we introduce the new faces $c, d, cd, de$, and $cde$. There are *two* minimal
new faces: $c$ and $d$.

There is in fact a close connection between $\Delta_c$ and $\Delta_d$. Given any simplicial
complex $\Delta$ with vertex set $V$, define the *cone* over $\Delta$, denoted $C(\Delta)$, to be the
simplicial complex with vertex set $V \cup \{v\}$, where $v$ is a new vertex not in $V$,
and with faces

$$C(\Delta) = \Delta \cup \{\{v\} \cup F : F \in \Delta\}.$$

Then $\Delta_d = C(\Delta_c)$. Moreover, it is not hard to see (Exercise 7) that a simplicial
complex $\Delta$ is shellable if and only if $C(\Delta)$ is shellable.

**12.12 Example.** For a somewhat more complicated example of a shellable simplicial complex, let $\Delta$ be the simplicial complex realized by the boundary of an

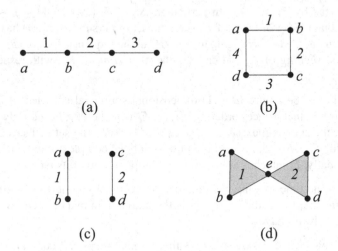

(a)                                           (b)

(c)                                           (d)

**Fig. 12.3** Some simplicial complexes

octahedron. Figure 12.4 shows a shelling of $\Delta$. This figure is a projection of the octahedron into the plane. All eight triangular regions, including the unbounded outside region with vertices $d, e, f$ represent faces of $\Delta$. The sets $G_i$ of minimal new faces are as follows:

$$G_1 = \emptyset, \quad G_2 = d, \quad G_3 = e, \quad G_4 = f$$
$$G_5 = de, \quad G_6 = ef, \quad G_7 = df, \quad G_8 = def.$$

**12.13 Example.** Figure 12.5 shows a more subtle example of a nonshellable simplicial complex $\Delta$. It has nine triangular facets. There is no "local" obstruction to shellabilty. That is, we cannot look at just a small part of $\Delta$ and conclude that it is nonshellable. We will explain why $\Delta$ is nonshellable in Corollary 12.16 (see the paragraph after its proof). In general, however, there is no simple way to tell whether a simplicial complex is shellable.

We now want to discuss a connection between $f$-vectors and shellability. Suppose that $F_1, \ldots, F_t$ is shelling of a $(d - 1)$-dimensional simplicial complex $\Delta$. Let $G_j$ and $\Delta_{j-1}$ have the meaning in Definition 12.10. Suppose that $\#G_j = m$. Thus some $m$-element subset $S$ of $F_j$ is the unique minimal face of $F_j$ not belonging to $\Delta_{j-1}$. This set $S$ is contained in $\binom{d-m}{i}$ $(m + i)$-element subsets $T$ of $F_j$, since $\#F_j = d$. Therefore, knowing the number of elements of $G_j$ tells us exactly how many new faces of each dimension we have adjoined to $\Delta$ at the $j$th shelling step.

There is an elegant and very useful way of organizing the above information. Given the $f$-vector $(f_0, f_1, \ldots, f_{d-1})$ of a $(d - 1)$-dimensional simplicial complex $\Delta$, define numbers $h_0, h_1, \ldots, h_d$ by the formula

$$\sum_{i=0}^{d} f_{i-1}(x - 1)^{d-i} = \sum_{i=0}^{d} h_i x^{d-i}, \tag{12.5}$$

**Fig. 12.4** A shelling order of the octahedron

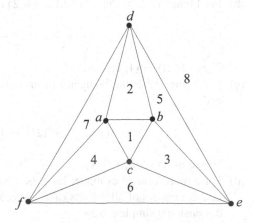

**Fig. 12.5** A nonshellable
simplicial complex

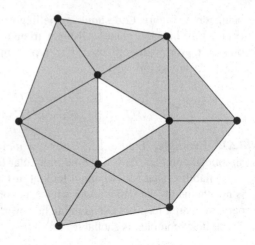

where as usual $f_{-1} = 1$ unless $\Delta = \emptyset$. We call the vector

$$h(\Delta) = (h_0, h_1, \ldots, h_d)$$

the *h-vector* of $\Delta$. It is clear from (12.5) that the $f$-vector and $h$-vector contain equivalent information—$f(\Delta)$ determines $h(\Delta)$ and *vice versa*.

### 12.14 Example.

(a) The $f$-vector of the simplicial complex of Figure 12.3a is $(3, 2)$. We compute that

$$(x - 1)^2 + 3(x - 1) + 2 = x^2 + x.$$

Hence $h(\Delta) = (1, 1, 0)$.

(b) For Figure 12.3c we have $f(\Delta) = (4, 2)$ and

$$(x - 1)^2 + 4(x - 1) + 2 = x^2 + 2x - 1.$$

Hence $h(\Delta) = (1, 2, -1)$.

(c) For Figure 12.2 (the boundary of an octahedron) we have $f(\Delta) = (6, 12, 8)$ and

$$(x - 1)^3 + 6(x - 1)^2 + 12(x - 1) + 8 = x^3 + 3x^2 + 3x + 1.$$

Hence $h(\Delta) = (1, 3, 3, 1)$.

(d) For a more general example, let $\Delta$ be generated by a single $d$-element face $F$, i.e., $\Delta$ consists of all subsets of $F$. A geometric realization of $\Delta$ is a $(d - 1)$-dimensional simplex. Now

$$f(\Delta) = \left( \binom{d}{1}, \binom{d}{2}, \binom{d}{3}, \ldots, \binom{d}{d} \right),$$

and

$$\sum_{i=0}^{d} \binom{d}{i} (x - 1)^{d-i} = x^d,$$

by the binomial theorem. Hence $h(\Delta) = (1, 0, 0, \ldots, 0)$.

There are some elementary properties of the $h$-vector worth noting:

- By taking coefficients of $x^d$ on both sides of (12.5), we see that $h_0 = 1$ unless $\Delta = \emptyset$.
- Taking coefficients of $x^{d-1}$ shows that $h_1 = f_0 - d$.
- If we set $x = 1$ in (12.5), then we obtain[1]

$$f_{d-1} = h_0 + h_1 + \cdots + h_d. \tag{12.6}$$

The left-hand side $f_{d-1}$ is the number of $(d-1)$-faces of $\Delta$. It would be nice if $h_i$ were the number of such faces with some property (depending on $i$). Example 12.14(b) shows that we can have $h_i < 0$, so in general $h_i$ does not have such a nice combinatorial interpretation. However, for *shellable* simplicial complexes $h_i$ has a simple interpretation given by Theorem 12.15 below.

- (for readers with some knowledge of topology) Putting $x = 0$ on both sides of (12.5) shows that

$$h_d = (-1)^{d-1}(-f_{-1} + f_0 - f_1 + f_2 - \cdots + (-1)^{d-1} f_{d-1}). \tag{12.7}$$

If $X$ is any topological space that possesses a finite triangulation, then for any triangulation $\Gamma$, say with $f$-vector $(f_0, \ldots, f_{d-1})$, the alternating sum $-f_{-1} + f_0 - f_1 + \cdots + (-1)^{d-1} f_{d-1}$ is independent of the triangulation and is known as the *reduced Euler characteristic* of $X$, denoted $\tilde{\chi}(X)$. We also write $\tilde{\chi}(\Gamma) = \tilde{\chi}(X)$ for any triangulation $\Gamma$ of $X$. We say that $\tilde{\chi}(\Gamma)$ is a *topological invariant* of $\Gamma$ since it depends only on the geometric realization $|\Gamma|$ as a topological space. Equation (12.7) therefore shows that

$$h_d = (-1)^{d-1} \tilde{\chi}(\Gamma). \tag{12.8}$$

Recall also that the ordinary *Euler characteristic* $\chi(X)$ is given by $f_0 - f_1 + f_2 - \cdots + (-1)^{d-1} f_{d-1}$ for any triangulation $\Gamma$ as above. Thus, if $\Gamma \neq \emptyset$, then

---

[1]Since we have $(x - 1)^0 = 1$ in the term indexed by $i = d$ on the left-hand side of (12.5), we need to interpret $0^0 = 1$ when we set $x = 1$. Although $0^0$ is an indeterminate form in calculus, in combinatorics it usually makes sense to set $0^0 = 1$.

$$\tilde{\chi}(X) = \chi(X) - 1$$

since $f_{-1} = 1$. Hence the difference between the reduced and ordinary Euler characteristics depends on whether or not we regard $\emptyset$ as a face.

We now come to the relationship between shellings and $h$-vectors.

**12.15 Theorem.** *Let* $F_1, \ldots, F_t$ *be a shelling of the simplicial complex* $\Delta$, *with restrictions* $G_1, \ldots, G_t$. *Then*

$$\sum_{i=0}^{d} h_i x^i = \sum_{j=1}^{t} x^{\#G_j}.$$

*In other words,* $h_i$ *is the number of restrictions with* $i$ *elements (independent of the choice of shelling).*

*Proof.* We noted after Example 12.13 that when we adjoin a facet $F_j$ to a shelling with restriction $G_j$ satisfying $\#G_j = m$, then we adjoin $\binom{d-m}{i}$ new faces with $m+i$ elements. Hence the contribution to the polynomial $\sum_{i=0}^{d} f_{i-1}(x-1)^{d-i}$ from adjoining $F_j$ (using the symmetry $\binom{d-m}{i} = \binom{d-m}{d-m-i}$ and the binomial theorem) is given by

$$\sum_{i=0}^{d-m} \binom{d-m}{i}(x-1)^{d-(m+i)} = \sum_{i=0}^{d-m} \binom{d-m}{i}(x-1)^i$$

$$= x^{d-m},$$

and the proof follows.                                                                                    □

**12.16 Corollary.** *A necessary condition for a (pure)* $(d-1)$-*dimensional simplicial complex* $\Delta$ *to be shellable is that* $h_i(\Delta) \geq 0$ *for all* $0 \leq i \leq d$. *Moreover, if* $\Delta$ *triangulates a topological space* $X$, *then a necessary condition for shellability is* $(-1)^{d-1}\tilde{\chi}(X) \geq 0$.

*Proof.* Assume that $\Delta$ is shellable. By Theorem 12.15 we have $h_i(\Delta) \geq 0$ for all $0 \leq i \leq d$. The second assertion then follows from (12.8).                                    □

Corollary 12.16 explains why the simplicial complex $\Delta$ of Figure 12.5 is not shellable. We have

$$h_3(\Delta) = (-1)^2(-1 + 9 - 18 + 9) = -1.$$

The geometric realization of $\Delta$ is a cylinder (or more accurately, homeomorphic to a cylinder). Since $h_3(\Delta) = -1$, it follows that *any* triangulation $\Gamma$ of a cylinder

$X$ satisfies $h_3(\Gamma) = -1 = \tilde{\chi}(X)$. Similarly, the two-dimensional torus $T$ satisfies $(-1)^2 \tilde{\chi}(T) = -1$, so no triangulation of $T$ can be shellable.

The condition of Corollary 12.16 is necessary but not sufficient for shellability. For instance, the disjoint union of two cycles (a one-dimensional simplicial complex) satisfies $h_d = 1$ but isn't shellable. (See Exercise 28.) For some more subtle examples, see Exercises 9 and 11.

## 12.2   The Face Ring

Our goal is a complete characterization of the $f$-vector of a shellable simplicial complex, analogous to the characterization of the $f$-vector of all simplicial complexes given by the Kruskal–Katona theorem (Theorem 12.6). The main tool will be a certain commutative ring associated with a simplicial complex $\Delta$ on the vertex set $V = \{x_1, \ldots, x_n\}$. To keep the presentation as simple as possible, we will develop the necessary ring theory to prove the main result of this chapter (Theorem 12.25), but no more. Most of our definitions, results, and proofs can be extended to a far greater context. We make a brief remark on one of these generalizations in Remark 12.26.

Let $K$ be a field. Any infinite field will do for our purposes. Think of the elements of the vertex set $V$ as indeterminates. Let $K[x_1, \ldots, x_n]$ or $K[V]$ denote the polynomial ring in the indeterminates $x_1, \ldots, x_n$. For any subset $S$ of $\{x_1, \ldots, x_n\}$, write

$$x_S = \prod_{x_i \in S} x_i. \tag{12.9}$$

Let $I_\Delta$ denote the ideal of $K[V]$ generated by all monomials $x_S$ such that $S \notin \Delta$. We call such a set $S$ a *nonface* of $\Delta$. If $S$ is a nonface and $T \supset S$, then clearly $T$ is a nonface. Hence $I_\Delta$ is generated by the *minimal nonfaces* of $\Delta$, that is, those nonfaces for which no proper subset is a nonface. A minimal nonface is also called a *missing face*.

**12.17 Example.** For the simplicial complexes of Figure 12.3 we have the following minimal generators of $I_\Delta$, i.e., the monomials corresponding to missing faces: (a) $ac, ad, bd$, (b) $ac, bd$, (c) $ac, ad, bc, bd$, (d) $ac, ad, bc, bd$. Note that (c) and (d) have the same missing faces. This is because (d) is a cone over (c). The cone vertex $e$ is attached to every face $F$ of (c) (i.e., $\{e\} \cup F$ is a face of (d)), so $e$ belongs to no missing face.

For Figure 12.1, the missing faces all have two elements except for $\{3, 5, 6\}$. For the octahedron of Figure 12.2, the missing faces are (writing as usual $11'$ for $\{1, 1'\}$, etc.) $11'$, $22'$, and $33'$.

The quotient ring $K[\Delta] := K[V]/I_\Delta$ is called the *face ring* (also called the *Stanley–Reisner ring*) of $\Delta$. It is the fundamental algebraic object of this chapter.

If face rings are to be useful in characterizing $f$-vectors, we need to connect the two together. For this aim, define the *support* supp($u$) of a monomial $u = x_1^{a_1} \cdots x_n^{a_n}$ by

$$\text{supp}(u) = \{x_i : a_i > 0\}.$$

Note that a $K$-basis for the ideal $I_\Delta$ consists of all monomials $u$ satisfying supp($u$) $\notin$ $\Delta$ [why?]. Hence a $K$-basis for $K[\Delta]$ consists of all monomials $u$ satisfying supp($u$) $\in \Delta$, including (unless $\Delta = \emptyset$) the monomial 1, whose support is $\emptyset$. (More precisely, we mean the images of these monomials in $K[\Delta]$ under the quotient map $K[V] \to K[\Delta]$, but in such situations we identify elements of $K[V]$ with their images in $K[\Delta]$.) For $i \geq 0$ define $K[\Delta]_i$ to be the span of all monomials $u$ of degree $i$ satisfying supp($u$) $\in \Delta$. Then

$$K[\Delta] = K[\Delta]_0 \oplus K[\Delta]_1 \oplus \cdots \text{ (vector space direct sum)}.$$

We define the *Hilbert series* of $K[\Delta]$ to be the power series

$$L(K[\Delta], \lambda) = \sum_{i \geq 0} (\dim_K K[\Delta]_i) \lambda^i,$$

where $\lambda$ is an indeterminate. Thus $L(K[\Delta], \lambda)$ is some kind of measurement of the "size" of $K[\Delta]$.

**12.18 Theorem.** *If* $\dim \Delta = d - 1$ *and* $h(\Delta) = (h_0, h_1, \ldots, h_d)$, *then*

$$L(K[\Delta], \lambda) = \frac{h_0 + h_1 \lambda + \cdots + h_d \lambda^d}{(1 - \lambda)^d}. \tag{12.10}$$

*Proof.* We have seen that a $K$-basis for $K[\Delta]$ consists of monomials whose support is a face $F$ of $\Delta$. Let $\mathcal{M}_F$ be the set of monomials with support $F$. Then

$$\sum_{u \in \mathcal{M}_F} \lambda^{\deg(u)} = \prod_{x_i \in F} \left( \sum_{a_i \geq 1} \lambda^{a_i} \right)$$

$$= \frac{\lambda^{\#F}}{(1 - \lambda)^{\#F}}. \tag{12.11}$$

In particular, when $F = \emptyset$ the two sides of (12.11) are equal to 1. Summing over all $F \in \Delta$ gives

$$L(K[\Delta], \lambda) = \sum_{F \in \Delta} \frac{\lambda^{\#F}}{(1 - \lambda)^{\#F}}$$

$$= \sum_{i=0}^{d} f_{i-1} \frac{\lambda^i}{(1-\lambda)^i}$$

$$= \frac{\sum_{i=0}^{d} f_{i-1} \lambda^i (1-\lambda)^{d-i}}{(1-\lambda)^d}.$$

Now

$$\sum_{i=0}^{d} f_{i-1} \lambda^i (1-\lambda)^{d-i} = \lambda^d \sum_{i=0}^{d} f_{i-1} \left( \frac{1}{\lambda} - 1 \right)^{d-i}$$

$$= \lambda^d \sum_{i=0}^{d} h_i \lambda^{-(d-i)} \quad \text{(by (12.5))}$$

$$= \sum_{i=0}^{d} h_i \lambda^i,$$

and the proof follows.                                                                                   □

The integer $d = 1 + \dim \Delta$ is called the *Krull dimension* of $K[\Delta]$, denoted $\dim K[\Delta]$. Do not confuse the vector space dimension $\dim_K$ with the Krull dimension dim! By (12.6) and (12.10) $\dim K[\Delta]$ is the order to which 1 is a pole of $L(K[\Delta], \lambda)$, i.e., the least integer $k$ for which $(1-\lambda)^k L(K[\Delta], \lambda)$ does not have a singularity at $\lambda = 1$. It is known (but not needed here) that $\dim K[\Delta]$ is also the most number of elements of $K[\Delta]$ that are algebraically independent over $K$, and is also the length $\ell$ of the longest chain $\mathfrak{p}_0 \subset \mathfrak{p}_1 \subset \cdots \subset \mathfrak{p}_\ell$ of prime ideals of $K[\Delta]$.

There is a special situation in which the $h_i$'s have a direct algebraic interpretation. In any commutative ring $R$, recall that an element $u$ is called a *non-zero-divisor* (or NZD) if whenever $y \in R$ and $uy = 0$, then $y = 0$. Now let $\theta \in K[\Delta]_1$ be an NZD. (Note that $\theta \in K[\Delta]_1$ means that $\theta$ is a linear combination $\sum_{i=1}^{n} \alpha_i x_i$ ($\alpha_i \in K$) of the vertices $x_1, \ldots, x_n$ of $\Delta$.) Since $\theta$ is an NZD, we have that for $i \geq 0$ the map $K[\Delta]_i \to K[\Delta]_{i+1}$ defined by $y \mapsto \theta y$ is injective (one-to-one). Hence

$$\dim_K \theta K[\Delta]_i = \dim_K K[\Delta]_i. \tag{12.12}$$

Let $(\theta)$ denote the ideal of $K[\Delta]$ generated by $\theta$. Since $\theta$ is homogeneous, the quotient ring $K[\Delta]/(\theta)$ has the vector space grading

$$K[\Delta]/(\theta) = (K[\Delta]/(\theta))_0 \oplus (K[\Delta]/(\theta))_1 \oplus \cdots, \tag{12.13}$$

where $(K[\Delta]/(\theta))_i$ is the image of $K[\Delta]_i$ under the quotient homomorphism $K[\Delta] \to K[\Delta]/(\theta)$.

If $A(\lambda) = \sum_{i \geq 0} a_i \lambda^i$ and $B(\lambda) = \sum_{i \geq 0} b_i \lambda^i$ are two power series with real coefficients, write $A(\lambda) \leq B(\lambda)$ to mean $a_i \leq b_i$ for all $i$.

**12.19 Lemma.** *Let* $\theta \in K[\Delta]_1$. *Then*

$$L(K[\Delta], \lambda) \leq \frac{L(K[\Delta]/(\theta), \lambda)}{1 - \lambda}, \tag{12.14}$$

*with equality if and only if* $\theta$ *in an NZD.*

*Proof.* If $\theta$ is an NZD, then by (12.12) we have

$$H(K[\Delta]/(\theta), i + 1) = H(K[\Delta], i + 1) - H(K[\Delta], i).$$

Multiplying both sides by $\lambda^{i+1}$ and summing on $i \geq -1$ gives

$$L(K[\Delta]/(\theta), \lambda) = L(K[\Delta], \lambda) - \lambda L(K[\Delta], \lambda),$$

so

$$L(K[\Delta], \lambda) = \frac{L(K[\Delta]/(\theta), \lambda)}{1 - \lambda}.$$

If $\theta$ is not an NZD, then we always have

$$\dim_K \theta K[\Delta]_i \leq \dim_K K[\Delta]_i,$$

and for at least one $i$ strict inequality holds. This is easily seen to imply that strict inequality holds in (12.14).                                                    □

By iteration of Lemma 12.19 we have the following result.

**12.20 Theorem.** *Let* $\theta_1, \ldots, \theta_j \in K[\Delta]_1$. *Then*

$$L(K[\Delta], \lambda) \leq \frac{L(K[\Delta]/(\theta_1, \ldots, \theta_j), \lambda)}{(1 - \lambda)^j},$$

*with equality if and only* $\theta_i$ *is an NZD in the ring* $K[\Delta]/(\theta_1, \ldots, \theta_{i-1})$ *for* $1 \leq i \leq j - 1$.

If $\theta_1, \ldots, \theta_j \in K[\Delta]_1$ has the property that $\theta_i$ is an NZD in the ring $K[\Delta]/(\theta_1, \ldots, \theta_{i-1})$ for $1 \leq i \leq j - 1$, then we say that $\theta_1, \ldots, \theta_j$ is a *regular sequence*. The number of elements of the largest regular sequence in $K[\Delta]_1$ is called the *depth* of $K[\Delta]$, denoted depth $K[\Delta]$. Let us remark that it can be shown that all maximal regular sequences have the same number of elements, though we do not need this fact here.

It is easy to see that a regular sequence $\theta_1, \ldots, \theta_j \in K[\Delta]_1$ is algebraically independent over $K$ (Exercise 22). In other words, there does not exist a polynomial

$0 \neq P(t_1, \ldots, t_k) \in K[t_1, \ldots, t_k]$ for which $P(\theta_1, \ldots, \theta_k) = 0$ in $K[\Delta]$. Let us point out that if the sequence $\theta_1, \ldots, \theta_j \in K[\Delta]$ is algebraically independent and moreover each $\theta_i$ is an NZD in $K[\Delta]$, then these conditions are *not* sufficient for $\theta_1, \ldots, \theta_j$ to be a regular sequence. For instance, if $\Delta$ has vertices $a, b, c$ and the single edge $ab$, then $a - c$ and $b - c$ are algebraically independent NZDs. However, in the ring $K[\Delta]/(a - c)$ we have $c \neq 0$ but $(b - c)c = 0$. In fact, we have depth $K[\Delta] = 1$, e.g., by Exercise 25. For another example, let $\Delta$ have vertices $a, b, c$ and edges $ab, bc$, and assume that $\mathrm{char}(K) \neq 2$. Now $a + b$ and $a - b$ are algebraically independent NZDs but not a regular sequence since in $K[\Delta]/(a + b)$ we have $c \neq 0$ and $c(a - b) = 0$. Unlike the previous example, this time we have depth $K[\Delta] = 2$. A regular sequence of length two is given by, for instance, $a - c, b$.

Suppose that $\theta_1, \ldots, \theta_d \in K[\Delta]_1$ is a regular sequence, where as usual $d = \dim K[\Delta] = \dim \Delta + 1$. Let $R = K[\Delta]/(\theta_1, \ldots, \theta_d)$. Thus $R$ inherits a grading $R = R_0 \oplus R_1 \oplus \cdots$ from $K[\Delta]$. Let $h(\Delta) = (h_0, h_1, \ldots, h_d)$ as usual. By Theorem 12.18 and the definition of regular sequence we have

$$L(R, \lambda) = h_0 + h_1 \lambda + \cdots + h_d \lambda^d, \tag{12.15}$$

a polynomial in $\lambda$. Hence $R$ is a finite-dimensional vector space, and $\dim_K R = \sum h_i = f_{d-1}$. Clearly $R_1$ cannot contain an NZD $\psi$, since, e.g., if $u$ is a nonzero element of $R$ of maximal degree (which must exist since $\dim_K R < \infty$), then $\psi u = 0$. Hence depth $K[\Delta] = d = \dim K[\Delta]$, the maximum possible. This motivates the following key definition.

**12.21 Definition.** Assume that $K$ is an infinite field. We say that the simplicial complex $\Delta$ is *Cohen–Macaulay* (with respect to the field $K$) and that the ring $K[\Delta]$ is a *Cohen–Macaulay ring* if $\dim K[\Delta] = \mathrm{depth}\, K[\Delta]$.

NOTE. Note that the above definition assumes that $K$ is infinite. There is a more algebraic definition of Cohen–Macaulay that coincides with our definition when $K$ is infinite but not always when $K$ is finite. For our purposes it doesn't hurt to assume that $K$ is infinite.

It follows from (12.15) that a Cohen–Macaulay simplicial complex $\Delta$ satisfies $h_i(\Delta) \geq 0$. However, two basic problems remain, as follows.

- What simplicial complexes are Cohen–Macaulay?
- What more can be said about the $h$-vector (or $f$-vector) of a Cohen–Macaulay simplicial complex?

The definitive answer to the first question is beyond the scope of this book, but for the benefit of readers with some knowledge of algebraic topology we discuss the answer in Remark 12.26. What we will prove is that shellable simplicial complexes are indeed Cohen–Macaulay. Regarding the second question, we will obtain a complete characterization of the $h$-vector of a Cohen–Macaulay simplicial complex, which will also characterize $h$-vectors of shellable simplicial complexes. This characterization is a multiset analogue of the Kruskal–Katona theorem (Theorem 12.6).

Let us first consider the second question. A *multicomplex* $\Gamma$ on a set $V$ is a multiset analogue of a simplicial complex whose vertex set is contained in $V$. More precisely, $\Gamma$ is a collection of multisets (sets with repeated elements, as discussed on page 1), such that every element of $\Gamma$ is contained in $V$, and if $M \in \Gamma$ and $N \subseteq M$, then $N \in \Gamma$. We will assume from now on that the underlying set $V$ is *finite*. For example (writing 112 for $\{1, 1, 2\}$, etc.), $\Gamma = \{\emptyset, 1, 2, 3, 11, 12, 112, 1112\}$ is *not* a multicomplex, since $1112 \in \Gamma$ and $111 \subseteq 1112$, but $111 \notin \Gamma$.

If $\Gamma$ is a multicomplex with $e_i$ elements of size $i$, then we call the sequence $e(\Gamma) = (e_0, e_1, \dots)$ the *e-vector* of $\Gamma$. Any integer vector $(e_0, e_1, \dots)$ which is the $e$-vector of some multicomplex is called an *e-vector*. Our $e$-vectors are also called *M-vectors* after F. S. Macaulay and *O-sequences*, where O stands for "order ideal of monomials," defined below.

NOTE ON TERMINOLOGY. It might seem more natural to let $f_i$ be the number of elements of $\Gamma$ of size $i + 1$, and define $(f_0, f_1, \dots)$ to be the $f$-vector of $\Gamma$ in complete analogy with simplicial complexes. Historically, the indexing of $f$-vectors is explained by $f_i$ being the number of faces of *dimension* (rather than cardinality) $i$. For multicomplexes, we have no need for the concept of the dimension of a face $F$ (and if we did, the "best" definition would be that $\dim F$ is one less than the number of *distinct* elements of $F$). Counting elements of multicomplexes by their cardinality is more natural for almost all purposes. In the literature our $e_i$ is often replaced with $h_i$, and our $e$-vector is called an $h$-vector. This is because $e$-vectors of multicomplexes do sometimes coincide with $h$-vectors of simplicial complexes (e.g., Theorem 12.25). Moreover, $e$-vectors of multicomplexes coincide with the sequence of Hilbert function values of standard graded $K$-algebras (not defined here, though $K[\Delta]$ and its quotients considered here are special cases), so one can think that $h$ stands for "Hilbert." To avoid any possible confusion we will use the new notation $e_i$ and terminology $e$-vector.

We now discuss a "multiplicative equivalent" of multicomplexes. If $u$ and $v$ are monomials in the variables $x_1, \dots, x_n$, we say that $u$ *divides* $v$ (written $u \mid v$) if there is a monomial $w$ for which $uw = v$. Equivalently, if $u = x_1^{a_1} \cdots x_n^{a_n}$ and $v = x_1^{b_1} \cdots x_n^{b_n}$, then $u \mid v$ if and only if $a_i \le b_i$ for all $i$. An *order ideal of monomials* in the variables $x_1, \dots, x_n$ is a collection $\mathfrak{o}$ of monomials in these variables such that if $v \in \mathfrak{o}$ and $u \mid v$, then $u \in \mathfrak{o}$. Equivalently, associate with the monomial $u = x_1^{a_1} \cdots x_n^{a_n}$ the multiset $M_u = \{1^{a_1}, \dots, n^{a_n}\}$ (i.e., $i$ has multiplicity $a_i$). Then a set $\mathcal{M}$ of monomials is an order ideal of monomials if and only if the collection $\{M_u : u \in \mathcal{M}\}$ is a multicomplex.

While we need the next result only for quotients of face rings by a regular sequence, it involves no extra work to prove it in much greater generality. For this purpose, we define a *homogeneous ideal* of the polynomial ring $K[x_1, \dots, x_n]$ to be an ideal $I$ generated by homogeneous polynomials.

**12.22 Theorem.** *Let $I$ be a homogeneous ideal of $K[x_1, \dots, x_n]$, and let $P = K[x_1, \dots, x_n]/I$. Then $P$ has a $K$-basis that is an order ideal of monomials in the variables $x_1, \dots, x_n$.*

*Proof.* We define *reverse lex order* on monomials of some fixed degree $m$ as follows: define

$$x_1^{a_1} \cdots x_n^{a_n} \overset{R}{<} x_1^{b_1} \cdots x_n^{b_n},$$

where $\sum a_i = \sum b_i$, if

$$(1^{a_1}, \ldots, n^{a_n}) \overset{R}{<} (1^{b_1}, \ldots, n^{b_n}),$$

where $i^j$ denotes a sequence $i$'s of length $j$, and $\overset{R}{<}$ is defined by 12.4. For instance, the reverse lex order on monomials of degree three in the variables $x_1, x_2, x_3$ is

$$x_1^3 \overset{R}{<} x_1^2 x_2 \overset{R}{<} x_1 x_2^2 \overset{R}{<} x_2^3 \overset{R}{<} x_1^2 x_3 \overset{R}{<} x_1 x_2 x_3 \overset{R}{<} x_2^2 x_3 \overset{R}{<} x_1 x_3^2 \overset{R}{<} x_2 x_3^2 \overset{R}{<} x_3^3.$$

For each $m \geq 0$ let $P_m$ denote the span (over the field $K$) of the homogeneous polynomials in $P$ of degree $m$. Because $I$ is generated by homogeneous polynomials we have the vector space direct sum

$$P = P_0 \oplus P_1 \oplus P_2 \oplus \cdots,$$

a direct generalization of (12.13).

For each degree $m$, let $B_m$ be the least $K$-basis for $P_m$ in reverse lex order. In other words, first choose the least monomial $u_1$ of degree $m$ in reverse lex order that is nonzero in $P$. Then choose the least monomial $u_2$ of degree $m$ in reverse lex order such that $\{u_1, u_2\}$ are linearly independent, etc. We eventually obtain a $K$-basis for $P_m$ by this process since every linearly independent subset of a vector space can be extended to a basis.

We claim that the set $B_0 \cup B_1 \cup B_2 \cup \cdots$ is a basis for $P$ which is an order ideal of monomials. Suppose not. Let $v \in B_j$, $u \mid v$, but $u \notin B_i$ where $i = \deg u$. Then $u$ is a linear combination of monomials $w \overset{R}{<} u$, say

$$u = \sum_{w \overset{R}{<} u} \alpha_w w.$$

Multiply both sides by $\frac{v}{u}$. It is easy to see that if $w \overset{R}{<} u$ then $w \cdot \frac{v}{u} \overset{R}{<} v$. Thus we have expressed $v$ as a linear combination of monomials $\frac{wv}{u} \overset{R}{<} v$, contradicting $v \in B_j$. $\qquad\square$

**12.23 Corollary.** *Let $\Delta$ be a Cohen–Macaulay simplicial complex. Then the h-vector of $\Delta$ is an e-vector.*

*Proof.* Let $\theta_1, \ldots, \theta_d$ be a regular sequence in $K[\Delta]_1$, and let

$$R = R_0 \oplus R_1 \oplus \cdots = K[\Delta]/(\theta_1, \ldots, \theta_d).$$

According to the definition of a Cohen–Macaulay simplicial complex and (12.15) we have $h_i(\Delta) = \dim_K R_i$. Thus if $D_i$ is any $K$-basis for $R_i$, then $\#D_i = h_i$. By Theorem 12.22 there is a $K$-basis $B_i$ for each $i$ such that $B_0 \cup B_1 \cup \cdots \cup B_d$ is an order ideal $\Gamma$ of monomials. Thus if $\Gamma$ has $e$-vector $(e_0, e_1, \dots)$, then $e_i = \#B_i = h_i$, and the proof follows.                                                                      $\square$

The next step is to find some simplicial complexes to which we can apply Corollary 12.23. The following theorem is the primary algebraic result of this chapter. Recall the notation $x_S = \prod_{x_i \in S} x_i$ of (12.9).

**12.24 Theorem.** *If $\Delta$ is a shellable simplicial complex on the vertex set $V = \{x_1, \dots, x_n\}$, then the face ring $K[\Delta]$ is Cohen–Macaulay for any infinite field $K$. Moreover, if $F_1, \dots, F_t$ is a shelling of $\Delta$ with restrictions $G_1, \dots, G_t$, then $x_{G_1}, \dots, x_{G_t}$ is a $K$-basis for $R = K[\Delta]/(\theta_1, \dots, \theta_d)$ for any regular sequence $\theta_1, \dots, \theta_d \in K[\Delta]_1$, and such a regular sequence always exists.*

*Proof.* Let $B = \{x_{G_1}, \dots, x_{G_t}\}$. Let $\theta_1, \dots, \theta_d \in K[\Delta]_1$ satisfy the following property.

(P) The restriction of $\theta_1, \dots, \theta_d$ to any facet (or face) $F$ spans the $K$-vector space $KF$ with basis $F$. In other words, if we define

$$\psi_i = \theta_i|_{x_j=0 \text{ if } x_j \notin F},$$

then $\psi_1, \dots, \psi_d$ span $KF$.

Note that $K$ being infinite guarantees that there is enough "room" to find such $\theta_1, \dots, \theta_d$. This is the reason why we require $K$ to be infinite. The argument below will show in particular that if $\theta_1, \dots, \theta_d \in K[\Delta]_1$ satisfies Property (P), then $\theta_1, \dots, \theta_d$ is a regular sequence.

Now let $R = K[\Delta]/(\theta_1, \dots, \theta_d)$. By Theorems 12.15 and 12.20 (in the case $j = d$) it follows that if $B$ spans $R$ (as a vector space over $K$) then $\theta_1, \dots, \theta_d$ is regular, and $B$ is a $K$-basis for $R$. Thus we need to show that $B$ spans $R$.

The proof is by induction on $t$.

First assume that $t = 1$. Then $\Delta$ is just a simplex and $K[\Delta] = K[x_1, \dots, x_d]$. Moreover, any $K$-basis $\theta_1, \dots, \theta_d$ is a regular sequence and $K[\Delta]/(\theta_1, \dots, \theta_d) = K$. The Hilbert series of the field $K$ is just 1. Finally, if $F$ is the unique facet of $\Delta$, then $F$ (regarded as a one-term sequence) is a shelling of $\Delta$ with $G_1 = \emptyset$. Since $x_\emptyset = 1$ is a basis for $K$, the theorem is true for $t = 1$.

Now assume the theorem for $t - 1$, and let $F_1, \dots, F_t$ be a shelling of $\Delta$.

*Claim.* $x_i x_{G_t} = 0$ in $R$ for all $1 \le i \le n$.

*Case 1.* Suppose that $x_i \notin F_t$. By definition of shelling the new faces $F$ obtained by adjoining $F_t$ to the shelling are given by $G_t \subseteq F \subseteq F_t$. Thus $\{x_i\} \cup G_t$ cannot be a new face, so $\{x_i\} \cup G_t \notin \Delta$. Hence $x_i x_{G_t} = 0$ in $K[\Delta]$, so also in $R$.

*Case 2.* Suppose that $x_i \in F_t$. Set

$$K[F_t] = K[\Delta]/(x_j : x_j \notin F_t) = K[x_j : x_j \in F_t],$$

a polynomial ring in the vertices of $F_t$. By property (P), the restrictions $\psi_1, \ldots, \psi_d$ of $\theta_1, \ldots, \theta_d$ to $F$ span the space $KF$. Hence there exists a linear combination of $\theta_1, \ldots, \theta_d$ of the form

$$\eta = x_i + \sum_{x_j \notin F_t} \alpha_j x_j, \quad \alpha_j \in K.$$

Then in the ring $R$ we have

$$x_i x_{G_t} = (x_i - \eta) x_{G_t} \quad (\text{since } \eta = 0 \text{ in } R)$$

$$= - \left( \sum_{x_j \notin F_t} \alpha_j x_j \right) x_{G_t}$$

$$= 0 \quad (\text{by Case 1}).$$

This completes the proof of the claim.

Now let $R' = R/(x_{G_t})$ and $\Delta_{t-1} = \langle F_1, \ldots, F_{t-1} \rangle$. By definition of $G_t$ we have

$$K[\Delta_{t-1}] = K[\Delta]/(x_{G_t}).$$

Condition (P) still holds for $K[\Delta_{t-1}]$ (since the facets of $\Delta_{t-1}$ are also facets of $\Delta$). Moreover,

$$R' = K[\Delta_{t-1}]/(\theta_1, \ldots, \theta_d).$$

By the induction hypothesis, $x_{G_1}, \ldots, x_{G_{t-1}}$ span $R'$. By the claim, the ideal $(x_{G_t})$ of $R$ is a vector space of dimension at most one. Hence $x_{G_1}, \ldots, x_{G_t}$ span $R$, and the proof follows for any sequence (necessarily regular by the argument above) $\theta_1, \ldots, \theta_d$ satisfying (P).

It remains to show that *every* regular sequence $\theta_1, \ldots, \theta_d \in K[\Delta]_1$ satisfies (P). This result is an easy exercise; a somewhat more general result is given by the "only if" part of Exercise 24. □

NOTE. Note the structure of the previous proof. We pick $\theta_1, \ldots, \theta_d \in K[\Delta]_1$ satisfying Property (P). Let $F_1, \ldots, F_t$ be a shelling of $\Delta$, and set $R = K[\Delta]/(\theta_1, \ldots, \theta_d)$. As we successively quotient $R$ by the monomials $x_{G_1}, \ldots, x_{G_t}$, the vector space dimension drops by at most one, and we end up with the ring 0. Hence $\dim_K R \leq t = f_{d-1}(\Delta)$. On the other hand, by Theorems 12.18 and 12.20 we have $\dim_K R \geq \sum h_i = t$. Hence $x_{G_1}, \ldots, x_{G_t}$ must be a $K$-basis for $R$.

We are finally ready for the main theorem of this chapter.

**12.25 Theorem.** *Let $h = (h_0, h_1, \ldots, h_d)$ be a sequence of integers. The following three conditions are equivalent.*

(a)  There exists a $(d - 1)$-dimensional Cohen–Macaulay simplicial complex (over any infinite field) $\Delta$ with $h(\Delta) = \boldsymbol{h}$.

(b)  There exists a $(d - 1)$-dimensional shellable simplicial complex $\Delta$ with $h(\Delta) = \boldsymbol{h}$.

(c)  The sequence $\boldsymbol{h}$ is an e-vector.

*Proof.* (b)$\Rightarrow$(a) Immediate from Theorem 12.24.

(a)$\Rightarrow$(c) This is Corollary 12.23.

(c)$\Rightarrow$(b) Given the e-vector $\boldsymbol{h}$, we need to construct a shellable simplicial complex $\Delta$ whose $h$-vector is $\boldsymbol{h}$. We will identify a (finite) multiset $M$ of positive integers with the increasing sequence of its elements. Given $0 \leq i \leq d$, let $\alpha_1, \alpha_2, \ldots, \alpha_{h_i}$ be the first $h_i$ terms of the reverse lex order on $i$-element multisets of positive integers. For instance, when $i = 3$ and $h_3 = 8$, we have

$$(\alpha_1, \ldots, \alpha_8) = (111, 112, 122, 222, 113, 123, 223, 133). \qquad (12.16)$$

If $\alpha_j = a_1 a_2 \cdots a_i$, define

$$\beta_j = 1, 2, 3, \ldots, d - i, a_1 + d - i + 1, a_2 + d - i + 2, \ldots, a_i + d. \qquad (12.17)$$

For the example of (12.16) and $d = 5$ we have

$$(\beta_1, \ldots, \beta_8) = (12456, 12457, 12467, 12567, 12458, 12468, 12568, 12478).$$

Now let $\sigma$ be the concatenation $\beta_1, \beta_2, \ldots, \beta_d$ of the sequences $\beta_1, \ldots, \beta_d$. For instance, if $\boldsymbol{h} = (1, 4, 2, 1)$ (so $d = 3$), then we get (where we separate the different $\beta_j$'s with semicolons, and we write in boldface the terms $a_1 + d - i + 1, a_2 + d - i + 2, \ldots, a_i + d$ of each $\beta_j$)

$$\sigma = (123; \ 124, \ 125, \ 126, \ 127; \ 134, \ 135; \ 234). \qquad (12.18)$$

We leave as an exercise (Exercise 29) to show that $\sigma$ is a shelling of a $(d - 1)$-dimensional simplicial complex $\Delta$ on the vertex set $\{1, 2, \ldots, h_1 + d\}$. Moreover, if we write $\sigma = (F_1, \ldots, F_m)$ (where $m = \sum h_i = f_{d-1}(\Delta)$) and if $F_k$ is given by the sequence on the right-hand side of (12.17), then the restriction $G_k$ is given by $G_k = \{a_1 + d - i + 1, a_2 + d - i + 2, \ldots, a_i + d\}$. This being the case, exactly $h_i$ restrictions $G_k$ have $i$ elements, so indeed $h(\Delta) = \boldsymbol{h}$. $\qquad \square$

As an example, the sequence $\sigma$ of (12.18) is a shelling $(F_1, \ldots, F_8)$ of a simplicial complex $\Delta$ with vertices $1, \ldots, 7$. The restrictions $G_1, \ldots, G_8$ are given by $\emptyset, 4, 5, 6, 7, 34, 35, 234$ (the elements in boldface).

**12.26 Remark.** We mentioned earlier that the complete characterization of Cohen–Macaulay simplicial complexes is beyond the scope of this book. For readers familiar with some algebraic topology (only the rudiments of simplicial homology are needed), we will state without proof the theorem of Gerald Reisner that provides this characterization. For this purpose, if $F \in \Delta$, then define the *link* of $F$, denoted $\mathrm{lk}_\Delta(F)$, by

$$\mathrm{lk}_\Delta(F) = \{G \in \Delta : F \cap G = \emptyset, \ F \cup G \in \Delta\}.$$

It is clear that $\mathrm{lk}_\Delta(F)$ is a subcomplex of $\Delta$. In particular, $\mathrm{lk}_\Delta(\emptyset) = \Delta$. For any simplicial complex $\Gamma$ we write $\widetilde{H}_i(\Delta; K)$ for the $i$th reduced homology group of $\Delta$ over the field $K$.

**12.27 Theorem.** *Let $K$ be an infinite field. The following two conditions on a simplicial complex $\Delta$ are equivalent.*

- *$\Delta$ is Cohen–Macaulay with respect to $K$.*
- *For every $F \in \Delta$ (including $F = \emptyset$), we have $\widetilde{H}_i(\mathrm{lk}_\Delta(F); K) = 0$ for all $i \neq \dim \mathrm{lk}_\Delta(F)$.*

It can be shown from Reisner's theorem that for fixed $K$ (or actually, for fixed characteristic of $K$), Cohen–Macaulayness is a topological property, i.e., it depends only on the geometric realization of $\Delta$ (as a topological space). For instance, all triangulations of spheres and balls (of any dimension) are Cohen–Macaulay over any (infinite) field. A triangulation of the real projective plane is Cohen–Macaulay with respect to $K$ if and only if $\mathrm{char}(K) \neq 2$.

Theorem 12.25 gives an elegant characterization of $h$-vectors (and hence $f$-vectors) of shellable simplicial complexes, but one ingredient is still missing—a "nice" characterization of $e$-vectors. Since an $e$-vector is a multiset analogue of an $f$-vector of a simplicial complex, it is not unreasonable to expect a characterization of $e$-vectors similar to the Kruskal–Katona theorem (Theorem 12.6) for ordinary $f$-vectors. We conclude this chapter by discussing such a characterization.

Given positive integers $n$ and $j$, let

$$n = \binom{n_j}{j} + \binom{n_{j-1}}{j-1} + \cdots + \binom{n_1}{1}$$

be the $j$-binomial expansion of $n$ (equation (12.1)). Now define

$$n^{\langle j \rangle} = \binom{n_j + 1}{j + 1} + \binom{n_{j-1} + 1}{j} + \cdots + \binom{n_1 + 1}{2}.$$

Thus instead of adding 1 to the bottom of each binomial coefficient as we did when we defined $n^{(j)}$, now we add 1 to the bottom *and* top. We now have the following exact analogue of the Kruskal–Katona theorem.

**12.28 Theorem.** *A vector $(e_0, e_1, \ldots, e_d) \in \mathbb{P}^{d+1}$ is an $e$-vector if and only if $e_0 = 1$ and*

$$e_{i+1} \leq e_i^{\langle i \rangle}, \quad 0 \leq i \leq d - 1. \tag{12.19}$$

The proof is analogous to that of the Kruskal–Katona theorem. Namely, we identify a finite multiset $M$ on $\mathbb{N}$ with the increasing sequence of its elements. For instance, the multiset $\{0, 0, 2, 3, 3, 3\}$ becomes the sequence 002333, and the sequence of 3-element multisets on $\mathbb{N}$ in reverse lex order begins

$$000 \ \ 001 \ \ 011 \ \ 111 \ \ 002 \ \ 012 \ \ 112 \ \ 022 \ \ 122 \ \ 222 \ \ 003 \cdots .$$

It can easily be checked that if $a_1 a_2 \cdots a_j$ is the $n$th term (beginning with term 0) in the reverse lex ordering of $j$-element multisets on $\mathbb{N}$, then $a_1, a_2 + 1, a_3 + 2, \ldots, a_j + j - 1$ is the $n$th term in the reverse lex ordering of $j$-element *subsets* of $\mathbb{N}$. Hence Theorem 12.7 applies equally well to multisets on $\mathbb{N}$.

We next have the following multiset analogue of Theorem 12.8. The proof is completely analogous to that of Theorem 12.8.

**12.29 Theorem.** *Given* $e = (e_0, e_1, \ldots, e_d) \in \mathbb{P}^d$ *with* $e_0 = 1$, *let* $\Omega_e$ *consist of the union over all* $i \geq 1$, *together with* $\emptyset$, *of the first* $e_i$ *of the* $i$-element multisets on $\mathbb{N}$ *in reverse lex order. The set* $\Omega_e$ *is a multicomplex if and only if* $e_{i+1} \leq e_i^{\langle i \rangle}$ *for* $1 \leq i \leq d - 1$.

Theorem 12.29 proves the "if" direction of Theorem 12.28. The proof of the "only if" direction is similar to that of the Kruskal–Katona theorem. A multicomplex $\Omega_e$ for some $e$-vector $e$ is called *compressed*. Given any multicomplex $\Gamma$, we transform it by a sequence of simple operations into a compressed multicomplex, at all steps preserving the $e$-vector. We omit the details, which are somewhat more complicated than in the simplicial complex case.

**12.30 Example.** Is $(e_0, e_1, e_2, e_3) = (1, 4, 5, 7)$ an $e$-vector? The first $e_i$ multisets on $\mathbb{N}$ in reverse lex order for $1 \leq i \leq 3$ are given by

$$0 \ \ 1 \ \ 2 \ \ 3$$
$$00 \ \ 01 \ \ 11 \ \ 02 \ \ 12$$
$$000 \ \ 001 \ \ 011 \ \ 111 \ \ 002 \ \ 012 \ \ 112$$

These multisets (together with $\emptyset$) form a multicomplex, so $(1, 4, 5, 7)$ is an $e$-vector. On the other hand, $(1, 4, 5, 8)$ is *not* an $e$-vector. We need to add the multiset 022, but 22 does not appear. We can also check this by writing $5 = \binom{3}{2} + \binom{2}{1}$. Then $5^{\langle 2 \rangle} = \binom{4}{3} + \binom{3}{2} = 7$, so if in an $e$-vector we have $e_2 = 5$ then $e_3 \leq 7$.

# Notes for Chapter 12

The embedding theorem of Menger discussed in Example 12.3 appears (in much greater generality) in Menger [93]. The statement that the simplicial complex whose facets are all $(d + 1)$-element subsets of a $(2d + 3)$-element set cannot be realized in $\mathbb{R}^{2d}$ is due to A. Flores and E. R. van Kampen. For a modern treatment, see Matoušek [91].

The Kruskal–Katona theorem (Theorem 12.6) was first stated by M.-P. Schützen-berger in a rather obscure journal [117]. The first published proofs were by Kruskal [79] and later independently by Katona [72]. A nice survey of this area is given by Greene and Kleitman [56], including a good presentation of a proof of the Kruskal–Katona theorem due to Clements and Lindström [25].

The first indication of a connection between commutative algebra and combi-natorial properties of simplicial complexes appears in a paper of Melvin Hochster [67]. The face ring of a simplicial complex first appeared in the Ph.D. thesis of Gerald Reisner (published version in [110]), which was supervised by Hochster, and independently in two papers of Stanley [121, 122]. For an exposition of the connections between combinatorics and commutative algebra, see Stanley [124].

The concept of shelling goes back to nineteenth century geometers, but perhaps the first substantial result on shellings is due to Bruggesser and Mani [15]. The characterization of $h$-vectors of shellable simplicial complexes (Theorem 12.25) is a special case of a result of Stanley [121]. Our proof here is based on that of Kind and Kleinschmidt [74].

The characterization of $e$-vectors (Theorem 12.28) is due to Macaulay [87], who gave a very complicated proof as part of his characterization of Hilbert series of graded algebras. It is interesting that Macaulay's theorem preceded the Kruskal–Katona theorem, though the latter is somewhat easier to prove. Simpler proofs of Macaulay's theorem were later given by Sperner [120], Whipple [144], and Clements and Lindström [25], among others.

Cohen–Macaulay rings are named after Cohen [26] and Macaulay [86], who were interested in them primarily because of their connection with "unmixedness" theorems. For a modern treatment, see the text of W. Bruns and J. Herzog [19].

# Exercises for Chapter 12

1. A simplicial complex $\Delta$ has 2,528 three-dimensional faces. Show that it has at most 6,454 four-dimensional faces, and that this result is best possible. ("Best possible" means that there exists some $\Delta$ with $f_3 = 2,528$ and $f_4 = 6,454$.)

2. Prove the assertion of the Note following Definition 12.10. That is, let $\Delta$ be a pure $(d-1)$-dimensional simplicial complex. Then a facet ordering $F_1, \ldots F_t$ is a shelling if and only if for all $2 \le i \le t$, the subcomplex $\langle F_1, \ldots, F_{i_1} \rangle \cap \langle F_i \rangle$ is a pure simplicial complex of dimension $d-2$.

3. (a) Find the number of shellings of a path of length $n$, i.e., the simplicial complex with $n+1$ vertices and $n$ edges forming a path.
   (b) Find the number of shellings of a cycle of length $n$.

4. Find explicitly every simplicial complex $\Delta$ with the property that *every* ordering of its facets is a shelling.

5. Suppose that $F_1, F_2, \ldots, F_t$ is a shelling of a simplicial complex $\Delta$. Is it always the case that the reverse order $F_t, F_{t-1}, \ldots, F_1$ is also a shelling of $\Delta$?

6. Show that if a simplicial complex $\Delta$ is shellable and $F \in \Delta$, then the link $\text{lk}_\Delta(F)$ (as defined in Remark 12.26) is also shellable.

7. Prove the assertion of Example 12.11(d) that a simplicial complex $\Delta$ is shellable if and only if the cone $C(\Delta)$ is shellable.

8. A *matroid complex* is a simplicial complex $\Delta$ on the vertex set $V$ such that for any $W \subseteq V$, the *restriction* $\Delta_W$ of $\Delta$ to $W$, i.e.,

$$\Delta_W = \{F \in \Delta : F \subseteq W\},$$

is pure.

(a) Let $V = \{x_1, \ldots, x_n\}$ be a set of distinct nonzero vectors in some vector space over a field. Define

$$\Delta = \{F \subseteq V : F \text{ is linearly independent}\}.$$

Clearly $\Delta$ is a simplicial complex on $V$. Show that $\Delta$ is a matroid complex.

(b) Show that a matroid complex is shellable.

9. (*) (for those who know a little topology) Let $X$ be the topological space obtained by identifying the three sides (edges) of a solid triangle as shown in Figure 12.6. (The edges are identified in the direction of the arrows.) This space is called the *topological dunce hat*. Show that if $\Delta$ is a simplicial complex whose geometric realization is homeomorphic to $\Delta$ then $\Delta$ is not shellable.

10. (*) Give an example of a shellable simplicial complex $\Delta$ with more than one facet, such that there is a unique facet $F$ that comes last in every shelling.

11. (a) Show that every triangulation $\Delta$ of a two-dimensional ball $X$ (i.e., the geometric realization of $\Delta$ is homeomorphic to $X$) is shellable.

(b) (very difficult) Find a triangulation of a three-dimensional ball that is not shellable.

(c) (even more difficult) Find a triangulation of a three-dimensional sphere that is not shellable.

12. A *partial shelling* of a pure $(d-1)$-dimensional complex is a sequence $F_1, \ldots, F_r$ of some *subset* of the facets such that this sequence is a shelling order for the simplicial complex $\langle F_1, \ldots, F_r \rangle$ which they generate. Clearly if $F_1, \ldots, F_t$ is a shelling of $\Delta$, then $F_1, \ldots, F_r$ is a partial shelling for all $1 \le r \le t$.

(a) (*) Give an example of a shellable simplicial complex that has a partial shelling that cannot be extended to a shelling.

(b) Let $\Delta_{n,k}$ be the simplicial complex on an $n$-element vertex set $V$ whose facets consist of all $k$-element subsets of $V$. Show that $\Delta_{n,k}$ is shellable.

(c) (unsolved) Can every partial shelling of $\Delta_{n,k}$ be extended to a shelling?

13. Let $n$ be the largest integer, if it exists, with the following property: if $\Delta$ is any shellable simplicial complex with at most $n$ facets, then for any facet $F$ there exists a shelling beginning with $F$. If no such $n$ exists, then set $n = \infty$.

**Fig. 12.6** The topological
dunce hat

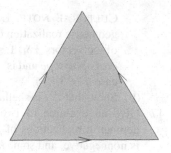

(a) (easy) Show that $n \geq 2$.

(b) (*) Show that $n \geq 9$.

(c) (*) Show that $n < \infty$.

14. Let $(f_0, f_1, \ldots, f_{d-1})$ be the $f$-vector of a $(d-1)$-dimensional simplicial complex $\Delta$. We will illustrate a certain procedure with the example $(6, 12, 8)$ (the $f$-vector of an octahedron). Write down the numbers $f_0, f_1, \ldots, f_{d-1}$ on a diagonal, and put 1 to the left of $f_0$:

$$1 \quad 6$$
$$12$$
$$8$$

Think of the 1 as being preceded by a string of 0's. Turn the array into a difference table by writing below each pair of consecutive numbers their difference:

$$1 \quad 6$$
$$1 \quad 5 \quad 12$$
$$1 \quad 4 \quad 7 \quad 8$$

Now write down one further row of differences:

$$1 \quad 6$$
$$1 \quad 5 \quad 12$$
$$\underline{1 \quad 4 \quad 7 \quad 8}$$
$$1 \quad 3 \quad 3 \quad 1$$

Show that this bottom row is the $h$-vector of $\Delta$.

15. Find the $f$-vector and $h$-vector of the simplicial complex whose geometric realization is the boundary of an icosahedron.

16. (a) Let $\Delta_d$ be the simplicial complex on the vertex set $V = \{x_1, \ldots, x_d, y_1, \ldots, y_d\}$ whose faces are those subsets of $V$ that do not contain both $x_i$ and $y_i$, for any $1 \leq i \leq d$. Compute the $h$-vector of $\Delta_d$.

CULTURAL NOTE. Let $\delta_i$ be the $i$th unit coordinate vector in $\mathbb{R}^d$. A geometric realization of $\Delta_d$ consists of the boundary of the convex hull $\mathcal{C}_d$ of the vectors $\pm\delta_i$, $1 \leq i \leq d$. The polytope $\mathcal{C}$ is called the $d$-dimensional *cross-polytope* and is a $d$-dimensional generalization of an octahedron, the case $d = 3$.

(b) Show that $\Delta_d$ is shellable.

17. Give an example of two simplicial complexes $\Delta_1$ and $\Delta_2$ such that the geometric realizations of $\Delta_1$ and $\Delta_2$ are homeomorphic, the $h$-vector of $\Delta_1$ is nonnegative, and some $h_i(\Delta_2) < 0$. What is the smallest possible dimension of $\Delta_1$ and $\Delta_2$?

18. (difficult from first principles) (*) Let $\Delta$ be a triangulation of a $(d - 1)$-dimensional sphere, and let $h(\Delta) = (h_0, h_1, \ldots, h_d)$. Show that $h_i = h_{d-i}$ for $0 \leq i \leq d$. This result is called the *Dehn–Sommerville equations* for spheres.

19. Let $h = (h_0, h_1, \ldots, h_d)$ and $k = (k_0, k_1, \ldots, k_d)$ be $e$-vectors. Define

$$h \wedge k = (\min\{h_0, k_0\}, \min\{h_1, k_1\}, \ldots, \min\{h_d, k_d\})$$

$$h \vee k = (\max\{h_0, k_0\}, \max\{h_1, k_1\}, \ldots, \max\{h_d, k_d\}).$$

Show that $h \wedge k$ and $h \vee k$ are $e$-vectors.

20. (*) Let $\Delta$ be the simplicial complex of Figure 12.7. Each triangle, including the "outer triangle" with vertices $a, b, c$ is a face, so $f(\Delta) = (6, 12, 8)$ and $\Delta$ triangulates a 2-sphere. Find the minimal nonfaces (or missing faces) of $\Delta$.

21. (*) Suppose that $\Gamma$ and $\Delta$ are simplicial complexes whose face rings $K[\Gamma]$ and $K[\Delta]$ are isomorphic as $K$-algebras (or even as rings). Show that $\Gamma$ and $\Delta$ are isomorphic.

22. Show that a regular sequence $\theta_1, \ldots, \theta_j \in K[\Delta]_1$ is algebraically independent over $K$.

**Fig. 12.7** A triangulation of the 2-sphere

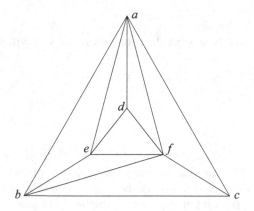

23. (a) Let $\theta_1, \ldots, \theta_j \in K[\Delta]_1$ be a regular sequence. Show that any permutation of this sequence is also a regular sequence.

    (b) Show that each $\theta_i$ is an NZD in $K[\Delta]$.

24. Let $\Delta$ be any $(d-1)$-dimensional simplicial complex, and let $\theta_1, \ldots, \theta_d \in K[\Delta]_1$. Show that the quotient ring $R = K[\Delta]/(\theta_1, \ldots, \theta_d)$ is a finite-dimensional vector space over $K$ if and only if $\theta_1, \ldots, \theta_d$ satisfy Property (P).

25. Show that the face ring $K[\Delta]$ of a simplicial complex $\Delta$ has depth one if and only if $\Delta$ is disconnected. Deduce that a disconnected simplicial complex of dimension at least one is not Cohen–Macaulay.

26. Let $\Gamma$ and $\Delta$ be simplicial complexes on disjoint vertex sets $V$ and $W$, respectively. Define the *join* $\Gamma * \Delta$ to be the simplicial complex on the vertex set $V \cup W$ with faces $F \cup G$, where $F \in \Gamma$ and $G \in \Delta$. (If $\Gamma$ consists of a single point, then $\Gamma * \Delta$ is the *cone* over $\Delta$. If $\Gamma$ consists of two disjoint points, then $\Gamma * \Delta$ is the *suspension* of $\Delta$.)

    (a) Compute the $h$-vector $h(\Gamma * \Delta)$ in terms of $h(\Gamma)$ and $h(\Delta)$.

    (b) Show that if $\Gamma$ and $\Delta$ are Cohen–Macaulay, then so is $\Gamma * \Delta$.

    (c) Generalizing Exercise 7, show that if $\Gamma$ and $\Delta$ are shellable, then so is $\Gamma * \Delta$.

27. Let $\Delta$ be a $(d-1)$-dimensional Cohen–Macaulay simplicial complex with $h$-vector $(h_0, h_1, \ldots, h_d)$. Suppose that $h_i = 1$ for some $1 \le i \le d-1$. What are the possible values of $h_d$?

28. Let $\Delta$ be a one-dimensional simplicial complex. Show that the following three conditions are equivalent: (a) $\Delta$ is Cohen–Macaulay, (b) $\Delta$ is shellable, and (c) $\Delta$ is connected.

29. Complete the proof of Theorem 12.25 by showing that the sequence $\sigma$ is a shelling of $\Delta$ with the stated restrictions $G_k$.

30. (*) Let $\Delta$ be a four-dimensional shellable simplicial complex with $f_0 = 13$, $f_1 = 67$, and $f_2 = 204$. What is the most number of facets that $\Delta$ can have? What if we drop the hypothesis of shellability?

31. (*) Let $\Delta$ be a $(d-1)$-dimensional Cohen–Macaulay simplicial complex with $h$-vector $(h_0, h_1, \ldots, h_d)$. Let $\Delta'$ be a $(d-1)$-dimensional Cohen–Macaulay subcomplex of $\Delta$ with $h$-vector $(h'_0, h'_1, \ldots, h'_d)$. Show that $h'_i \le h_i$ for all $0 \le i \le d$.

32. (*) Let $\Delta$ be a $(d-1)$-dimensional simplicial complex on the vertex set $V$. We say that $\Delta$ is *balanced* if we can write $V$ as a disjoint union $V = V_1 \cup V_2 \cup \cdots \cup V_d$ such that for every $F \in \Delta$ and every $1 \le i \le d$ we have $\#(F \cap V_i) \le 1$. In particular, if $\Delta$ is pure, then $\#(F \cap V_i) = 1$ when $F$ is a facet. (Sometimes $\Delta$ is required to be pure in the definition of balanced.) Suppose that $(h_0, h_1, \ldots, h_d)$ is the $h$-vector of a Cohen–Macaulay balanced simplicial complex $\Delta$. Show that $(h_1, h_2, \ldots, h_d)$ is the $f$-vector of a balanced simplicial complex. You may want to use the result of Exercise 24.

# Chapter 13
# Miscellaneous Gems of Algebraic Combinatorics

## 13.1 The 100 Prisoners

An evil warden is in charge of 100 prisoners (all with different names). He puts a row of 100 boxes in a room. Inside each box is the name of a different prisoner. The prisoners enter the room one at a time. Each prisoner must open 50 of the boxes, one at a time. If any of the prisoners does not see his or her own name, then they are all killed. The prisoners may have a discussion before the first prisoner enters the room with the boxes, but after that there is no further communication. A prisoner may not leave a message of any kind for another prisoner. In particular, all the boxes are shut once a prisoner leaves the room. If all the prisoners choose 50 boxes at random, then each has a success probability of 1/2, so the probability that they are not killed is $2^{-100}$, not such good odds. Is there a strategy that will increase the chances of success? What is the best strategy?

It's not hard to see that the prisoners can achieve a success probability of greater than $2^{-100}$. For instance, suppose that the first prisoner opens the first 50 boxes and the second prisoner opens the last 50. If the first prisoner succeeds (with probability 1/2), then the first prisoner's name is guaranteed not to be in one of the boxes opened by the second prisoner, so the second prisoner's probability of success is 50/99. Each pair of prisoners can do this strategy, increasing the overall success probability to $(25/99)^{50}$, still an extremely low number. Can they do significantly better? The key to understanding this problem is the realization that the prisoners do not have to decide in advance on which boxes they will open. A prisoner can decide which box to open next based on what he has seen in the boxes previously opened.

**13.1 Theorem.** *There exists a strategy with a success probability of*

$$1 - \sum_{j=51}^{100} \frac{1}{j} = 0.3118278207 \cdots .$$

© Springer International Publishing AG, part of Springer Nature 2018
R. P. Stanley, *Algebraic Combinatorics*, Undergraduate Texts in Mathematics,
https://doi.org/10.1007/978-3-319-77173-1_13

*Proof.* The prisoners assign themselves the numbers $1, 2, \ldots, 100$ by whatever method they prefer. Each prisoner is assigned a different number. The prisoners memorize everyone's number. They regard the boxes, which are lined up in a row, as being numbered $1, 2, \ldots, 100$ from left to right. A prisoner with number $k$ first goes to box $k$. If the prisoner sees his name, then he breathes a temporary sigh of relief, and the next prisoner enters. Otherwise the first prisoner will see the name of some other prisoner, say with number $n_1$. He then opens box $n_1$ and repeats the procedure, so whenever he opens a box $B$ that doesn't contain his own name, the next box that he opens has the number of the prisoner whose name appears in box $B$.

What is the probability of success of this strategy? Suppose that box $i$ contains the name of the prisoner numbered $\pi(i)$. Thus $\pi$ is a permutation of $1, 2, \ldots, 100$. The boxes opened by prisoner $i$ are those containing the names of prisoners with numbers $\pi(i)$, $\pi^2(i)$, $\pi^3(i)$, etc. If $k$ is the length of the cycle of $\pi$ containing $i$, then the prisoner will see his name after opening the $k$th box. This will happen whenever $k \leq 50$. Thus all prisoners see their names if and only if every cycle of $\pi$ has length at most 50. If $\pi$ does not have this property, then it has exactly one cycle of length $r > 50$. There are $\binom{100}{r}$ ways to choose the elements of the cycle and $(r - 1)!$ ways to arrange them in a cycle. There are then $(100 - r)!$ ways to arrange the other elements of $\pi$. Thus the number of permutations $\pi \in \mathfrak{S}_{100}$ with a cycle of length $r > 50$ is

$$\binom{100}{r}(r - 1)!(100 - r)! = \frac{100!}{r}.$$

(There are more clever ways to see this.) Hence the probability of success, i.e., the probability that $\pi$ has *no* cycle of length more than 50, is

$$1 - \frac{1}{100!} \sum_{r=51}^{100} \frac{100!}{r} = 1 - \sum_{r=51}^{100} \frac{1}{r},$$

as claimed.                                                                    $\square$

If we apply the above argument to $2n$ prisoners rather than 100, then we get a success probability of

$$1 - \sum_{r=n+1}^{2n} \frac{1}{r} = 1 - \sum_{r=1}^{2n} \frac{1}{r} + \sum_{r=1}^{n} \frac{1}{r}.$$

From calculus we know that there is a constant $\gamma = 0.577215665 \cdots$, known as *Euler's constant*, for which

$$\lim_{n \to \infty} \left( \sum_{r=1}^{n} \frac{1}{r} - \log n \right) = \gamma.$$

Thus if there are $n$ prisoners, then it follows that as $n \to \infty$, the success probability of the prisoners is

$$\lim_{n \to \infty} (1 - \log 2n + \log n) = 1 - \log 2 = 0.3068528194 \cdots.$$

It seems quite amazing that no matter how many prisoners there are, they can always achieve a success probability of over 30%!

NOTE. It can be shown that the above strategy is in fact *optimal*, i.e., no strategy achieves a higher probability of success. The proof, however, is not so easy.

## 13.2 Oddtown

The village of Oddtown has a population of $n$ people. Inhabitants of Oddtown like to form clubs. Every club has an odd number of members, and every pair of clubs share an even number of members (possibly none).

**13.2 Theorem.** *There are at most $n$ clubs.*

*Proof.* Let $k$ be the number of clubs. Define a matrix $M = (M_{ij})$ over the two-element field $\mathbb{F}_2$ as follows. The rows of $M$ are indexed by the clubs $C_i$ and the columns by the inhabitants $x_j$ of Oddtown. Set

$$M_{ij} = \begin{cases} 1, & x_j \in C_i \\ 0, & \text{otherwise.} \end{cases}$$

The matrix $M$ is called the *incidence matrix* corresponding to the clubs and their members.

In general, let $S$ be a subset of $[n]$, and let $\chi_S \in \mathbb{Z}^n$ be the *characteristic vector* of $S$, i.e., $\chi_S = (a_1, \ldots, a_n)$ where

$$a_i = \begin{cases} 1, & i \in S \\ 0, & i \notin S. \end{cases}$$

If $T$ is another subset of $[n]$, then the key observation is that the scalar (dot) product of $\chi_S$ and $\chi_T$ is given by $\chi_S \cdot \chi_T = \#(S \cap T)$. Hence if we now work over $\mathbb{F}_2$, then

$$\chi_S \cdot \chi_T = \begin{cases} 1, & \text{if } \#(S \cap T) \text{ is odd} \\ 0, & \text{if } \#(S \cap T) \text{ is even.} \end{cases} \tag{13.1}$$

Let $A = MM^t$, a $k \times k$ matrix. By (13.1) and the assumption that every club has an odd number of members, we see that main diagonal elements of $A$ are 1. Similarly the off-diagonal elements of $A$ are 0, so $A = I_k$, the $k \times k$ identity matrix. Hence $\operatorname{rank}(A) = k$.

Recall that if $B$ is a $k \times m$ matrix and $C$ is an $m \times n$ matrix (over some field), then rank$(BC) \leq$ rank$(B)$ (as well as rank$(BC) \leq$ rank$(C)$), since for any matrix $D$, rank$(D) = \dim$ image$(D)$. Hence, since $M$ has $n$ columns,

$$n \geq \text{rank}(M) \geq \text{rank}(MM^t) = \text{rank}(A) = k. \qquad \Box$$

While Theorem 13.2 can be proved without linear algebra, the proof is not easy.

## 13.3   Complete Bipartite Partitions of $K_n$

Figure 13.1 shows the six edges of the complete graph $K_4$ partitioned (according to the edge label) into the edge sets of the three complete bipartite graphs $K_{3,1}$, $K_{2,1}$, and $K_{1,1}$. Clearly we can extend this construction, achieving a partition of the edges $E(K_n)$ of $K_n$ into the edge sets of $n - 1$ complete bipartite graphs. Specifically, let $E_1$ be the set of edges incident to a fixed vertex $v$. Thus $E_1$ is the edge set of a complete bipartite graph $K_{n-1,1}$. Remove $E_1$ from $E(K_n)$ and proceed by induction, obtaining a partition of $E(K_n)$ into the edges of $K_{n-1,1}$, $K_{n-2,1}, \ldots, K_{1,1}$. The question thus arises as to whether $E(K_n)$ can be partitioned into *fewer* than $n - 1$ edge sets of complete bipartite graphs.

**13.3 Theorem.** *If $E(K_n)$ is the disjoint union of the edge sets of $m$ complete bipartite graphs, then $m \geq n - 1$.*

*Proof.* Take the vertex set of $K_n$ to be $[n]$. Let $E(K_n) = E(B_1) \cup E(B_1) \cup \cdots \cup E(B_m)$ (disjoint union), where $B_k$ is a complete bipartite graph with vertex bipartition $(X_k, Y_k)$ (so $X_k \cap Y_k = \emptyset$). For $1 \leq k \leq n$, define an $n \times n$ matrix $A_k$ by

$$(A_k)_{ij} = \begin{cases} 1, & i \in X_k, \ j \in Y_k \\ 0, & \text{otherwise.} \end{cases}$$

All nonzero rows of $A_k$ are equal, so rank $A_k = 1$. Let $S = \sum_{k=1}^{m} A_k$. For $i \neq j$, exactly one of the $2m$ numbers $(A_k)_{ij}$ and $(A_k)_{ji}$, $1 \leq k \leq m$, is equal to 1, since

**Fig. 13.1** A decomposition of the edges of $K_4$ into three complete bipartite graphs

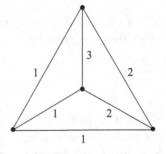

every edge $ij$ of $K_n$ appears in one $E(B_k)$ with either $i \in X_k$ and $j \in Y_k$, or else $j \in X_k$ and $i \in Y_k$. Hence

$$S + S^t = J - I,$$

where as usual $J$ is the $n \times n$ all 1's matrix, and $I$ is the $n \times n$ identity matrix.

*Claim.* If $T$ is any real matrix satisfying $T + T^t = J - I$, then rank $T \geq n - 1$.

Suppose to the contrary that rank $T \leq n - 2$. Then $T$ has (at least) two linearly independent eigenvectors $x, y$ such that $Tx = Ty = 0$ [why?]. Since $J$ has rank one, the space $\langle x, y \rangle$ spanned by $x$ and $y$ contains a nonzero vector $z$ satisfying $Jz = 0$ [why?]. Then from $T + T^t = J - I$ and $Tz = 0$ we get $-z = T^t z$. Take the dot product with $z^t$ on the left. We get

$$-|z|^2 = z^t T^t z$$

$$= (z^t T^t z)^t \text{ (since a } 1 \times 1 \text{ matrix is symmetric)}$$

$$= z^t T z \text{ (since in general } (AB)^t = B^t A^t)$$

$$= 0 \text{ (since } Tz = 0),$$

contradicting $z \neq 0$. Hence the claim is proved, so in particular rank $S \geq n - 1$. But in general rank$(A + B) \leq$ rank $A +$ rank $B$ [why?]. Therefore from $S = \sum_{k=1}^{m} A_k$ and rank $A_k = 1$ we get rank $S \leq m$. It follows that $m \geq n - 1$, completing the proof. $\qquad\square$

## 13.4 The Nonuniform Fisher Inequality

A *balanced incomplete block design* (BIBD) with parameters $(v, k, \lambda, r, b)$ is a $v$-element set $X$ and a collection $\mathcal{A}$ of $k$-element subsets (blocks), with $\#\mathcal{A} = b$, such that any two points $x, y \in X$ lie in exactly $\lambda$ blocks, and each point is in exactly $r$ blocks. We also assume that $k < v$, which is the reason for the word "incomplete." We can draw a BIBD as a bipartite graph with vertex bipartition $(X, \mathcal{A})$. There is an edge from $x \in X$ to $A \in \mathcal{A}$ if $x \in A$. Thus the degree of each vertex $x \in X$ is $r$, and the degree of each vertex $A \in \mathcal{A}$ is $k$. It follows that $vr = kb$ (the total number of edges of the graph). We can also count the number of two-element sets of edges that are incident to the same vertex of $\mathcal{A}$. On the one hand, since each vertex in $\mathcal{A}$ has degree $k$ this number is $b\binom{k}{2}$. On the other hand, each pair of points in $X$ are mutually adjacent to $\lambda$ points in $\mathcal{A}$, so we get $\lambda\binom{v}{2} = b\binom{k}{2}$. A little manipulation shows that the two equalities $vr = kb$ and $\lambda\binom{v}{2} = b\binom{k}{2}$ are equivalent to

$$vr = kb, \quad \lambda(v - 1) = r(k - 1),$$

the usual form in which they are written.

R.A. Fisher showed in 1940 that $b \geq v$. This inequality was generalized by R.C. Bose in 1949. The most convenient way to state Bose's inequalities, known as the *nonuniform Fisher inequality*, is to reverse the roles of points and blocks. Thus consider the elements $x$ of $X$ to be sets whose elements are the blocks $A \in \mathcal{A}$ that contain them. In other words, we have a collection $C_1, \ldots, C_v$ of $r$-element sets whose union contains $b$ points $x_1, \ldots, x_b$. Each point is in exactly $k$ of the sets. Finally, $\#(C_i \cap C_j) = \lambda$ for all $i \neq j$.

**13.4 Theorem.** *Let $C_1, \ldots, C_v$ be distinct subsets of a $b$-element set $X$ such that for all $i \neq j$ we have $\#(C_i \cap C_j) = \lambda$ for some $1 \leq \lambda < b$ (independent of $i$ and $j$). Then $v \leq b$.*

*Proof. Case 1.* Some $\#C_i = \lambda$. Then all other $C_j$'s contain $C_i$ and are disjoint otherwise, so

$$v \leq \underbrace{1}_{\text{from } C_i} + \underbrace{b - \lambda}_{\text{from all } C_j \neq C_i} \leq b.$$

*Case 2.* All $\#C_i > \lambda$. Let $\gamma_i = \#C_i - \lambda > 0$. Let $M$ be the incidence matrix of the set system $C_1, \ldots, C_v$, i.e., the rows of $M$ correspond to the $C_i$'s and the columns to the elements $x_1, \ldots, x_b$ of $X$, with

$$M_{ij} = \begin{cases} 1, & x_j \in C_i \\ 0, & x_j \notin C_i. \end{cases}$$

Let $A = MM^t$. The hypotheses imply that $A = \lambda J + G$, where $J$ as usual is the all 1's matrix (of size $v$), and $G$ is the diagonal matrix $\text{diag}(\gamma_1, \ldots, \gamma_v)$.

*Claim.* $\text{rank}(A) = v$ (i.e., $A$ is invertible). We would then have

$$v = \text{rank}(A) \leq \text{rank}(M) \leq b,$$

the last inequality because $M$ has $b$ columns.

As in the proof of Theorem 4.7, a real symmetric matrix $B$ is positive semidefinite if it has nonnegative eigenvalues. Equivalently, by basic linear algebra, $uBu^t \geq 0$ for all row vectors $u$ of length $v$. Moreover $B$ is positive definite (and so has positive eigenvalues) if $uBu^t > 0$ for all $u \neq 0$.

Now we easily compute that

$$u(\lambda J + G)u^t = \lambda(u_1 + \cdots + u_v)^2 + \gamma_1 u_1^2 + \cdots + \gamma_v u_v^2 > 0$$

for all $u \neq 0$. Thus $A = \lambda J + G$ is positive definite and hence of full rank $v$.  $\square$

**Fig. 13.2** The $3 \times 4$ grid
graph

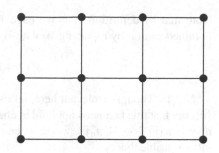

## 13.5   Odd Neighborhood Covers

Consider an $m \times n$ grid graph. The case $m = 3$, $n = 4$ is shown in Figure 13.2. At
each vertex are a turned on light bulb and also a switch that changes the state of its
bulb and those of its neighbors (adjacent vertices). Can all the lights be turned off?

This problem was open for many years until in 1989 K. Sutner, then a graduate
student, showed using automata theory that the answer is yes for *any* (finite) graph!
More explicitly, let $G$ be a finite graph with a turned on light bulb at each vertex.
At each vertex is a switch that changes the state of that vertex and all its neighbors.
Then it is possible to turn off all the lights. We will give a modification of a simpler
proof due to Y. Caro based on linear algebra.

Without loss of generality we may assume that $G$ is simple. If $v \in V(G)$, then
the *neighborhood* $N(v)$ of $v$ is the set consisting of $v$ and all vertices adjacent to $v$.
A little thought shows that we need to prove the following result.

**13.5 Theorem.** *There exists a subset $S \subseteq V = V(G)$ such that $\#(S \cap N(v))$ is odd
for all $v \in V$. (It follows that switching at the vertices $v \in S$ turns all the lights off.)*

*Proof.* Let $V(G) = \{v_1, \ldots, v_p\}$. Let $A$ be the adjacency matrix of $G$ over the field
$\mathbb{F}_2$, and let $y = (1, 1, \ldots, 1) \in \mathbb{F}_2^p$. Write row$(A + I)$ for the row space of the
matrix $A + I$, and let $\gamma_v$ denote the row of $A + I$ indexed by $v \in V$. Note that
switching at $S$ turns all the lights off if and only if $\sum_{v \in S} \gamma_v = y$. Hence we need to
show that $y \in \text{row}(A + I)$ [why?].

Let us recall from linear algebra some standard facts about orthogonal subspaces.
Let $K$ be a field, and for $u, v \in K^n$ let $u \cdot v$ be the usual dot product (2.1) of $u$ and $v$,
so $u \cdot v \in K$. If $W$ is a subspace of $K^n$, then define the *orthogonal subspace* $W^\perp$ by

$$W^\perp = \{u \in K^n : u \cdot v = 0 \text{ for all } v \in W\}.$$

(In Chapter 11 we discussed the case $K = \mathbb{R}$.) Let $d = \dim W$. Since $W^\perp$ is the
set of solutions to $d$ linearly independent homogeneous linear equations [why?], we
have

$$\dim W + \dim W^\perp = n. \tag{13.2}$$

Note that by definition of $^\perp$ we have $W \subseteq (W^\perp)^\perp$. By (13.2) and the equation obtained from it by replacing $W$ with $W^\perp$, we get $\dim W = \dim (W^\perp)^\perp$. Hence

$$(W^\perp)^\perp = W. \tag{13.3}$$

NOTE. Though irrelevant here, let us point out that if $K \subseteq \mathbb{R}$ then $W \cap W^\perp = \{0\}$, but that this fact need not hold in characteristic $p \neq 0$. Over $\mathbb{C}$ we should define $u \cdot v = u_1 \bar{v}_1 + \cdots + u_n \bar{v}_n$, where $^-$ denotes complex conjugation, in order to get the most sensible theory.

Now by (13.3) the vector $y = (1, 1, \ldots, 1)$ (or any vector in $\mathbb{F}_2^n$) lies in the row space of $A + I$ if and only if it is orthogonal to every vector in $\mathrm{row}(A + I)^\perp = \ker(A+I)$. Thus we need to show that if $(A+I)v^t = 0$, then $v \cdot y = 0$. Equivalently, if $yv^t \neq 0$ then $(A+I)v^t \neq 0$. Note that (a) $yv^t \neq 0$ means that $v$ has an odd number of 1's, and (b) $(A + I)v^t$ is the sum of the rows of $A + I$ indexed by the positions of the 1's in $v$. Thus we need to show that $A + I$ does not have an odd number of rows summing to 0.

Suppose that $v_1, \ldots, v_k$ are vertices indexing rows of $A + I$ summing to 0. Let $H$ be the subgraph *induced* by $v_1, \ldots, v_k$, i.e., $H$ consists of the vertices $v_1, \ldots, v_k$ and all edges of $G$ between two of these vertices. Let $b_{ij}$ be the $(i, j)$-entry of $A+I$. Since $\sum_{i=1}^k b_{ij} = 0$ for $1 \leq j \leq n$, and each $b_{ii} = 1$, it follows that every vertex of $H$ has odd degree. Since [why?]

$$\sum_{v \in V(H)} \deg v = 2 \cdot \#E(H),$$

we have that $k = \#V(H)$ is even, completing the proof.      $\square$

## 13.6   Circulant Hadamard Matrices

For our next "gem of algebraic combinatorics," we will provide some variety by leaving the realm of linear algebra and looking at some simple algebraic number theory.

An $n \times n$ matrix $H$ is a *Hadamard matrix* if its entries are $\pm 1$ and its rows are orthogonal. Equivalently, its entries are $\pm 1$ and $HH^t = nI$. In particular [why?],

$$\det H = \pm n^{n/2}. \tag{13.4}$$

It is easy to see that if $H$ is an $n \times n$ Hadamard matrix then $n = 1, n = 2$, or $n = 4m$ for some integer $m \geq 1$. (See Exercise 12.19.) It is conjectured that the converse is true, i.e., for every such $n$ there exists an $n \times n$ Hadamard matrix.

An $n \times n$ matrix $A = (b_{ij})$ is a *circulant* or *circulant matrix* if it has the form $b_{ij} = a_{i-j}$ for some $a_0, a_1, \ldots, a_{n-1}$, where the subscript $i - j$ is taken modulo $n$.

For instance,

$$A = \begin{bmatrix} a & b & c & d \\ d & a & b & c \\ c & d & a & b \\ b & c & d & a \end{bmatrix}$$

is a circulant. Let $A = (a_{i-j})$ be an $n \times n$ circulant, and let $\zeta = e^{2\pi i/n}$, a primitive $n$th root of unity. It is straightforward to compute that for $0 \le j < n$ the column vector $[1, \zeta^j, \zeta^{2j}, \ldots, \zeta^{(n-1)j}]^t$ is an eigenvector of $A$ with eigenvalue $a_0 + \zeta^j a_1 + \zeta^{2j} a_2 + \cdots + \zeta^{(n-1)j} a_{n-1}$. Hence

$$\det A = \prod_{j=0}^{n-1} (a_0 + \zeta^j a_1 + \zeta^{2j} a_2 + \cdots + \zeta^{(n-1)j} a_{n-1}). \tag{13.5}$$

Note that the matrix

$$\begin{bmatrix} -1 & 1 & 1 & 1 \\ 1 & -1 & 1 & 1 \\ 1 & 1 & -1 & 1 \\ 1 & 1 & 1 & -1 \end{bmatrix}$$

is both a Hadamard matrix and a circulant.

**Conjecture.** Let $H$ be an $n \times n$ circulant Hadamard matrix. Then $n = 1$ or $n = 4$.

The first significant work on this conjecture is due to R.J. Turyn. He showed that there does not exist a circulant Hadamard matrix of order $8m$, and he also excluded certain other orders of the form $4(2m + 1)$. Turyn's proofs use the machinery of algebraic number theory. Here we will give a proof for the special case $n = 2^k$, $k \ge 3$, where the algebraic number theory can be "dumbed down" to elementary commutative algebra and field theory. (Only in Theorem 13.14 do we use a little Galois theory, which can be avoided with a bit more work.) It would be interesting to find similar proofs for other values of $n$.

**13.6 Theorem.** *There does not exist a circulant Hadamard matrix $H$ of order $2^k$, $k \ge 3$.*

NOTE. It is curious that the numbers $2^k$ ($k \ge 3$) are the easiest multiples of 4 to show are *not* the orders of circulant Hadamard matrices, while on the other hand the numbers $2^k$ ($k \ge 1$) are the easiest numbers to show *are* the orders of Hadamard matrices. To see that $2^k$ is the order of a Hadamard matrix $H$, first note that the case $k = 1$ is trivial. It is routine to show that if $H_1$ is a Hadamard matrix of order $a$ and $H_2$ is a Hadamard matrix of order $b$, then the tensor (or Kronecker) product $A \otimes B$ is a Hadamard matrix of order $ab$. It follows that there exists a Hadamard matrix of order $2^k$, $k \ge 1$.

From now on we assume $n = 2^k$ and $\zeta = e^{2\pi i/2^k}$. Clearly $\zeta$ is a zero of the polynomial $p_k(x) = x^{2^{k-1}} + 1$. We will be working in the ring $\mathbb{Z}[\zeta]$, the smallest subring of $\mathbb{C}$ containing $\mathbb{Z}$ and $\zeta$. Write $\mathbb{Q}(\zeta)$ for the quotient field of $\mathbb{Z}[\zeta]$, i.e., the field obtained by adjoining $\zeta$ to $\mathbb{Q}$.

**13.7 Lemma.** *The polynomial $p_k(x)$ is irreducible over $\mathbb{Q}$.*

*Proof.* If $p_k(x)$ is reducible then so is $p_k(x + 1)$. A standard fact about polynomial factorization is *Gauss' lemma*, namely, an integral polynomial that factors over $\mathbb{Q}$ also factors over $\mathbb{Z}$. If $p(x), q(x) \in \mathbb{Z}[x]$, write $p(x) \equiv q(x) \,(\mathrm{mod}\,2)$ to mean that the coefficients of $p(x) - q(x)$ are even. Now [why?]

$$p_k(x + 1) \equiv (x + 1)^{2^{k-1}} + 1 \equiv x^{2^{k-1}} \;(\mathrm{mod}\,2).$$

Hence any factorization of $p_k(x + 1)$ over $\mathbb{Z}$ into two factors of degree at least one has the form $p_k(x + 1) = (x^r + 2a)(x^s + 2b)$, where $r + s = 2^{k-1}$ and $a, b$ are polynomials of degrees less than $r$ and $s$, respectively. Hence the constant term of $p_k(x + 1)$ is divisible by 4, a contradiction. $\qquad\square$

It follows by elementary field theory that every element $u \in \mathbb{Z}[\zeta]$ can be uniquely written in the form

$$u = b_0 + b_1\zeta + b_2\zeta^2 + \cdots + b_{n/2-1}\zeta^{n/2-1}, \quad b_i \in \mathbb{Z}.$$

The basis for our proof of Theorem 13.6 is the two different ways to compute $\det H$ given by (13.4) and (13.5), yielding the formula

$$\prod_{j=0}^{n-1} (a_0 + \zeta^j a_1 + \zeta^{2j} a_2 + \cdots + \zeta^{(n-1)j} a_{n-1}) = \pm n^{n/2} = \pm 2^{k2^{k-1}}. \qquad (13.6)$$

Thus we have a factorization in $\mathbb{Z}[\zeta]$ of $2^{k2^{k-1}}$. Algebraic number theory is concerned with factorization of algebraic integers (and ideals) in algebraic number fields, so we have a vast amount of machinery available to show that no factorization (13.6) is possible (under the assumption that each $a_j = \pm 1$). Compare Kummer's famous approach toward Fermat's Last Theorem (which led to his creation of algebraic number theory), in which he considered the equation $x^n + y^n = z^n$ as $\prod_{\tau^n=1}(x + \tau y) = z^n$ when $n$ is odd.

We are continuing to assume that $H = (a_{j-i})$ is an $n \times n$ circulant Hadamard matrix. We will denote the eigenvalues of $H$ by

$$\gamma_j = a_0 + a_1\zeta^j + a_2\zeta^{2j} + \cdots + a_{n-1}\zeta^{(n-1)j}, \quad 0 \le j \le n - 1.$$

**13.8 Lemma.** *For $0 \leq j \leq n - 1$ we have*

$$|\gamma_j| = \sqrt{n}.$$

*Thus all the factors appearing on the left-hand side of (13.6) have absolute value $\sqrt{n}$.*

*First proof* (naive). Let $H_i$ denote the $i$th row of $H$, let $\cdot$ denote the usual dot product, and let $\bar{\ }$ denote complex conjugation. Then

$$\gamma_j \bar{\gamma}_j = (a_0 + a_1 \zeta^j + \cdots + a_{n-1} \zeta^{(n-1)j})(a_0 + a_1 \zeta^{-j} + \cdots + a_{n-1} \zeta^{-(n-1)j})$$

$$= H_1 \cdot H_1 + (H_1 \cdot H_2) \zeta^j + (H_1 \cdot H_3) \zeta^{2j} + \cdots + (H_1 \cdot H_n) \zeta^{(n-1)j}.$$

By the Hadamard property we have $H_1 \cdot H_1 = n$, while $H_1 \cdot H_k = 0$ for $2 \leq k \leq n$, and the proof follows.  $\square$

*Second proof* (algebraic). The matrix $\frac{1}{\sqrt{n}} H$ is a real orthogonal matrix. By linear algebra, all its eigenvalues have absolute value 1. Hence all eigenvalues $\gamma_j$ of $H$ have absolute value $\sqrt{n}$.  $\square$

**13.9 Lemma.** *We have*

$$2 = (1 - \zeta)^{n/2} u, \tag{13.7}$$

*where $u$ is a unit in $\mathbb{Z}[\zeta]$.*

*Proof.* Put $x = 1$ in

$$x^{n/2} + 1 = \prod_{\substack{j=0 \\ j \text{ odd}}}^{n-1} (x - \zeta^j)$$

to get $2 = \prod_j (1 - \zeta^j)$. Since

$$1 - \zeta^j = (1 - \zeta)(1 + \zeta + \cdots + \zeta^{j-1}),$$

it suffices to show that $1 + \zeta + \cdots + \zeta^{j-1}$ is a unit when $j$ is odd. Let $j\bar{j} \equiv 1 \pmod{n}$. Note that $\bar{j}$ exists since $j$ and $n$ are relatively prime. Then

$$(1 + \zeta + \cdots + \zeta^{j-1})^{-1} = \frac{1 - \zeta}{1 - \zeta^j}$$

$$= \frac{1 - (\zeta^j)^{\bar{j}}}{1 - \zeta^j}$$

$$= 1 + \zeta^j + \zeta^{2j} + \cdots + \zeta^{(\bar{j}-1)j} \in \mathbb{Z}[\zeta],$$

as desired.  $\square$

**13.10 Lemma.** *We have* $\mathbb{Z}[\zeta]/(1 - \zeta) \cong \mathbb{F}_2$.

*Proof.* Let $R = \mathbb{Z}[\zeta]/(1 - \zeta)$. The integer 2 is not a unit in $\mathbb{Z}[\zeta]$, e.g., because $1/2$ is not an algebraic integer (the zero of a *monic* polynomial $f(x) \in \mathbb{Z}[x]$). Thus by Lemma 13.9, $1 - \zeta$ is also not a unit. Hence $R \neq 0$ (where 0 is short for $\{0\}$).

For all $j$ we have $\zeta^j = 1$ in $R$ since $\zeta = 1$ in $R$. Hence all elements of $R$ can be written as ordinary integers $m$. But $0 = 2$ in $R$ by Lemma 13.9, so the only elements of $R$ are 0 and 1. $\qquad\square$

**13.11 Lemma.** *For all* $0 \leq j \leq n - 1$ *there is an integer* $h_j \geq 0$ *such that*

$$a_0 + a_1 \zeta^j + a_2 \zeta^{2j} + \cdots + a_{n-1} \zeta^{(n-1)j} = v_j (1 - \zeta)^{h_j},$$

*where* $v_j$ *is a unit in* $\mathbb{Z}[\zeta]$.

*Proof.* Since 2 is a multiple of $1 - \zeta$ by Lemma 13.9, we have by (13.6) that

$$\prod_{j=0}^{n-1} (a_0 + a_1 \zeta^j + a_2 \zeta^{2j} + \cdots + a_{n-1} \zeta^{(n-1)j}) = 0$$

in $\mathbb{Z}[\zeta]/(1 - \zeta)$. Since $\mathbb{Z}[\zeta]/(1 - \zeta)$ is an integral domain by Lemma 13.10, some factor $a_0 + a_1 \zeta^j + \cdots + a_{n-1} \zeta^{(n-1)j}$ is divisible by $1 - \zeta$. Divide this factor and the right-hand side of (13.6) by $1 - \zeta$, and iterate the procedure. We continue to divide a factor of the left-hand side and the right-hand side by $1 - \zeta$ until the right-hand side becomes the unit $u$. Hence each factor of the original product has the form $v(1-\zeta)^h$, where $v$ is a unit. $\qquad\square$

**13.12 Corollary.** *Either* $\gamma_0/\gamma_1 \in \mathbb{Z}[\zeta]$ *or* $\gamma_1/\gamma_0 \in \mathbb{Z}[\zeta]$. *(In fact, both* $\gamma_0/\gamma_1 \in \mathbb{Z}[\zeta]$ *and* $\gamma_1/\gamma_0 \in \mathbb{Z}[\zeta]$, *as will soon become apparent, but we don't need this fact here.)*

*Proof.* By the previous lemma, each $\gamma_j$ has the form $v_j (1 - \zeta)^{h_j}$. If $h_0 \geq h_1$ then $\gamma_0/\gamma_1 \in \mathbb{Z}[\zeta]$; otherwise $\gamma_1/\gamma_0 \in \mathbb{Z}[\zeta]$. $\qquad\square$

We now need to appeal to a result of Kronecker on elements of $\mathbb{Z}[\zeta]$ of absolute value one. For completeness we include a proof of this result, beginning with a lemma. Recall that if $\theta$ is an algebraic number (the zero of an irreducible polynomial $f(x) \in \mathbb{Q}[x]$), then a *conjugate* of $\theta$ is any zero of $f(x)$.

**13.13 Lemma.** *Let* $\theta$ *be an algebraic integer such that* $\theta$ *and all its conjugates have absolute value one. Then* $\theta$ *is a root of unity.*

*Proof.* Suppose the contrary. Let $\deg \theta = d$, i.e., $[\mathbb{Q}(\theta) : \mathbb{Q}] := \dim_{\mathbb{Q}} \mathbb{Q}(\theta) = d$. Now $\theta, \theta^2, \theta^3, \ldots$ are all distinct and hence infinitely many of them have the property that no two are conjugate. Each $\theta^j \in \mathbb{Z}[\theta]$ and so is the root of a monic integral polynomial of degree at most $d$, since the set of algebraic integers forms a ring. If $\theta_1, \theta_2, \ldots, \theta_d$ are the conjugates of $\theta$, then all the conjugates of $\theta^j$ are among $\theta_1^j, \theta_2^j, \ldots, \theta_d^j$. Hence each $\theta^j$ satisfies the hypothesis that all its conjugates have absolute value 1 (and $\theta^j$ is an algebraic integer). Thus the $r$th elementary symmetric function $e_r$ in $\theta^j$ and its conjugates has at most $\binom{d}{r}$ terms, each of absolute value 1, so $|e_r| \le \binom{d}{r}$. Moreover, $e_r \in \mathbb{Z}$ since $\theta^j$ is an algebraic integer. It follows that there are only finitely many possible polynomials that can be an irreducible monic polynomial with a zero equal to some $\theta^j$, contradicting the fact that there are infinitely many $\theta^j$'s for which no two are conjugate.                      $\square$

**13.14 Theorem** (Kronecker). *Let $\tau$ be any root of unity and $\alpha \in \mathbb{Z}[\tau]$ with $|\alpha| = 1$. Then $\alpha$ is a root of unity.*

*Proof.* Since $\alpha \in \mathbb{Z}[\tau]$, we see that $\alpha$ is an algebraic integer. We use the basic fact from Galois theory that the Galois group of the extension field $\mathbb{Q}(\tau)/\mathbb{Q}$ is abelian. Let $\beta$ be a conjugate of $\alpha$, so $\beta = w(\alpha)$ for some automorphism $w$ of $\mathbb{Q}(\tau)$. Apply $w$ to the equation $\alpha\bar{\alpha} = 1$. Since complex conjugation is an automorphism of $\mathbb{Q}(\tau)$ it commutes with $w$, so we obtain $\beta\bar{\beta} = 1$. Hence all the conjugates of $\alpha$ have absolute value one, so $\alpha$ is a root of unity by the previous lemma.                      $\square$

For our next result, we need the standard algebraic fact that if $\tau = e^{2\pi i/m}$, a primitive $m$th root of unity, then $[\mathbb{Q}(\tau) : \mathbb{Q}] = \phi(m)$ (the Euler $\phi$-function). Equivalently, the unique monic polynomial $\Phi_m(x)$ whose zeros are the primitive $m$th roots of unity is irreducible. This polynomial is by definition given by

$$\Phi_m(x) = \prod_{\substack{1 \le j \le m \\ \gcd(j,m)=1}} (x - \tau^j)$$

and is called a *cyclotomic polynomial*. Lemma 13.7 is the case $m = n \, (= 2^k)$.

**13.15 Lemma.** *If $\tau \in \mathbb{Z}[\zeta]$ is a root of unity, then $\tau = \zeta^r$ for some $r \in \mathbb{Z}$.*

*Proof.* Suppose not. It is easy to see that then either $\tau$ is a primitive $2^m$th root of unity for some $m > k$, or else $\tau^s$ is a primitive $p$th root of unity for some odd prime $p$ and some $s \ge 1$. In the former case

$$[\mathbb{Q}(\tau) : \mathbb{Q}] = \phi(2^m) = 2^{m-1} > 2^{k-1} = \phi(2^k) = [\mathbb{Q}(\zeta) : \mathbb{Q}],$$

a contradiction. In the latter case, $\tau^s \zeta$ is a primitive $pn$-th root of unity, so

$$[\mathbb{Q}(\tau^s \zeta) : \mathbb{Q}] = \phi(pn) = (p - 1)\phi(n) > \phi(n) = [\mathbb{Q}(\zeta) : \mathbb{Q}],$$

again a contradiction.                      $\square$

We now have all the ingredients to complete the proof of Theorem 13.6. Note that we have yet to use the hypothesis that $a_i = \pm 1$. By Lemma 13.8 we have

$$|\gamma_1/\gamma_0| = |\gamma_0/\gamma_1| = 1.$$

Hence by Corollary 13.12, Theorem 13.14 and Lemma 13.15 we have $\gamma_0 = \zeta^{-r}\gamma_1$ for some $r$. Expand $\gamma_0$ and $\zeta^{-r}\gamma_1$ uniquely as integer linear combinations of $1, \zeta, \zeta^2, \ldots, \zeta^{\frac{n}{2}-1}$:

$$\gamma_0 = a_0 + a_1 + \cdots + a_{n-1} = \pm\sqrt{n}$$

$$\zeta^{-r}\gamma_1 = \zeta^{-r}((a_0 - a_{n/2}) + (a_1 - a_{n/2+1})\zeta + \cdots)$$

$$= (a_r - a_{n/2+r}) + (a_{r+1} - a_{n/2+r+1})\zeta + \cdots.$$

Equating coefficients of $\zeta^0$ yields $\pm\sqrt{n} = a_r - a_{n/2+r}$. Since each $a_i = \pm 1$, we must have $n \leq 4$, completing the proof.

## 13.7  P-Recursive Functions

A function $f : \mathbb{N} \to \mathbb{C}$ is called *polynomially recursive*, or *P-recursive* for short, if there exist polynomials $P_0(n), \ldots, P_d(n) \in \mathbb{C}[n]$, with $P_d(n) \neq 0$, such that

$$P_d(n)f(n+d) + P_{d-1}(n)f(n+d-1) + \cdots + P_0(n)f(n) = 0 \qquad (13.8)$$

for all $n \geq 0$.

For instance, the Fibonacci sequence $F_n$ is $P$-recursive since $F_{n+2} - F_{n+1} - F_n = 0$ for all $n \geq 0$. Here $d = 2$ and $P_2(n) = 1$, $P_1(n) = P_0(n) = -1$. This situation is quite special since the polynomials $P_i(n)$ are *constants*. Another $P$-recursive function is $f(n) = n!$, since $f(n+1) - (n+1)f(n) = 0$ for all $n \geq 0$.

Let $\mathcal{P}$ denote the set of all $P$-recursive functions $f : \mathbb{N} \to \mathbb{C}$. Our goal in this section is to prove that $\mathcal{P}$ is a $\mathbb{C}$-*algebra*, which amounts to showing that for any $f, g \in \mathcal{P}$ and $\alpha, \beta \in \mathbb{C}$, we have

$$\alpha f + \beta g \in \mathcal{P}, \qquad fg \in \mathcal{P}.$$

There is one technical problem that needs to be dealt with before proceeding to the proof. We would like to conclude from (13.8) that

$$f(n+d) = -\frac{1}{P_d(n)}(P_{d-1}(n)f(n+d-1) + \cdots + P_0(n)f(n)). \qquad (13.9)$$

This formula, however, is problematical when $P_d(n) = 0$. This can happen only for finitely many $n$, so (13.9) is valid for $n$ sufficiently large. Thus we want to deal with functions $f(n)$ only for $n$ sufficiently large. To this end, define $f \sim g$ if $f(n) = g(n)$ for all but finitely many $n$. Clearly $\sim$ is an equivalence relation; the equivalence classes are called *germs* at $\infty$ of functions $f : \mathbb{N} \to \mathbb{C}$. The germ containing $f$ is denoted $[f]$. Write $\mathcal{G}$ for the set of all germs.

**13.16 Lemma.** *(a) If $f$ is $P$-recursive and $f \sim g$, then $g$ is $P$-recursive. In other words, the property of $P$-recursiveness is compatible with the equivalence relation $\sim$.*

*(b) Write $\mathbb{C}^{\mathbb{N}}$ for the complex vector space of all functions $f : \mathbb{N} \to \mathbb{C}$. Let $\alpha, \beta \in \mathbb{C}$ and $f_1, f_2, g_1, g_2 \in \mathbb{C}^{\mathbb{N}}$. If $f_1 \sim f_2$ and $g_1 \sim g_2$, then $\alpha f_1 + \beta g_1 \sim \alpha f_2 + \beta g_2$ and $f_1 g_1 \sim f_2 g_2$. In other words, linear combinations and multiplication are compatible with the equivalence relation $\sim$. Thus the set $\mathcal{G}$ has the structure of a $\mathbb{C}$-algebra, i.e., a complex vector space and a ring (with obvious compatibility properties such as $(\alpha f)g = f(\alpha g) = \alpha(fg)$).*

*Proof.* (a) Suppose that $f(n) = g(n)$ for all $n > n_0$. Let (13.8) be the recurrence satisfied by $f$. Multiply both sides by $\prod_{j=0}^{n_0}(n - j)$. We then get a recurrence relation satisfied by $g$. Hence $g$ is $P$-recursive.

(b) This is clear. $\qquad\qquad\qquad\qquad\qquad\qquad\qquad\qquad\qquad\qquad\qquad\qquad\qquad\qquad$ $\square$

Let $\mathbb{C}[n]$ denote the ring of complex polynomials in $n$. Let $\mathbb{C}(n)$ denote the quotient field of $\mathbb{C}[n]$, i.e., the field of all rational functions $P(n)/Q(n)$, where $P, Q \in \mathbb{C}[n]$. Suppose that $f \in \mathbb{C}^{\mathbb{N}}$ and $R \in \mathbb{C}(n)$. Then $f(n)R(n)$ is defined for $n$ sufficiently large (i.e., when the denominator of $R(n)$ is nonzero). Thus we can define the germ $[f(n)R(n)] \in \mathcal{G}$ to be the germ of any function that agrees with $f(n)R(n)$ for $n$ sufficiently large. It is easy to see that this definition of scalar multiplication makes $\mathcal{G}$ into a vector space over the field $\mathbb{C}(n)$. We now come to the key characterization of $P$-recursive functions (or their germs).

**13.17 Lemma.** *A function $f \in \mathbb{C}^{\mathbb{N}}$ is $P$-recursive if and only if the vector space $\mathcal{V}_f$ over $\mathbb{C}(n)$ spanned by the germs $[f(n)], [f(n + 1)], [f(n + 2)], \ldots$ is finite-dimensional.*

*Proof.* Suppose that $f(n)$ satisfies (13.8). Let $\mathcal{V}_f'$ be the vector space over $\mathbb{C}(n)$ spanned by $[f(n)], [f(n + 1)], [f(n + 2)], \ldots, [f(n+d-1)]$, so $\dim_{\mathbb{C}(n)} \mathcal{V}_f' \leq d$. Equation (13.9) shows that $[f(n+d)] \in \mathcal{V}_f'$. Substitute $n + 1$ for $n$ in (13.9). We get that $[f(n+d+1)]$ is in the span (over $\mathbb{C}(n)$) of $[f(n+1)], [f(n+2)], \ldots, [f(n+d)]$. Since these $d$ germs are all in $\mathcal{V}_f'$, we get that $[f(n + d + 1)] \in \mathcal{V}_f'$. Continuing in this way, we get by induction on $k$ that $f(n + d + k) \in \mathcal{V}_f'$ for all $k \geq 0$, so $\mathcal{V}_f' = \mathcal{V}_f$. Thus $\mathcal{V}_f$ is finite-dimensional.

Conversely, assume that $\dim_{\mathbb{C}(n)} \mathcal{V}_f < \infty$. Then for some $d$, the germs $[f(n)]$, $[f(n+1)], \ldots, [f(n+d)]$ are linearly dependent over $\mathbb{C}(n)$. Write down this linear dependence relation and clear denominators to get a recurrence (13.8) satisfied by $f$. Hence $f$ is $P$-recursive. $\qquad\qquad\qquad\qquad\qquad\qquad\qquad\qquad$ $\square$

We now have all the ingredients necessary for the main result of this section.

**13.18 Theorem.** *Let $f, g \in \mathcal{P}$ and $\alpha, \beta \in \mathbb{C}$. Then:*

(a) $\alpha f + \beta g \in \mathcal{P}$

(b) $fg \in \mathcal{P}$.

*Proof.* (a) By Lemma 13.17 it suffices to show that $\dim \mathcal{V}_{\alpha f + \beta g} < \infty$. Now by definition, the *sum* $\mathcal{V}_f + \mathcal{V}_g$ is the vector space consisting of all linear combinations $\gamma[u] + \delta[v]$, where $[u] \in \mathcal{V}_f$ and $[v] \in \mathcal{V}_g$ and $\gamma, \delta \in \mathbb{C}(n)$. In particular, $\mathcal{V}_f + \mathcal{V}_g$ contains all the germs $\alpha[f(n+k)] + \beta[g(n+k)] = [\alpha f(n+k) + \beta g(n+k)], k \geq 0$. Hence

$$\mathcal{V}_{\alpha f + \beta g} \subseteq \mathcal{V}_f + \mathcal{V}_g.$$

Now if $V$ and $W$ are subspaces of some vector space, then $V + W$ is spanned by the union of a basis for $V$ and basis for $W$. In particular, if $V$ and $W$ are finite-dimensional, then $\dim(V + W) \leq \dim V + \dim W$. Hence

$$\dim \mathcal{V}_{\alpha f + \beta g} \leq \dim(\mathcal{V}_f + \mathcal{V}_g) \leq \dim \mathcal{V}_f + \dim \mathcal{V}_g < \infty,$$

as was to be proved.

(b) The proof is analogous to (a), except that instead of the sum $V + W$ we need the *tensor product* $V \otimes_K W$ over the field $K$. Recall from linear algebra that $V \otimes_K W$ may be thought of (somewhat naively) as the vector space spanned by all symbols $v \otimes w$, where $v \in V$ and $w \in W$, subject to the conditions

$$(v_1 + v_2) \otimes w = v_1 \otimes w + v_2 \otimes w$$

$$v \otimes (w_1 + w_2) = v \otimes w_1 + v \otimes w_2$$

$$\alpha v \otimes w = v \otimes \alpha w = \alpha(v \otimes w),$$

where $\alpha$ is a scalar. A standard and simple consequence is that if $V$ has the basis $\{v_1, \ldots, v_m\}$ and $W$ has the basis $\{w_1, \ldots, w_n\}$, then $V \otimes_K W$ has the basis $v_i \otimes w_j$, for $1 \leq i \leq m$ and $1 \leq j \leq n$. In particular,

$$\dim(V \otimes_K W) = (\dim V)(\dim W).$$

Recall the basic "universality" property of the tensor product $V \otimes W = V \otimes_K W$: there is a bilinear map $\Psi : V \times W \to V \otimes W$ such that for any vector space $Y$ and bilinear map $\Phi : V \times W \to Y$, there is a unique linear map $\varphi : V \otimes W \to Y$ for which $\Phi = \varphi \Psi$. In particular, there is a unique linear transformation $\varphi : \mathcal{V}_f \otimes_{\mathbb{C}(n)} \mathcal{V}_g \to \mathcal{G}$ satisfying

$$[f(n+i)] \otimes g[(n+j)] \overset{\varphi}{\mapsto} [f(n+i)g(n+j)].$$

The image of $\varphi$ contains all germs $[f(n+i)g(n+i)]$, so $\mathcal{V}_{fg} \subseteq \text{image}(\varphi)$. Thus

$$\dim \mathcal{V}_{fg} \leq \dim(\mathcal{V}_f \otimes_{\mathbb{C}(n)} \mathcal{V}_g) = (\dim \mathcal{V}_f)(\dim \mathcal{V}_g) < \infty,$$

and the proof follows.                                                                    $\square$

## 13.8 Affine Monoids

In this section we give a simple application of commutative algebra to a topic on the interface of algebra (but not commutative algebra *per se*) and combinatorics. It is not difficult to prove our main result (Theorem 13.21) without using commutative algebra, but it is nonetheless of interest to see the connection with algebra. In Chapter 12 we gave a more substantial application of commutative algebra which has no known more elementary proof.

Recall that a *semigroup* is a set with an associative binary operation. A *monoid* is a semigroup with an identity element. For example, the set

$$\mathbb{N}^d = \{(a_1, \ldots, a_d) : a_i \in \mathbb{N}\},$$

with the usual operation of componentwise addition, is a monoid. The identity element is $(0, 0, \ldots, 0)$. This monoid is in fact *commutative* since $\alpha + \beta = \beta + \alpha$ for all $\alpha, \beta \in \mathbb{N}^d$. We will be concerned here with *submonoids* $M$ of $\mathbb{N}^d$, that is, subsets of $\mathbb{N}^d$ which are closed under addition and which contain $(0, 0, \ldots, 0)$.

A monoid $M$ is *finitely generated* if there exists a finite subset $G$ of $M$ such that every element of $M$ is a (finite) sum (allowing any number of repetitions) of elements of $G$. Thus $\mathbb{N}^d$ is finitely generated, since we can take $G$ to be the set of $d$ unit coordinate vectors (vectors whose components are all 0 except for one 1). However, it is not true in general that every submonoid of $\mathbb{N}^d$ is finitely generated. For instance, let

$$M = \{(i, j) \in \mathbb{N}^2 : i > 0\} \cup \{(0, 0)\}. \tag{13.10}$$

Then a subset $G$ of $M$ generates $M$ if and only if $G$ contains the infinitely many elements $(1, j)$, $j \geq 0$. Let us mention that, on the other hand, every submonoid of $\mathbb{N}$ is finitely generated (Exercise 13.30).

We will be concerned here with special submonoids of $\mathbb{N}^d$. First note that the set of all solutions $(a_1, \ldots, a_d)$ in nonnegative integers to a system of homogeneous linear equations with integer coefficients in the unknowns $x_1, \ldots, x_d$ is a submonoid of $\mathbb{N}^d$.

**13.19 Example.** (a) Let $M$ be the monoid of all solutions $(a_1, a_2, a_3, a_4) \in \mathbb{N}^4$ to the equations

$$x_1 + x_2 - x_3 - x_4 = 0$$

$$x_1 - 2x_4 = 0.$$

The reader should try to show that $M$ is generated by $(2, 0, 1, 1)$ and $(0, 1, 1, 0)$.
(b) Let $M$ be the monoid of all solutions $(a_1, \ldots, a_6) \in \mathbb{N}^6$ of the equation

$$x_1 + x_2 + x_3 - x_4 - x_5 - x_6 = 0.$$

Can the reader see why the smallest set of generators for $M$ has nine elements?

(c) Let $M$ be the monoid of all solutions $(a_1, a_2, a_3) \in \mathbb{N}^3$ of the equations

$$2x_1 + x_2 - x_3 = 0$$
$$-x_1 + x_2 + 2x_3 = 0.$$

Then $M = \{(0, 0, 0)\}$. A quick way to see this is to add the two equations: $x_1 + 2x_2 + x_3 = 0$.

In complete analogy to the concept of isomorphism of groups, rings, etc., define two monoids $M$ and $N$ to be *isomorphic* if there is a bijection $f : M \to N$ such that $f(uv) = f(u)f(v)$ for all $u, v \in M$ (where we have written the monoid operation multiplicatively). Naturally the map $f$ is called an *isomorphism*.

Define a submonoid $M$ of $\mathbb{N}^d$ to be *linear* if it is the set of solutions $(a_1, \ldots, a_d)$ in nonnegative integers to a system of homogeneous linear equations with integer coefficients in the unknowns $x_1, \ldots, x_d$.

We say that an arbitrary monoid $M$ is *normal* if it is isomorphic to a linear monoid. For the reason behind this terminology, see Exercise 13.43. The following simple result shows that some monoids more general than linear monoids are in fact normal.

**13.20 Theorem.** *Let $M$ be the set of solutions $(a_1, \ldots, a_d) \in \mathbb{N}^d$ to a set of homogeneous linear equations with integer coefficients, homogeneous linear congruences modulo a positive integer, and homogeneous linear inequalities with integer coefficients. Then $M$ is normal.*

As an example of a monoid covered by the above theorem, we could take all vectors $(a_1, \ldots, a_7) \in \mathbb{N}^7$ satisfying the five conditions

$$3x_1 - 4x_2 - x_3 + 7x_4 + 5x_6 - 4x_7 = 0$$
$$x_1 + x_3 - 9x_6 - 4x_7 = 0$$
$$x_1 + x_2 - 3x_6 \equiv 0 \,(\mathrm{mod}\,5)$$
$$4x_2 - x_3 + 2x_4 + 2x_5 - 3x_7 \geq 0$$
$$x_3 - 3x_5 \geq 0.$$

*Proof.* Let the given congruences be $L_i(x_1, \ldots, x_d) \equiv 0 \,(\mathrm{mod}\,n_i)$, $1 \leq i \leq r$, so each $L_i(x_1, \ldots, x_d)$ has the form

$$L_i(x_1, \ldots, x_d) = \sum_{m=1}^{d} a_{im} x_m, \quad a_{im} \in \mathbb{Z}.$$

We can replace each $a_{im}$ with a *nonnegative* integer $b_{im}$ satisfying $a_{im} \equiv b_{im} \,(\mathrm{mod}\,n_i)$ without affecting the solutions to our system of equations, congruences, and inequalities. Let

$$L_i'(x_1, \ldots, x_d) = \sum_{m=1}^{d} b_{im} x_m.$$

Introduce a new variable $y_i$, and replace the congruence $L_i'(x_1, \ldots, x_d) \equiv 0 \pmod{n_i}$ with the equation

$$L_i'(x_1, \ldots, x_d) - n_i y_i = 0. \tag{13.11}$$

Thus if $(x_1, \ldots, x_d, y) = (a_1, \ldots, a_d, b)$ is a solution to (13.11) in integers with each $a_i \geq 0$, then also $b \geq 0$. Similarly for each inequality $M_j(x_1, \ldots, x_d) \geq 0$, introduce a new variable $z_j$ (called a *slack variable*), and replace the inequality with the equation

$$M_j(x_1, \ldots, x_d) - z_j = 0.$$

It is clear [why?] that the monoid of solutions in nonnegative integers to the original set of equations, congruences, and inequalities is isomorphic to the monoid of solutions in nonnegative integers to the new system of equations only, so the proof follows.                                                                           □

We now come to the main result of this section. For the proof, recall that a commutative ring $R$ with 1 is said to be *noetherian* if there does not exist an infinite strictly ascending chain $I_1 \subset I_2 \subset \cdots$ of ideals of $R$. (This condition is called the *ascending chain condition* or *ACC*.) Equivalently, every ideal of $R$ is finitely-generated. A fundamental (and not so difficult to prove) result of commutative algebra, called the *Hilbert basis theorem*, asserts that if $R$ is noetherian, then so is the polynomial ring $R[x]$. In particular, if $K$ is a field (and hence noetherian since its only ideals are $\{0\}$ and $K$), then so is the polynomial ring $K[x_1, \ldots, x_d]$. Moreover, it is not hard to show that if $I$ is an ideal of a noetherian ring and $\Gamma$ is a set of generators of $I$, then some finite subset of $\Gamma$ generates $I$.

**13.21 Theorem.** *If $M$ is a normal monoid, then $M$ is finitely-generated.*

*Proof.* We may assume that $M$ is linear, since if a monoid $N$ is finitely generated then any monoid isomorphic to $N$ is also finitely generated. Suppose that $M \subseteq \mathbb{N}^d$. Let $K$ be a field, and let $R = K[x_1, \ldots, x_d]$, the polynomial ring over $K$ in the indeterminates $x_1, \ldots, x_d$. If $\alpha = (\alpha_1, \ldots, \alpha_d) \in \mathbb{N}^d$, then we use the "multivariate notation"

$$x^\alpha = x_1^{\alpha_1} \cdots x_d^{\alpha_d}.$$

Let $I$ be the ideal of $R$ generated by all monomials $x^\alpha$ where $\alpha \in M$. By the Hilbert basis theorem $I$ is finitely generated as an ideal of $R$; in fact, some finite set $\Gamma$ of monomials $x^\alpha$, $\alpha \in R$, generates $I$. However, this does not imply that $M$ is finitely generated as a monoid. (If it did, then every submonoid of $\mathbb{N}^d$ would be finitely generated, but we have seen in (13.10) that this is not the case.) The key property (whose proof is trivial) that we need of linear submonoids $M$ of $\mathbb{N}^d$ is the following.

   (P) If $\alpha, \beta \in M$ and $\beta - \alpha \in \mathbb{N}^d$, then $\beta - \alpha \in M$.

We claim that $M$ is generated by those elements $\alpha$ for which $x^\alpha \in \Gamma$. For any $\beta = (\beta_1, \ldots, \beta_d) \in M$ write

$$s(\beta) = \beta_1 + \cdots + \beta_d.$$

We prove by induction on $s(\beta)$ that any $\beta \in M$ is a nonnegative linear combination of elements of $\Gamma$.

The assertion is clearly true for $s(\beta) = 0$, since then $\beta = (0, 0, \ldots, 0)$. Now let $\beta \in M$ with $s(\beta) > 0$, and assume the induction hypothesis for every $\gamma \in M$ with $s(\gamma) < s(\beta)$. Since $\beta \in M$ we have $x^\beta \in I$, so we can write

$$x^\beta = \sum_{x^\alpha \in \Gamma} f_\alpha(x) x^\alpha, \tag{13.12}$$

where $f_\alpha(x) \in R$. In order for the term $x^\beta$ to appear on the right-hand side of (13.12), some $f_\alpha(x)$ must have a term $x^\gamma$ (with some nonzero coefficient) such that $\beta = \alpha + \gamma$. By Property (P) above, we have $\gamma \in M$. Note that $s(\gamma) < s(\beta)$, so by the induction hypothesis $\gamma$ is a nonnegative integer linear combination of elements of $\Gamma$. Since $\beta = \alpha + \gamma$ where $\alpha \in \Gamma$, the same is true for $\beta$. Hence the proof follows by induction.                                                                                      □

# Notes for Chapter 13

The 100 prisoners problem was first considered by Miltersen. It appeared in a paper with Gál [50]. Further information on the history of this problem, together with a proof of optimality of the prisoners' strategy, is given by Curtin and Warshauer [27].

The Oddtown theorem is due to Berlekamp [7]. Theorem 13.3 on decomposing $K_n$ into complete bipartite subgraphs is due to Graham and Pollak [54, 55]. For Fisher's original proof of the inequality $v \le b$ for BIBD's and Bose's nonuniform generalization, see [41] and [12]. Sutner's original proof of the odd neighborhood theorem (Theorem 13.5) appears in [133], while the simpler proof of Caro may be found in [20]. The odd neighborhood problem is also known as the *Lights Out Puzzle*. For a host of other applications of linear algebra along the lines of Sections 13.2–13.5, see the unpublished manuscript [4] of Babai and Frankl, and the book [92] of Matoušek.

The circulant Hadamard matrix conjecture was first mentioned in print by Ryser [113, p. 134], though its precise origin is obscure. The work of Turyn mentioned in the text appears in [137, 138]. Some more recent progress is due to Leung and Schmidt [81].

While $P$-recursive functions and their cousins the $D$-finite series of Exercise 12.26 were known to nineteenth century analysts, the first systematic treatment of them did not appear until the paper of Stanley [125] in 1980, which includes a statement and proof of Theorem 13.18. For an exposition, see Stanley [131, §6.4].

Theorem 13.21 is known as *Gordan's lemma*, named after the German mathematician Paul Gordan (1837–1912). See [53]. For further information on submonoids of $\mathbb{N}^d$, see Bruns and Gubeladze [18, Chapter 2].

## Exercises for Chapter 13

1. Suppose that we have $2n$ prisoners and the same evil warden as in Section 13.1. Let $0 < \alpha < 1$. Now the prisoners open $2\alpha n$ of the boxes (more precisely, the closest integer to $2\alpha n$). For what value of $\alpha$ will the strategy used in the proof of Theorem 13.1 yield a 50% chance of success in the limit as $n \to \infty$?

2. Suppose that we have the same 100 prisoners and evil warden as in Section 13.1. This time, however, each prisoner must open 99 boxes. If any prisoner sees his or her name, then they are all killed. Find the best strategy for the prisoners and the resulting probability $p$ of success. Note that $10^{-200} \le p \le 10^{-2}$, the upper bound because the first prisoner has a success probability of $1/100$. (Unlike the situation in Section 13.1, once the best strategy is found for the present problem the proof of optimality is easy.)

3. (a) This time the evil warden puts a red hat or a blue hat on the head of each of the 100 prisoners. Each prisoner sees all the hats except for his own. The prisoners simultaneously guess the color of their hat. If any prisoner guesses wrong, then all are killed. What strategy minimizes the probability that all are killed?

   (b) Now the prisoners have hats as before, but only the prisoners who guess wrong are killed. What is the largest integer $m$ such that there is some strategy *guaranteeing* that at least $m$ prisoners survive?

4. (*) Our poor 100 prisoners have distinct real numbers written on their foreheads. They can see every number but their own. They each choose (independently, without communication) a red or blue hat and put it on their heads. The warden lines them up in increasing order of the numbers on their foreheads. If any two consecutive prisoners have the same color hat, then all are killed. What is the best strategy for success?

5. (a) (*) Suppose that $n$ people live in Reverse Oddtown. Every club contains an even number of persons, and any two clubs share an odd number of persons. Show that no more than $n$ clubs can be formed.

   (b) (rather difficult) Show that if $n$ is even, then at most $n - 1$ clubs can be formed.

6. a. Suppose that $n$ people live in Eventown. Every club contains an even number of persons, every two clubs share an even number of persons, and no two clubs have identical membership. Show that the maximum number of clubs is $2^{\lfloor n/2 \rfloor}$.

   b. (rather difficult) Suppose that fewer than $2^{\lfloor n/2 \rfloor}$ clubs have been formed using the Eventown rules of (a). Show that another club can be formed without breaking the rules.

7. (Bipartite Oddtown) A town of $n$ citizens has $m$ red clubs $R_1, \ldots, R_m$ and $m$ blue clubs $B_1, \ldots, B_m$. Assume that $\#(R_i \cap B_i)$ is odd for every $i$, and that $\#(R_i \cap B_j)$ is even for every $i \neq j$. Prove that $m \leq n$.

8. Triptown has $n$ persons and some clubs. Any two distinct clubs share a number of members that is a multiple of 3. Let $j$ be the number of clubs $C$ for which $\#C$ is not a multiple of 3. Show that $j \leq n$.

9. Strangetown has $n$ people and $k$ clubs $C_1, \ldots, C_k$. Each club has an even number of members, and any two clubs have an even number of members in common, with the following exception: $C_i$ and $C_{k+1-i}$ have an odd number of members in common for $1 \leq i \leq k$. (If $k = 2m+1$ and $i = m+1$, then this means that club $C_{m+1}$ has an odd number of members.) Show that there are at most $n$ clubs.

10. Weirdtown has $n$ people and $k$ clubs $C_1, \ldots, C_k$. Each club has an even number of members, and any two clubs have an even number of members in common, with the following exception: if $1 \leq i \leq k-1$, then $C_i$ and $C_{i+1}$ have an odd number of members in common. As a function of $n$, what is the largest possible value of $k$?

11. (a) An $n \times n$ real matrix is *skew-symmetric* if $A^t = -A$. Let $A$ be such a matrix with $n$ odd. Show that $\det A = 0$. (This is a standard result from linear algebra.)

   (b) Let $G$ be a simple graph with an odd number of vertices. The *deleted neighborhood* $N'(v)$ of a vertex $v$ is the set of all vertices adjacent to $v$. Show that there is a nonempty subset $S$ of the vertices such that $S$ intersects every deleted neighborhood $N'(v)$ in an even number of elements.

12. (*) Let $M_n$ denote the vector space of all real $n \times n$ matrices, so $\dim M_n = n^2$. Let $V$ be a subspace of $M_n$ such that every eigenvalue of every matrix $A \in V$ is real. Show that $\dim V \leq \binom{n+1}{2}$.

13. Let the edge set $E(K_n)$ of the complete graph $K_n$ be a union of the edge sets of complete bipartite graphs $B_1, \ldots, B_m$ such that every edge of $K_n$ is covered an odd number of times. Show that $m \geq (n-1)/2$. (The minimum value of $m$ is not known.)

14. A *complete tripartite graph* with vertex tripartition $(X_1, X_2, X_3)$ is the graph on the disjoint vertex sets $X_i$ with an edge between any two vertices not in the same set $X_i$. (We allow one of the $X_i$ to be empty, so for instance $K_2$ is a complete tripartite graph.) Thus if $\#X_i = p_i$ then the complete tripartite graph has $p_1 p_2 + p_1 p_3 + p_2 p_3$ edges. Suppose that the edge set $E(K_n)$ is partitioned into $m$ disjoint edge sets of complete tripartite graphs. What is the minimum value of $m$?

15. (*) Let $A_1, \ldots, A_n$ be distinct subsets of an $n$-set $X$. Give a linear algebra proof that for some $x \in X$, all the sets $A_i - x$ (short for $A_i - \{x\}$) are distinct. (There is a fairly simple combinatorial proof, but that is not what is asked for.) NOTE ON NOTATION: $A_i - x = A_i$ if $x \notin A_i$.

16. Show that the number of "switching sets" $S$ in Theorem 13.5 has the form $2^n$ for some $n \geq 0$.

17. (a) Let $G$ be a simple graph with $p$ vertices such that exactly $2^{p-1}$ subsets of the vertices are switching sets, i.e., they turn off all the light bulbs in the scenario of Section 13.5. Show that $G$ is a complete graph $K_p$. Give a proof based on linear algebra.

   (b) Describe the $2^{p-1}$ switching sets for $K_p$.

   (c) (more difficult) Same as above, but with exactly $2^{p-2}$ switching sets. Show that $G$ is a disjoint union of two complete graphs.

18. Given $v = (v_1, \ldots, v_n) \in \mathbb{Z}^n$, let $f(v)$ be the number of values of $1 \le i \le n$ for which the sum $v_i + v_{i+1} + v_{i+2}$ is even, where the subscripts are taken modulo $n$ so that they always lie in the set $\{1, 2, \ldots, n\}$. For which positive integers $n$ is the following true: for all $0 \le k \le n$, exactly $\binom{n}{k}$ vectors among the $2^n$ vectors $v \in \{0, 1\}^n$ satisfy $f(v) = k$?

19. (*) Show that a Hadamard matrix $H$ has order 1, 2, or $n = 4m$ for some integer $m \ge 1$.

20. For what values of $n$ do there exist $n + 1$ vertices of an $n$-dimensional cube such that any two of them are the same distance apart? For instance, it's clearly impossible for $n = 2$, while for $n = 3$ the vertices can be 000, 110, 101, 011. Your answer should involve Hadamard matrices.

21. (a) Show that if $H$ is an $n \times n$ Hadamard matrix all of whose rows have the same number of 1's, then $n$ is a square.

   (b) Show also that all columns of $H$ have the same number of 1's.

22. (*) Clearly $2^n$ and $n!$ are $P$-recursive functions, so by Theorem 13.18 so is $f(n) = 2^n + n!$. Find a recurrence of the form (13.8) satisfied by $f(n)$.

23. (a) (difficult) Let $f(n) = \sum_{k=0}^{n} \binom{n}{k}^3$. Show that

$$(n+2)^2 f(n+2) - (7n^2+21n+16) f(n+1) - 8(n+1)^2 f(n) = 0, \quad n \ge 0.$$

   (b) (*) Fix a positive integer $d$. Show that the function $f_d(n) = \sum_{k=0}^{n} \binom{n}{k}^d$ is $P$-recursive.

24. (*) (difficult) Let $f(n)$ be the number of paths from $(0, 0)$ to $(n, n)$, where each step in the path is $(1, 0)$, $(0, 1)$, or $(1, 1)$. For instance, $f(1, 1) = 3$, corresponding to the paths (where we abbreviate $(1, 0)$ as 10, etc.) $00 \to 10 \to 11$, $00 \to 01 \to 11$, and $00 \to 11$. Show that

$$(n + 2) f(n + 2) - 3(2n + 3) f(n + 1) + (n + 1) f(n) = 0, \quad n \ge 0.$$

25. (a) Let $\alpha_1, \ldots, \alpha_k$ be distinct nonzero complex numbers, and let $Q_1(n)$, $\ldots$, $Q_k(n)$ be distinct nonzero complex polynomials. Define

$$f(n) = Q_1(n)\alpha_1^n + Q_2(n)\alpha_2^n + \cdots + Q_k(n)\alpha_k^n, \quad n \ge 0.$$

   Show that $f(n)$ is $P$-recursive.

   (b) (difficult) Show that the least degree $d$ of a recurrence (13.8) satisfied by $f(n)$ is equal to $k$.

26. (a) Let $\mathbb{C}[[x]]$ denote the ring of all power series $\sum_{n\geq0} a_n x^n$ over $\mathbb{C}$. It is easy to see that $\mathbb{C}[[x]]$ is an integral domain. A *Laurent series* is a series $\sum_{n\in\mathbb{Z}} b_n x^n$, i.e., any integer exponents are allowed. Show that the quotient field of $\mathbb{C}[[x]]$, denoted $\mathbb{C}((x))$, consists of all Laurent series of the form $\sum_{n\geq n_0} b_n x^n$ for some $n_0 \in \mathbb{Z}$, i.e., all Laurent series with finitely many negative exponents. Equivalently, $\mathbb{C}((x))$ is obtained from $\mathbb{C}[[x]]$ by inverting the single element $x$. Show also that $\mathbb{C}((x))$ contains the field $\mathbb{C}(x)$ of all rational functions $P(x)/Q(x)$, where $P, Q \in \mathbb{C}[x]$, in the sense that there is a Laurent series $F(x) \in \mathbb{C}((x))$ satisfying $Q(x)F(x) = P(x)$.

(b) A Laurent series $y \in \mathbb{C}((x))$ is called *D-finite* (short for *differentiably finite*) if there exist polynomials $p_0(x), \ldots, p_d(x)$, not all 0, such that

$$p_d(x)y^{(d)} + \cdots + p_1(x)y' + p_0(x)y = 0,$$

where $y^{(d)}$ denotes the $d$th derivative of $y$ with respect to $x$. Show that a power series $\sum_{n\geq0} f(n)x^n$ is D-finite if and only if the function $f(n)$ is *P-recursive*.

(c) Show that $y \in \mathbb{C}((x))$ is D-finite if and only if the vector space over $\mathbb{C}(x)$ spanned by $y, y', y'', \ldots$ is finite-dimensional (whence the terminology "*D-finite*").

(d) Show that the set $\mathcal{D}$ of D-finite Laurent series $f \in \mathbb{C}((x))$ forms a subalgebra of $\mathbb{C}((x))$, i.e., $\mathcal{D}$ is closed under complex linear combinations and under product.

(e) (more difficult) Show that $\mathcal{D}$ is not a subfield of $\mathbb{C}((x))$.

27. (*) A Laurent series $y \in \mathbb{C}((x))$ is called *algebraic* if there exist polynomials $p_0(x), p_1(x), \ldots, p_d(x) \in \mathbb{C}[x]$, not all 0, such that

$$p_d(x)y^d + p_{d-1}(x)y^{d-1} + \cdots + p_1(x)y + p_0(x) = 0. \tag{13.13}$$

Show that an algebraic series $y$ is D-finite.

28. (very difficult) Show that a nonzero Laurent series $f(x) \in \mathbb{C}((x))$ and its reciprocal $1/f(x)$ are both D-finite if and only if $f'(x)/f(x)$ is algebraic.

29. Fix $k \geq 1$ and write $x = (x_1, \ldots, x_k)$. Suppose that

$$F(x) = \sum_{n_1,\ldots,n_k\geq0} a_{n_1,\ldots,n_k} x_1^{n_1} \cdots x_k^{n_k}$$

is a power series in $x$ with complex coefficients $a_{n_1,\ldots,n_k}$. Define the *diagonal* of $F(x)$ to be the power series $\sum_{n\geq0} a_{n,n,\ldots,n}x^n$ in the single variable $x$.

(a) (difficult) Suppose that $F(x)$ represents a rational function. In other words, there are polynomials $P(x), Q(x) \in \mathbb{C}[x]$ such that $Q(x)F(x) = P(x)$. Show that if $k = 2$ then the diagonal of $F(x)$ is algebraic.

(b) Show that the diagonal of the rational function $1/(1 - x - y - xy)$ is equal to $1/\sqrt{1 - 6x + x^2}$.

(c)  (difficult) Let

$$F(x, y, z) = \frac{1}{1 - x - y - z} = \sum_{k,m,n \geq 0} \frac{(k + m + n)!}{k! \, m! \, n!} x^k y^m z^n.$$

Show that the diagonal series $\sum_{n \geq 0} \frac{(3n)!}{n!^3} x^n$ is not algebraic.

(d)  (very difficult) Show that the diagonal of any power series over $\mathbb{C}$ in finitely many variables that represents a rational function is $D$-finite.

30.  Show that every submonoid of $\mathbb{N}$ (under the operation of addition) is finitely-generated.

31.  Let $a$ and $b$ be relatively prime positive integers, and let $M$ be the submonoid of $\mathbb{N}$ generated by $a$ and $b$.

(a)  Show that

$$\sum_{j \in M} x^j = \frac{1 - x^{ab}}{(1 - x^a)(1 - x^b)}.$$

(b)  Show that if $j \geq ab - a - b + 1$ then $j \in M$.

(c)  Show that $ab - a - b \notin M$.

(d)  Let $0 \leq i \leq ab - a - b$. Show that exactly one of $i$ and $ab - a - b - i$ belongs to $M$. Note that this generalizes (c) since $0 \in M$.

32.  Does the monoid $\mathbb{N}^2$ have a submonoid isomorphic to $\mathbb{N}^3$?

33.  (a)  Show that any submonoid of $\mathbb{N}^d$ has a unique minimal generating set $G$, i.e., every set of generators contains $G$.

(b)  In contrast to (a), give an example of a commutative monoid $M$ with more than one element with the following property: if $G$ generates $M$ and $S$ is any finite subset of $G$, then $G - S$ generates $M$.

34.  Let $n \geq 1$. Show that there exists a submonoid of $\mathbb{N}$ whose minimal generating set $G$ (as defined in the previous problem) has exactly $n$ elements.

35.  (moderately difficult) (*) Write $x = (x_1, \ldots, x_d)$. Let $L_1(x), \ldots, L_m(x)$ be homogeneous linear forms with integer coefficients. Show that the following two conditions are equivalent:

• The only solution $x = (a_1, \ldots, a_d) \in \mathbb{N}^d$ of the equations $L_1(x) = 0, \ldots, L_m(x) = 0$ is $(0, 0, \ldots, 0)$.

• Some integer linear combination of $L_1(x), \ldots, L_m(x)$ has every coefficient positive.

An example is given by Example 13.19(c).

36.  (*) Let $d \geq 1$. Does there exist an infinite increasing chain $M_1 \subset M_2 \subset \cdots$ of submonoids of $\mathbb{N}^d$?

37.  Let $M$ be the submonoid of $\mathbb{N}^d$ consisting of all vectors $(a_1, \ldots, a_d) \in \mathbb{N}^d$ such that $\sum a_i$ is even. Describe the unique minimal set $G$ of generators of $M$.

38. (a) Show that every normal submonoid of $\mathbb{N}$ is linear.
    (b) Given an example of a finitely-generated submonoid $M$ of $\mathbb{N}^2$ such that $M$ is normal, but it fails to satisfy Property (P) (from the proof of Theorem 13.21).
39. Let $S$ be a set of monomials in the variables $x_1, \ldots, x_d$. Suppose that no monomial in $S$ is divisible by another monomial is $S$. Show that $S$ is finite. Try to give a simple proof using commutative algebra.
40. Prove the assertion preceding Theorem 13.21 that if $R$ is a noetherian ring (commutative with 1) and $\Gamma$ generates an ideal $I$ of $R$, then some finite subset of $\Gamma$ generates $R$.
41. Give a proof of Theorem 13.21 that does not use commutative algebra.
42. (difficult) Show that a finitely generated commutative monoid $M$ is normal if and only if the following condition holds: if $\alpha, \beta, \gamma \in M, n \in \mathbb{P}$, and $n\alpha + \beta = n\gamma$, then $\beta = n\delta$ for some $\delta \in M$. Note that this result gives an *intrinsic* characterization of normal monoids, i.e., a characterization defined solely in terms of $M$ and not how it is embedded in a larger monoid.
43. Let $R$ be a commutative integral domain (with 1). Recall from commutative algebra that $R$ is *normal* if it is integrally closed in its quotient field $\bar{R}$, i.e., if $\alpha \in \bar{R}$ satisfies a *monic* polynomial equation with coefficients in $R$, then $\alpha \in R$. For instance, the ring $R = K[x^2, x^3, x^4, \ldots]$ (where $K$ is a field) is not normal since $x = \frac{x^3}{x^2} \in \bar{R}$, $x \notin R$, and $x$ is a zero of the polynomial $f(t) = t^2 - x^2 \in R[t]$. Let $M$ be a finitely generated submonoid of $\mathbb{N}^d$, and let $R$ be the subalgebra of $K[x_1, \ldots, x_d]$ generated (or, in this situation, spanned) by the monomials $x^\alpha$, $\alpha \in M$. Show that $R$ is normal if and only if $M$ is a normal monoid. This result explains the terminology "normal monoid."

# Hints and Comments for Some Exercises

## Chapter 1

**1.5** Consider $A(H_n)^2$ and use Exercise 1.3.

**1.6** (a) First count the number of sequences $V_{i_0}, V_{i_1}, \ldots, V_{i_\ell}$ for which there exists a closed walk with vertices $v_0, v_1, \ldots, v_\ell = v_0$ (in that order) such that $v_j \in V_{i_j}$.

**1.11** Consider the rank of $A(\Gamma)$, and also consider $A(\Gamma)^2$. The answer is very simple and does not involve irrational numbers.

**1.12** (b) Consider $A(G)^2$.

## Chapter 2

**2.2** Give an argument analogous to the proof of Theorem 8.8.9.

**2.3** This result is known as the "middle levels conjecture" and was proved by Torsten Mütze in 2016.

**2.5** (c) Mimic the proof for the graph $C_n$, using the definition

$$\langle \chi_u, \chi_v \rangle = \sum_{w \in \mathbb{Z}_n} \chi_u(w) \overline{\chi_v(w)},$$

where an overhead bar denotes complex conjugation.

## Chapter 3

**3.4** You may find Example 3.1 useful.

**3.8** It is easier *not* to use linear algebra.

**3.9** See previous hint.

© Springer International Publishing AG, part of Springer Nature 2018
R. P. Stanley, *Algebraic Combinatorics*, Undergraduate Texts in Mathematics,
https://doi.org/10.1007/978-3-319-77173-1

**3.11** First show (easy) that if we start at a vertex $v$ and take $n$ steps (using our random walk model), then the probability that we traverse a fixed closed walk $W$ is equal to the probability that we traverse $W$ in reverse order.

**3.13** See hint for Exercise 3.8.

# Chapter 4

**4.4** (b) One way to do this is to count in two ways the number of $k$-tuples $(v_1, \ldots, v_k)$ of linearly independent elements from $\mathbb{F}_q^n$: (1) first choose $v_1$, then $v_2$, etc., and (2) first choose the subspace $W$ spanned by $v_1, \ldots, v_k$, and then choose $v_1$, $v_2$, etc.

**4.4** (c) The easiest way is to use (b).

**4.7** (b) This result was proved by R. P. Dilworth in 1950 and is known as *Dilworth's theorem*.

# Chapter 5

**5.5** (a) Show that $N_n \cong B_n/G$ for a suitable group $G$.

**5.9** (a) Use Corollary 2.4 with $n = \binom{p}{2}$.

**5.13** Use Exercise 5.12.

# Chapter 6

**6.2** (b) Not really a hint, but the result is equivalent [why?] to the case $r = m$, $s = n, t = 2$, and $x = 1$ of Exercise 36.

**6.3** Consider $\mu = (8, 8, 4, 4)$.

**6.5** First consider the case where $S$ has $\zeta$ elements equal to 0 (so $\zeta = 0$ or 1), $\nu$ elements that are negative, and $\pi$ elements that are positive, so $\nu + \zeta + \pi = 2m + 1$.

# Chapter 7

**7.3** *Hint.* Dark red and burgundy are the same color!

For those who are interested, the cycle indicator $Z_G$ of the full symmetry group $G$ of the dodecahedron acting on the vertices is

$$\frac{1}{120} \left( z_1^{20} + 15z_2^{10} + 20z_1^2 z_3^6 + 24z_5^4 + 15z_1^4 z_2^8 + z_2^{10} + 20z_2 z_6^3 + 24z_{10}^2 \right).$$

The number of inequivalent vertex colorings using two *distinct* colors is $Z_G(2, 2, \ldots, 2) = 9436$.

**7.19** (a) Use Pólya's theorem.

# Chapter 8

**8.3** Encode a maximal chain by an object that we already know how to enumerate.

**8.7** Partially order by diagram inclusion the set of all partitions whose diagrams can be covered by nonoverlapping dominos, thereby obtaining a subposet $Y_2$ of Young's lattice $Y$. Show that $Y_2 \cong Y \times Y$.

**8.14** Use induction on $n$.

**8.17** (a) One way to do this is to use the generating function $\sum_{n \geq 0} Z_{\mathfrak{S}_n}(z_1, z_2, \dots)$ $x^n$ for the cycle indicator of $\mathfrak{S}_n$ (Theorem 7.13). Another method is to find a recurrence for $B(n + 1)$ in terms of $B(0), \dots, B(n)$ and then convert this recurrence into a generating function

**8.18** Consider the generating function

$$G(q, t) = \sum_{k, n \geq 0} \kappa(n \to n + k \to n) \frac{t^k q^n}{(k!)^2}$$

and use (8.25).

**8.20** (b) Consider the square of the adjacency matrix of $Y_{j-1, j}$.

**8.24** Use Exercise 8.14.

**8.33** Use an inclusion–exclusion argument (a topic not covered in the text).

# Chapter 9

**9.1** There is a simple proof based on the formula $\kappa(K_p) = p^{p-2}$, avoiding the Matrix-Tree Theorem.

**9.3** Consider the matrices $L - nI$ and $L - (n - m)I$.

**9.2** (c) Use the fact that the rows of $L$ sum to 0, and compute the trace of $L$.

**9.7** (b) Use Exercise 1.3.

**9.8** (a) For the most elegant proof, use the fact that commuting $p \times p$ matrices $A$ and $B$ can be simultaneously triangularized, i.e., there exists an invertible matrix $X$ such that both $XAX^{-1}$ and $XBX^{-1}$ are upper triangular.

**9.8** (d) Use Exercise 9.10(a).

**9.9** Let $G^*$ be the full dual graph of $G$, i.e., the vertices of $G^*$ are the faces of $G$, including the outside face. For every edge $e$ of $G$ separating two faces $R$ and $S$ of $G$, there is an edge $e^*$ of $G^*$ connecting the vertices $R$ and $S$. Thus $G^*$ will have some multiple edges, and $\#E(G) = \#E(G^*)$. First show combinatorially that $\kappa(G) = \kappa(G^*)$. (See Corollary 11.19.)

**9.10** (a) Use the Binet–Cauchy theorem.

**9.12** (a) The Laplacian matrix $L = L(G)$ acts on the space $\mathbb{R}V(G)$, the real vector space with basis $V(G)$. Consider the subspace $W$ of $\mathbb{R}V(G)$ spanned by the elements $v + \varphi(v)$, $v \in V(G)$.

**9.13** (a) Let $s(n, q, r)$ be the number of $n \times n$ symmetric matrices of rank $r$ over $\mathbb{F}_q$. Find a recurrence satisfied by $s(n, q, r)$, and verify that this recurrence is satisfied by

$$
s(n, q, r) = \begin{cases} \displaystyle\prod_{i=1}^{t} \frac{q^{2i}}{q^{2i} - 1} \cdot \prod_{i=0}^{2t-1} (q^{n-i} - 1), \ 0 \le r = 2t \le n \\[4mm] \displaystyle\prod_{i=1}^{t} \frac{q^{2i}}{q^{2i} - 1} \cdot \prod_{i=0}^{2t} (q^{n-i} - 1), \ \ 0 \le r = 2t + 1 \le n. \end{cases}
$$

**9.14** Any of the three proofs of the Appendix to Chapter 9 can be carried over to the present exercise.

## Chapter 10

**10.4** (a) Use a suitable generalization of the deBruijn graph.

(b) Use techniques similar to those in Theorem 7.7.10.

**10.5** (b) Use the Perron–Frobenius theorem (Theorem 3.3).

**10.8** (a) Consider $A^{\ell}$.

**10.8** (f) There is an example with nine vertices that is not a de Bruijn graph.

**10.8** (c) Let $E$ be the (column) eigenvector of $A(D)$ corresponding to the largest eigenvalue. Consider $AE$ and $A^{t}E$, where $^{t}$ denotes transpose.

## Chapter 11

**11.2**(d) Let $e$ be an edge of $G$ that is not an isthmus. Let $G - e$ denote $G$ with the edge $e$ removed, and let $G/e$ denote $G$ with the edge $e$ contracted to a point (so $G/e$ has one less vertex than $G$). Find a simple recurrence satisfied by $C_G(n)$ in terms of $C_{G-e}(n)$ and $C_{G/e}(n)$, and then use induction.

**11.5** Use the unimodularity of the basis matrices $C_T$ and $B_T$.

**11.6** Use the formula (11.5) for the resistance of a parallel connection.

**11.8** (a) Mimic the proof of Theorem 9.8 (the Matrix-Tree Theorem).

**11.8** (b) Consider $ZZ^t$.

## Chapter 12

**12.9.** Compute the reduced Euler characteristic of $X$ and consider the last step of a possible shelling.

**12.10.** Modify the dunce hat of the previous exercise.

**12.12.** There are two examples with $f$-vector $(6, 14, 9)$ and none smaller.

**12.13.** (b) Consider the hint for Problem 12.12.

(c) Use Exercise 12.10.

**12.18.** The following property of triangulations $\Delta$ of spheres is sufficient for solving this exercise. For every $F \in \Delta$ we have

$$
\tilde{\chi}(\text{lk}_\Delta(F)) = (-1)^{\dim \text{lk}_\Delta(F)}.
$$

Here $\tilde{\chi}$ denotes reduced Euler characteristic, and lk denotes link, as defined in Remark 12.26.

**12.20.** There are *five* minimal nonfaces.

**12.21.** One approach is to consider minimal ideals $I$ of $K[\Gamma]$, say, for which $K[\Gamma]/I$ is a polynomial ring, i.e., isomorphic to $K[y_1, \ldots, y_r]$ for indeterminates $y_1, \ldots, y_r$.

**12.30.** *Answer:* 588. Without shellability, the answer is 589.

**12.31.** Let $I$ be the ideal of $K[\Delta]$ generated by all monomials $x^F$, where $F \notin \Delta'$. Clearly $K[\Delta']$ is isomorphic to $K[\Delta]/I$ (as a $K$-algebra). Let $\theta_1, \ldots, \theta_d \in K[\Delta]_1$ satisfy Property (P), and consider the natural map

$$f : K[\Delta]/(\theta_1, \ldots, \theta_d) \to (K[\Delta]/I)/(\theta_1, \ldots, \theta_d).$$

**12.32.** Consider $\theta_i \in K[\Delta]_1$ defined by $\theta_i = \sum_{x_j \in V_i} x_j$. Also consider the product $x_j \theta_i$ in the ring $K[\Delta]/(\theta_1, \ldots, \theta_d)$ for all $1 \le i \le d$ and $x_j \in V_i$.

# Chapter 13

**13.4** The best strategy involves the concept of odd and even permutations.

**13.5** For the easiest solution, don't use linear algebra but rather use the original Oddtown theorem.

**13.12** What are the eigenvalues of skew-symmetric matrices?

**13.15** Consider the incidence matrix $M$ of the sets and their elements. Consider two cases: $\det M = 0$ and $\det M \neq 0$.

**13.19** Consider the first three rows of $H$. Another method is to use row operations to factor a large power of 2 from the determinant.

**13.22** It is easiest to proceed directly and not use the proof of Theorem 13.18.

**13.23** (b) Consider the power series $\sum_{n \ge 0} \frac{x^n}{n!^d}$.

**13.24** Use Exercise 13.29(b).

**13.27** Differentiate with respect to $x$ (13.13) satisfied by $y$.

**13.35** This result is known as *Stiemke's theorem* (after E. Stiemke, 1892–1915) and is a forerunner of the duality theorem of linear programming.

**13.36** The answer depends on $d$.

# Bibliography

1. I. Anderson, *Combinatorics of Finite Sets* (Oxford University Press, Oxford, 1987); Corrected republication by Dover, New York, 2002
2. G.E. Andrews, *The Theory of Partitions* (Addison-Wesley, Reading, 1976)
3. G.E. Andrews, K. Eriksson, *Integer Partitions* (Cambridge University Press, Cambridge, 2004)
4. L. Babai, P. Frankl, *Linear Algebra Methods in Combinatorics*, preliminary version 2 (1992), 216 pp.
5. E.A. Beem, Craige and Irene Schensted don't have a car in the world, in *Maine Times* (March 12, 1982), pp. 20–21
6. E.A. Bender, D.E. Knuth, Enumeration of plane partitions. J. Comb. Theor. **13**, 40–54 (1972)
7. E.R. Berlekamp, On subsets with intersections of even cardinality. Can. Math. Bull. **12**, 363–366 (1969)
8. O. Bernardi, On the spanning trees of the hypercube and other products of graphs. Electron. J. Comb. **19**(P52), 16 pp (2012)
9. H. Bidkhori, S. Kishore, Counting the spanning trees of a directed line graph. Preprint [arXiv:0910.3442]
10. E.D. Bolker, The finite Radon transform, in *Integral Geometry*, Brunswick, Maine, 1984. Contemporary Mathematics, vol. 63 (American Mathematical Society, Providence, 1987), pp. 27–50
11. C.W. Borchardt, Ueber eine der Interpolation entsprechende Darstellung der Eliminations–Resultante. J. Reine Angew. Math. (Crelle's J.) **57**, 111–121 (1860)
12. R.C. Bose, A note on Fisher's inequality for balanced incomplete block designs. Ann. Math. Stat. 619–620 (1949)
13. F. Brenti, Log-concave and unimodal sequences in algebra, combinatorics, and geometry: an update, in *Jerusalem Combinatorics '93*. Contemporary Mathematics, vol. 178 (American Mathematical Society, Providence, 1994)
14. A.E. Brouwer, W.H. Haemers, *Spectra of Graphs* (Springer, New York, 2012)
15. H. Bruggesser, P. Mani, Shellable decompositions of cells and spheres. Math. Scand. **29**, 197–205 (1971)
16. W. Burnside, *Theory of Groups of Finite Order* (Cambridge University Press, Cambridge, 1897)
17. W. Burnside, *Theory of Groups of Finite Order*, 2nd edn. (Cambridge University Press, Cambridge, 1911); Reprinted by Dover, New York, 1955
18. W. Bruns, J. Gubeladze, *Polytopes, Rings, and K-Theory* (Springer, Dordrecht, 2009)

© Springer International Publishing AG, part of Springer Nature 2018

R. P. Stanley, *Algebraic Combinatorics*, Undergraduate Texts in Mathematics, https://doi.org/10.1007/978-3-319-77173-1

19. W. Bruns, J. Herzog, *Cohen-Macaulay Rings*, revised edn. Cambridge Studies in Advanced Mathematics, vol. 39 (Cambridge University Press, Cambridge/New York, 1998)
20. Y. Caro, Simple proofs to three parity theorems. Ars Combin. **42**, 175–180 (1996)
21. N. Caspard, B. Leclerc, B. Monjardet, *Finite Ordered Sets*. Encyclopedia of Mathematics and Its Applications, vol. 144 (Cambridge University Press, Cambridge, 2012)
22. A.L. Cauchy, Mémoire sur diverses propriétés remarquables des substitutions régulaires ou irrégulaires, et des systèmes de substitutions conjuguées (suite). C. R. Acad. Sci. Paris **21**, 972–987 (1845); Oeuvres Ser. 1 **9**, 371–387
23. A. Cayley, Note sur une formule pour la réversion des séries. J. Reine Angew. Math. (Crelle's J.) **52**, 276–284 (1856)
24. A. Cayley, A theorem on trees. Q. J. Math. **23**, 376–378 (1889); *Collected Papers*, vol. 13 (Cambridge University Press, Cambridge, 1897), pp. 26-28
25. G.F. Clements, B. Lindström, A generalization of a combinatorial theorem due to Macaulay. J. Comb. Theory **7**, 230–238 (1969)
26. I.S. Cohen, On the structure and ideal theory of complete local rings. Trans. Am. Math. Soc. **59**, 54–106 (1946)
27. E. Curtin, M. Warshauer, The locker puzzle. Math. Intelligencer **28**, 28–31 (2006)
28. D.M. Cvetković, M. Doob, H. Sachs, *Spectra of Graphs: Theory and Applications*, 3rd edn. (Johann Ambrosius Barth, Heidelberg/Leipzig, 1995)
29. D.M. Cvetković, P. Rowlinson, S. Simić, in *An Introduction to the Theory of Graph Spectra*. London Mathematical Society. Student Texts, vol. 75 (Cambridge University Press, Cambridge, 2010)
30. N.G. de Bruijn, A combinatorial problem. Proc. Koninklijke Nederlandse Akademie v. Wetenschappen **49**, 758–764 (1946); Indagationes Math. **8**, 461–467 (1946)
31. N.G. de Bruijn, Pólya's theory of counting, in *Applied Combinatorial Mathematics*, ed. by E.F. Beckenbach (Wiley, New York, 1964); Reprinted by Krieger, Malabar, FL, 1981
32. N.G. de Bruijn, Acknowledgement of priority to C. Flye Sainte-Marie on the counting of circular arrangements of $2^n$ zeros and ones that show each $n$-letter word exactly once, Technische Hogeschool Eindhoven, T.H.-Report 75-WSK-06, 1975
33. M.R. DeDeo, E. Velasquez, The Radon transform on $\mathbb{Z}_n^k$. SIAM J. Discrete Math. **18**, 472–478 (electronic) (2004/2005)
34. P. Diaconis, R.L. Graham, The Radon transform on $\mathbb{Z}_2^k$. Pac. J. Math. **118**, 323–345 (1985)
35. P. Diaconis, R.L. Graham, *Magical Mathematics* (Princeton University Press, Princeton, 2012)
36. P.G. Doyle, J.L. Snell, *Random Walks and Electrical Networks*. Carus Mathematical Monographs (Book 22) (Mathematical Association of America, Washington, DC, 1984); Version 3.02 at arXiv:math/0001057
37. E.B. Dynkin, Some properties of the weight system of a linear representation of a semisimple Lie group (in Russian). Dokl. Akad. Nauk SSSR (N.S) **71**, 221–224 (1950)
38. E.B. Dynkin, The maximal subgroups of the classical groups. Am. Math. Soc. Transl. Ser. 2 **6**, 245–378 (1957); Translated from Trudy Moskov. Mat. Obsc. **1**, 39–166
39. K. Engel, *Sperner Theory*. Encyclopedia of Mathematics and Its Applications, vol. 65 (Cambridge University Press, Cambridge, 1997)
40. P. Fishburn, *Interval Orders and Interval Graphs: A Study of Partially Ordered Sets* (Wiley, New York, 1985)
41. R.A. Fisher, An examination of the different possible solutions of a problem in incomplete blocks. Ann. Eugen. **10**, 52–75 (1940)
42. C. Flye Sainte-Marie, Solution to question nr. 48. l'Intermédiaire des Mathématiciens **1**, 107–110 (1894)
43. S. Fomin, Duality of graded graphs. J. Algebr. Combin. **3**, 357–404 (1994)
44. S. Fomin, Schensted algorithms for dual graded graphs. J. Algebr. Combin. **4**, 5–45 (1995)
45. J.S. Frame, G. de B. Robinson, R.M. Thrall, The hook graphs of $S_n$. Can. J. Math. **6** 316–324 (1954)

46. D.S. Franzblau, D. Zeilberger, A bijective proof of the hook-length formula. J. Algorithms **3**, 317–343 (1982)
47. F.G. Frobenius, Über die Congruenz nach einem aus zwei endlichen Gruppen gebildeten Doppelmodul. J. Reine Angew. Math. (Crelle's J.) **101**, 273–299 (1887); Reprinted in *Gesammelte Abhandlungen*, vol. 2 (Springer, Heidelberg, 1988), pp. 304–330
48. F.G. Frobenius, Über die Charaktere der symmetrischen Gruppe, *Sitzungsber. Kön. Preuss. Akad. Wissen. Berlin* (1900), pp. 516–534; *Gesammelte Abh. III*, ed. by J.-P. Serre (Springer, Berlin, 1968), pp. 148–166
49. W.E. Fulton, *Young Tableaux*. Student Texts, vol. 35 (London Mathematical Society/Cambridge University Press, Cambridge, 1997)
50. A. Gál, P.B. Miltersen, The cell probe complexity of succinct data structures, in *Proceedings of the 30th International Colloquium on Automata, Languages and Programming (ICALP)* (2003), pp. 332–344
51. M. Gardner, Squaring the square, in *The 2nd Scientific American Book of Mathematical Puzzles and Diversions* (Simon and Schuster, New York, 1961)
52. I.J. Good, Normally recurring decimals. J. Lond. Math. Soc. **21**, 167–169 (1947)
53. P. Gordan, Über die Auflosung linearer Gleichungen mit reellen Coefficienten. Math. Ann. **6**, 23–28 (1873)
54. R.L. Graham, H.O. Pollak, On the addressing problem for loop switching. Bell Syst. Tech. J. **50**, 2495–2519 (1971)
55. R.L. Graham, H.O. Pollak, On embedding graphs in squashed cubes, in *Lecture Notes in Mathematics*, vol. 303 (Springer, New York, 1973), pp. 99-110
56. C. Greene, D.J. Kleitman, Proof techniques in the theory of finite sets, in *Studies in Combinatorics*, ed. by G.-C. Rota. M.A.A. Studies in Mathematics, vol. 17 (Mathematical Association of America, Washington, DC, 1978), pp. 22–79
57. C. Greene, A. Nijenhuis, H.S. Wilf, A probabilistic proof of a formula for the number of Young tableaux of a given shape. Adv. Math. **31**, 104–109 (1979)
58. J.I. Hall, E.M. Palmer, R.W. Robinson, Redfield's lost paper in a modern context. J. Graph Theor. **8**, 225–240 (1984)
59. F. Harary, E.M. Palmer, *Graphical Enumeration* (Academic, New York, 1973)
60. F. Harary, R.W. Robinson, The rediscovery of Redfield's papers. J. Graph Theory **8**, 191–192 (1984)
61. G.H. Hardy, J.E. Littlewood, G. Pólya, *Inequalities*, 2nd edn. (Cambridge University Press, Cambridge, 1952)
62. L.H. Harper, Morphisms for the strong Sperner property of Stanley and Griggs. Linear Multilinear Algebra **16**, 323–337 (1984)
63. T.W. Hawkins, The origins of the theory of group characters. Arch. Hist. Exact Sci. **7**, 142–170 (1970/1971)
64. T.W. Hawkins, Hypercomplex numbers, Lie groups, and the creation of group representation theory. Arch. Hist. Exact Sci. **8**, 243–287 (1971/1972)
65. T.W. Hawkins, New light on Frobenius' creation of the theory of group characters. Arch. Hist. Exact Sci. **12**, 217–243 (1974)
66. A.P. Hillman, R.M. Grassl, Reverse plane partitions and tableaux hook numbers. J. Comb. Theor. A **21**, 216–221 (1976)
67. M. Hochster, Rings of invariants of tori, Cohen-Macaulay rings generated by monomials, and polytopes. Ann. Math. **96**, 318–337 (1972)
68. R.A. Horn, C.R. Johnson, *Matrix Analysis* (Cambridge University Press, Cambridge, 1985)
69. J.W.B. Hughes, Lie algebraic proofs of some theorems on partitions, in *Number Theory and Algebra*, ed. by H. Zassenhaus (Academic, New York, 1977), pp. 135–155
70. A. Hurwitz, Über die Anzahl der Riemannschen Flächen mit gegebenen Verzweigungspunkten. Math. Ann. **55**, 53–66 (1902)
71. A. Joyal, Une théorie combinatoire des séries formelles. Adv. Math. **42**, 1–82 (1981)
72. G.O.H. Katona, A theorem of finite sets, in *Proc. Tihany Conf. 1966*, Budapest (1968)

73. N.D. Kazarinoff, R. Weitzenkamp, Squaring rectangles and squares. Am. Math. Mon. **80**, 877–888 (1973)

74. B. Kind, P. Kleinschmidt, Schälbare Cohen-Macaulay-Komplexe und ihre Parametrisierung. Math. Z. **167**, 173–179 (1979)

75. G. Kirchhoff, Über die Auflösung der Gleichungen, auf welche man bei der Untersuchung der Linearen Vertheilung galvanischer Ströme geführt wird. Ann. Phys. Chem. **72**, 497–508 (1847)

76. C. Krattenthaler, Bijective proofs of the hook formulas for the number of standard Young tableaux, ordinary and shifted. Electronic J. Combin. **2**, R13, 9 pp. (1995)

77. D.E. Knuth, Permutations, matrices, and generalized Young tableaux. Pac. J. Math. **34**, 709–727 (1970)

78. I. Krasikov, Y. Roditty, Balance equations for reconstruction problems. Arch. Math. (Basel) **48**, 458–464 (1987)

79. J. Kruskal, The number of simplices in a complex, in *Mathematical Optimization Techniques* (University of California Press, Berkeley/Los Angeles, 1963), pp. 251–278

80. J.P.S. Kung, Radon transforms in combinatorics and lattice theory, in *Combinatorics and Ordered Sets*, Arcata, CA, 1985. Contemporary Mathematics, vol. 57 (American Mathematical Society, Providence, 1986), pp. 33–74

81. K.H. Leung, B. Schmidt, New restrictions on possible orders of circulant Hadamard matrices. Designs Codes Cryptogr. **64**, 143–151 (2012)

82. E.K. Lloyd, J. Howard Redfield: 1879–1944. J. Graph Theor. **8**, 195–203 (1984)

83. L. Lovász, A note on the line reconstruction problem. J. Comb. Theor. Ser. B **13**, 309–310 (1972)

84. L. Lovász, Random walks on graphs: a survey, in *Combinatorics. Paul Erdős is Eighty*, vol. 2, Bolyai Society Mathematical Studies, vol. 2 (Keszthely, Hungary, 1993), pp. 1–46

85. D. Lubell, A short proof of Sperner's lemma. J. Comb. Theor. **1**, 299 (1966)

86. F.S. Macaulay, *The Algebraic Theory of Modular Systems* (Cambridge University Press, Cambridge, 1916); Reprinted by Cambridge University Press, Cambridge/New York, 1994

87. F.S. Macaulay, Some properties of enumeration in the theory of modular systems. Proc. Lond. Math. Soc. **26**, 531–555 (1927)

88. P.A. MacMahon, Memoir on the theory of the partitions of numbers — Part I. Philos. Trans. R. Soc. Lond. A **187**, 619–673 (1897); *Collected Works*, vol. 1, ed. by G.E. Andrews (MIT, Cambridge, 1978), pp. 1026–1080

89. P.A. MacMahon, Memoir on the theory of the partitions of numbers — Part IV. Philos. Trans. R. Soc. Lond. A **209**, 153–175 (1909); *Collected Works*, vol. 1, ed. by G.E. Andrews (MIT, Cambridge, 1978), pp. 1292–1314

90. P.A. MacMahon, *Combinatory Analysis*, vols. 1, 2 (Cambridge University Press, Cambridge, 1915/1916); Reprinted in one volume by Chelsea, New York, 1960

91. J. Matoušek, *Using the Borsuk-Ulam Theorem: Lectures on Topological Methods in Combinatorics and Geometry*. Universitext (Springer, Berlin, 2003)

92. J. Matoušek, *Thirty-Three Miniatures* (American Mathematical Society, Providence, 2010)

93. K. Menger, Untersuchungen über allgemeine Metrik. Math. Ann. **100**, 75–163 (1928)

94. J.W. Moon, *Counting Labelled Trees*. Canadian Mathematical Monographs, vol. 1 (Canadian Mathematical Congress, 1970)

95. V. Müller, The edge reconstruction hypothesis is true for graphs with more than $n \log_2 n$ edges. J. Comb. Theor. Ser. B **22**, 281–283 (1977)

96. P.M. Neumann, A lemma that is not Burnside's. Math. Sci. **4**, 133–141 (1979)

97. J.-C. Novelli, I. Pak, A.V. Stoyanovskii, A new proof of the hook-length formula. Discrete Math. Theor. Comput. Sci. **1**, 053–067 (1997)

98. K.M. O'Hara, Unimodality of Gaussian coefficients: a constructive proof. J. Comb. Theor. Ser. A **53**, 29–52 (1990)

99. W.V. Parker, The matrices $AB$ and $BA$. Am. Math. Mon. **60**, 316 (1953); Reprinted in *Selected Papers on Algebra*, ed. by S. Montgomery et al. (Mathematical Association of America, Washington), pp. 331–332

100. J. Pitman, Coalescent random forests. J. Comb. Theor. Ser. A **85**, 165–193 (1999)
101. G. Pólya, Kombinatorische Anzahlbestimmungen für Gruppen, Graphen und chemische Verbindungen. Acta Math. **68**, 145–254 (1937)
102. G. Pólya, R.C. Read, *Combinatorial Enumeration of Groups, Graphs, and Chemical Compounds* (Springer, New York, 1987)
103. M. Pouzet, Application d'une propriété combinatoire des parties d'un ensemble aux groupes et aux relations. Math. Zeit. **150**, 117–134 (1976)
104. M. Pouzet, I.G. Rosenberg, Sperner properties for groups and relations. Eur. J. Combin. **7**, 349–370 (1986)
105. R.A. Proctor, A solution of two difficult combinatorial problems with linear algebra. Am. Math. Mon. **89**, 721–734 (1982)
106. H. Prüfer, Neuer Beweis eines Satzes über Permutationen. Arch. Math. Phys. **27**, 742–744 (1918)
107. J. Radon, Über die Bestimmung von Funktionen durch ihre Integralwerte längs gewisser Mannigfaltigkeiten. Berichte über die Verhandlungen der Königlich Sächsischen Gesellschaft der Wissenschaften zu Leipzig **69**, 262–277 (1917); Translation by P.C. Parks, On the determination of functions from their integral values along certain manifolds. IEEE Trans. Med. Imaging **5**, 170–176 (1986)
108. J.H. Redfield, The theory of group reduced distributions. Am. J. Math. **49**, 433–455 (1927)
109. J.H. Redfield, Enumeration by frame group and range groups. J. Graph Theor. **8**, 205–223 (1984)
110. G. Reisner, Cohen-Macaulay quotients of polynomial rings. Adv. Math. **21**, 30–49 (1976)
111. J.B. Remmel, Bijective proofs of formulae for the number of standard Young tableaux. Linear Multilinear Algebra **11**, 45–100 (1982)
112. G. de B. Robinson, On the representations of $S_n$. Am. J. Math. **60**, 745–760 (1938)
113. H.J. Ryser, *Combinatorial Mathematics* (Mathematical Association of America, Washington, 1963)
114. B.E. Sagan, *The Symmetric Group*, 2nd edn. (Springer, New York, 2001)
115. C.E. Schensted, Longest increasing and decreasing subsequences. Can. J. Math. **13**, 179–191 (1961)
116. J. Schmid, A remark on characteristic polynomials. Am. Math. Mon. **77**, 998–999 (1970); Reprinted in *Selected Papers on Algebra*, ed. by S. Montgomery et al. (Mathematical Association of America, Washington), pp. 332-333
117. M.-P. Schützenberger, A characteristic property of certain polynomials of E.F. Moore and C.E. Shannon. RLE Quarterly Progress Report, No. 55 (Research Laboratory of Electronics, M.I.T., 1959), pp. 117–118
118. C.A.B. Smith, W.T. Tutte, On unicursal paths in a network of degree 4. Am. Math. Mon. **48**, 233–237 (1941)
119. E. Sperner, Ein Satz über Untermengen einer endlichen Menge. Math. Z. **27**(1), 544–548 (1928)
120. E. Sperner, Über einen kombinatorischen Satz von Macaulay und seine Anwendung auf die Theorie der Polynomideale. Abh. Math. Sem. Univ. Hamburg **7**, 149–163 (1930)
121. R. Stanley, Cohen-Macaulay rings and constructible polytopes. Bull. Am. Math. Soc. **81**, 133–135 (1975)
122. R. Stanley, The upper bound conjecture and Cohen-Macaulay rings. Stud. Appl. Math. **54**, 135–142 (1975)
123. R. Stanley, Weyl groups, the hard Lefschetz theorem, and the Sperner property. SIAM J. Algebr. Discrete Meth. **1**, 168–184 (1980)
124. R. Stanley, *Combinatorics and Commutative Algebra*, 2nd edn. Progress in Mathematics, vol. 41 (Birkhäuser, Boston/Basel/Berlin, 1996)
125. R. Stanley, Differentiably finite power series. Eur. J. Combin. **1**, 175–188 (1980)
126. R. Stanley, Quotients of Peck posets. Order **1**, 29–34 (1984)
127. R. Stanley, Differential posets. J. Am. Math. Soc. **1**, 919–961 (1988)

128. R. Stanley, Unimodal and log-concave sequences in algebra, combinatorics, and geometry, in *Graph Theory and Its Applications: East and West*. Annals of the New York Academy of Sciences, vol. 576 (1989), pp. 500–535

129. R. Stanley, Variations on differential posets, in *Invariant Theory and Tableaux*, ed. by D. Stanton. The IMA Volumes in Mathematics and Its Applications, vol. 19 (Springer, New York, 1990), pp. 145–165

130. R. Stanley, *Enumerative Combinatorics*, vol. 1, 2nd edn. (Cambridge University Press, Cambridge, 2012)

131. R. Stanley, *Enumerative Combinatorics*, vol. 2 (Cambridge University Press, New York, 1999)

132. B. Stevens, G. Hurlbert, B. Jackson (eds.), Special issue on "Generalizations of de Bruijn cycles and Gray codes". Discrete Math. **309** (2009)

133. K. Sutner, Linear cellular automata and the Garden-of-Eden. Math. Intelligencer **11**, 49–53 (1989)

134. J.J. Sylvester, On the change of systems of independent variables. Q. J. Math. **1**, 42–56 (1857); *Collected Mathematical Papers*, vol. 2 (Cambridge, 1908), pp. 65–85

135. J.J. Sylvester, Proof of the hitherto undemonstrated fundamental theorem of invariants. Philos. Mag. **5**, 178–188 (1878); *Collected Mathematical Papers*, vol. 3 (Chelsea, New York, 1973), pp. 117–126

136. W.T. Trotter, *Combinatorics and Partially Ordered Sets: Dimension Theory*. Johns Hopkins Studies in the Mathematical Sciences, vol. 6 (Johns Hopkins University Press, Baltimore, 1992)

137. R.J. Turyn, Character sums and difference sets. Pac. J. Math. **15**, 319–346 (1965)

138. R.J. Turyn, Sequences with small correlation, in *Error Correcting Codes*, ed. by H.B. Mann (Wiley, New York, 1969), pp. 195–228

139. W.T. Tutte, The dissection of equilateral triangles into equilateral triangles. Proc. Camb. Philos. Soc. **44**, 463–482 (1948)

140. W.T. Tutte, Lectures on matroids. J. Res. Natl. Bur. Stand. Sect. B **69**, 1–47 (1965)

141. W.T. Tutte, The quest of the perfect square. Am. Math. Mon. **72**, 29–35 (1965)

142. T. van Aardenne-Ehrenfest, N.G. de Bruijn, Circuits and trees in oriented linear graphs. Simon Stevin (Bull. Belgian Math. Soc.) **28**, 203–217 (1951)

143. M.A.A. van Leeuwen, The Robinson-Schensted and Schützenberger algorithms, Part 1: new combinatorial proofs, Preprint no. AM-R9208 1992, Centrum voor Wiskunde en Informatica, 1992

144. F. Whipple, On a theorem due to F.S. Macaulay. J. Lond. Math. Soc. **28**, 431–437 (1928)

145. E.M. Wright, Burnside's lemma: a historical note. J. Comb. Theor. B **30**, 89–90 (1981)

146. A. Young, Qualitative substitutional analysis (third paper). Proc. Lond. Math. Soc. (2) **28**, 255–292 (1927)

147. D. Zeilberger, Kathy O'Hara's constructive proof of the unimodality of the Gaussian polynomials. Am. Math. Mon. **96**, 590–602 (1989)

# Index

© Springer International Publishing AG, part of Springer Nature 2018
R. P. Stanley, *Algebraic Combinatorics*, Undergraduate Texts in Mathematics,
https://doi.org/10.1007/978-3-319-77173-1

Printed in the United States
By Bookmasters